Structural analysis
using computers

Structural analysis using computers

W. M. JENKINS

Professor Emeritus, School of Engineering
The Hatfield Polytechnic

LONGMAN

CO-PUBLISHED IN THE UNITED STATES WITH
JOHN WILEY & SONS, INC., NEW YORK

Longman Scientific & Technical,
Longman Group UK Ltd,
Longman House, Burnt Mill, Harlow,
Essex CM20 2JE, England
and Associated Companies throughout the world.

Copublished in the United States with
John Wiley & Sons, Inc., 605 Third Avenue, New York, NY 10158

First published 1990

British Library Cataloguing in Publication Data

Jenkins, W. M. (William McLaren)
Structural analysis using computers
1. Structures. Analysis. Applications of computers
I. Title
624.1'71'0285

ISBN 0-582-02007-7

Library of Congress Cataloging-in-Publication Data

Jenkins, W. M. (William McLaren)
Structural analysis using computers / W.M. Jenkins.
p. cm.
Bibliography: p.
Includes index.
ISBN 0-470-21474-0
1. Structural analysis (Engineering)—Data processing. I. Title.
TA647.J46 1990
624.1'71'0285—dc20 89-12273
CIP

Produced by Longman Group (FE) Limited
Printed in Hong Kong

To Carol

CONTENTS

PREFACE

Computers have been used in structural analysis since the 1950s. Their potential for carrying out the more laborious parts of analysis was quickly appreciated and there began a steady development of methods designed to utilize the speed and numerical processing capacity of computers.

Initially considerable attention was directed to the formulation of methods of analysis using matrix algebra. It was clear that these methods had much to offer in terms of compact organization of the arithmetical processes and were eminently suitable for the digital computer, especially if a high-level matrix interpretative language was available. In these circumstances the computer could be used to generate the required matrices and then go on to carry out the necessary operations of addition, multiplication and inversion to provide values of structural displacements and stress resultants.

The basic concepts and theorems of structural analysis were subjected to considerable review and restatement in matrix terms. The concepts of flexibility and stiffness were developed and were the foundations of the modern methods of analysis. A general matrix formulation was then possible and this was to prove capable of handling structures of a degree of complexity undreamt of. The new methods were able to analyse not only skeletal structures comprised of struts, ties, beams and columns but also structural continua such as plates and shells. Thus the method of finite elements was developed simultaneously with growth of computing power.

Before the advent of the computer, structural engineers had of necessity to avoid structural analysis formulations which led to the solution of more than three or four simultaneous equations. Indeed methods such as moment distribution were developed to avoid this difficulty. The practical removal of this constraint was the signal for a quantum leap forward in analytical power. Moment distribution was very extensively developed and was of considerable value since it solved the rotational equilibrium equations at the joints iteratively and also brought a useful 'feel' for the structure because the method had an obvious physical interpretation. Distribution methods of analysis played an important role in the structural engineering curriculum from the 1940s to the 1970s. The methods now have a more limited role and in some cases have disappeared from the teaching syllabuses in structural analysis. Unfortunately the structural appreciation which they provided will be lost and must be replaced in some way. One of the purposes of this book is to try to find this educational replacement.

In the space of some forty years the computer has transformed the subject of structural analysis, and the widespread availability of computers, particularly the microcomputer, now enables virtually all analysis to be carried out by machine. The implications of this for the teaching of structural theory and analysis are profound and have not yet been fully grasped. Many teachers of the subject are reluctant to discard the old emphasis on 'hand' methods of analysis, arguing that, in spite of computers, students need to develop and consolidate their understanding by these well-trodden paths. The ability to carry out long involved calculations by hand, often so laboriously developed in the past, is no longer required. All the necessary understanding and analytical ability can be developed using the computer and it is the purpose of this book to try to assist in this development.

It is now possible to give increased attention and emphasis to an improved understanding of basic structural theory and the synthesis and design of structures. It is important for the student to understand how structures function and are described in mathematical terms for the purposes of computation. There is a need for accurate presentation of data and an informed interpretation of results. Above all the student must have sufficient knowledge of what the computer is doing to have confidence in the results it produces and be able to tell when something has gone wrong. Methods of verification need to be available and ideally the computer should assist in checking its own results wherever possible. The engineer also needs to be able to apply approximate methods to assist in this important task of verifying computer output. The awareness of orders of magnitude of, for example, bending moments, deflections and stresses is a necessary attribute of the engineer. A chapter of the book will be devoted specifically to this purpose.

This book is written in three parts. The first states and explains the basic principles of the theory of structures. Where numerical examples are needed to exemplify the development, these are simple and uncomplicated from the computational point of view. The computer will not figure prominently at this stage; nevertheless the development of the material is carried out with the computer in mind. For example, when we come to choose a system of orthogonal axes with attendant sign conventions, we will prepare ourselves for work with three-dimensional structures since most computer systems will be described in this way.

The pre-eminence of the stiffness method is recognized by a progressive development in the early chapters, with one chapter devoted to the derivation of stiffness and related matrices and the process of assembling the structure stiffness from that of the elements.

Part II is a workshop consisting of a systematic and progressive analysis of structures from an ordered catalogue. The structures are chosen to illustrate and test principles and to encourage the reader to adopt an exploratory approach. It is the intention that suitable computer programs will be used throughout. Any conveniently available programs can be used; however, programs are provided in Part III and can be used for this purpose. Specifications and listings of the programs are provided. The programs are unsophisticated and are written in a fairly simple style of BASIC so that they are not severely machine-dependent. They are annotated sufficiently to enable adaptation to different microcomputers.

The book is intended for all students of structural theory and analysis at degree and Higher National Certificate/Diploma levels. It should provide for the needs of the degree student over two years of study. Final-year studies generally require support from texts in elasticity, plasticity, finite element theory, stability and vibrations. The HNC/D student will be able to cover all his needs and be able to omit some of the more advanced material. The book should be suitable for group teaching or for individual learning. In group teaching the lecturer is encouraged to concentrate on the principles and leave the practice to the students and the computers. A lively and interesting style of teaching will ensue if this attitude is adopted. Questions and observations on the structures in Part II are by no means exhaustive and perceptive students will uncover many aspects of structural behaviour to add to those mentioned. The reader is expected to have a reasonable grasp of the basic principles of structural mechanics and some facility in matrix algebra.

The author is greatly indebted to the many colleagues and students he has worked with over many years, for their willingness to share ideas in a common purpose. The material for the book has developed from the author's teaching experience in the Universities of London and Bradford and in the Teesside and Hatfield Polytechnics. Many ideas originated in the normal course of discussions with academic colleagues and in work with students for which the author is very grateful.

Whilst the subject of the book is the analysis of structures, considerable attention is paid to topics normally coming under the heading of structural theory or strength of materials. This is done deliberately to try to make the treatment self-contained. There is virtually no treatment of structural design as such;

however, mention is made where appropriate of British Standards for design, for example BS 5950 in Chapter 7 in connection with column design and in Chapter 10 in connection with plastic design. The treatment is however not restricted to British Standards.

Finally, the author appreciates the help given by his friend Harold Walker with his word processor and records his thanks to his family for their patience and forbearance.

W M JENKINS
July 1988

NOTATION

P, Q	forces
H, V	horizontal and vertical forces
N	axial force
V, S	shearing force at a section
T	torque (twisting moment)
M	bending moment
$\boldsymbol{R}$	external nodal forces
$\boldsymbol{W}_e$	equivalent nodal forces
$\boldsymbol{S}$	member forces in local coordinates
$\boldsymbol{Q}$	member forces in global coordinates
$\boldsymbol{r}$, $\boldsymbol{s}$, $\boldsymbol{q}$	displacements corresponding to $\boldsymbol{R}$, $\boldsymbol{S}$ and $\boldsymbol{Q}$
A	cross-sectional area
$\boldsymbol{A}$	displacement transformation matrix
I	second moment of area of cross-section
J	torsion constant of cross-section
$\boldsymbol{T}$	force transformation matrix
X, Y, Z	coordinate axes
$\boldsymbol{D}$	elasticity matrix
E	modulus of elasticity
EI	flexural rigidity
EA	extensional rigidity
G	shear modulus of elasticity
GJ	torsional rigidity
L, l	length
u, v, w	linear coordinate displacements
θ	rotations
ω	sectorial coordinate
n_s	degree of statical indeterminacy
n_k	degree of kinematical indeterminacy
σ	normal stress
τ	shear stress
$\boldsymbol{\sigma}$	stress matrix
ε	direct strain
γ	shear strain
$\boldsymbol{\varepsilon}$	strain matrix
ν	Poisson's ratio
U	strain energy

PART

I

The theory of structures

CHAPTER

1

Introduction

1.1 The role of structural analysis

An engineering structure is a device intended to support loads. It is usually an assemblage of parts or elements such as ties, struts, beams, columns and plates, and the result is often described as a structural 'frame'. The loads may be 'imposed', for example those caused by the traffic on a bridge, or they may be gravity loads resulting from the weight of material to be supported, including the self-weight of the structure. Although the weight of material to be supported can usually be accurately predicted, imposed loads are much more difficult to predict especially if they are dynamic in character as is the case with bridges. Although it appears that an element of uncertainty has already been introduced in estimating the actual loads which will be applied to a given structure, in fact a good deal is known about the real loads resulting from most practical situations. What is important of course is to be able to predict the most severe loading which is likely to occur during the expected life of the structure. Load estimation for the purposes of structural design is controlled by Codes of Practice and is usually based on statistical predictions.

The role of structural analysis is to apply the principles of the theory of structures to determine the behaviour of a structure under the design loads in order to assess the adequacy of the design. The analysis will reveal the magnitudes of the forces and stresses set up in the structure thus enabling an assessment of strength to be made; adjustments to the design can then be made if the structure is in some sense unsatisfactory. The analysis will also yield magnitudes of the displacements occurring in the structure and these can be examined for acceptability.

It will be clear that the role of structural analysis is to assist the design process by providing information about what will happen to the structure when the loads are applied. The closer the two operations of analysis and design are connected, the more effective the design will be. Ideally the complete design process should include whatever analysis is required and an integrated process is very attractive. Fortunately the computer enables this integration to be achieved to a significant extent, thus enabling improvements to be made in the design by iteration.

In certain circumstances, structural analysis can be generalized in so-called 'closed form' solutions. For example the maximum bending moment and deflection in a uniformly loaded beam can be expressed by standard formulae. These standard results are very useful in the structural engineer's 'tool-kit' and indeed are frequently incorporated in computer programs; however, structural analysis carried out on computers is invariably numerical in style. Widespread use is made of ordered sets of quantities, that is matrices, and matrix algebra is a useful vehicle for organizing the actual calculations needed. In the computer these matrices will be stored as ordered 'arrays' of numbers and the program will instruct the computer in the processing of the arrays. Once the program has been verified and is operating reliably, then the structural designer can concentrate his attention on the design process using the analysis primarily as an aid to design.

There are other aspects of structural engineering understanding which are developed by studying the analysis of structures. An engineer with a good grasp of analysis will better understand the behaviour of his structure and will be capable of designing the more complex structures arising in modern engineering. It is essential for the structural designer to gain this understanding and to be fully aware of what the computer is doing. It is necessary for the student to carry out some analysis without the computer but using the same data and the same matrices that the computer will use in order to establish confidence. This of course will have to be confined to problems of a manageable size and will often appear to be using a sledge-hammer to crack a nut! The gain in confidence in the methods will amply repay the effort.

1.2 Some fundamental concepts

1.2.1 Force

This is the generic name used for the physical actions applied to structures and developing in structures. In reality forces in nature are usually distributed; however, in analysis it is convenient to regard discrete forces acting at a point or along an axis in a structure. The term 'force' is a general one and includes all kinds of physical actions such as axial force, bending moment, shear force and twisting moment. Thus forces and couples are collectively termed 'forces'.

Force is a vector quantity having both magnitude and direction. In structural analysis it is frequently necessary to resolve forces to obtain equivalent actions in specific directions. Consider forces P and Q in Fig. 1.1(a). We can 'resolve' these forces into the directions X, Y as follows:

Figure 1.1 Resolution of forces and moments

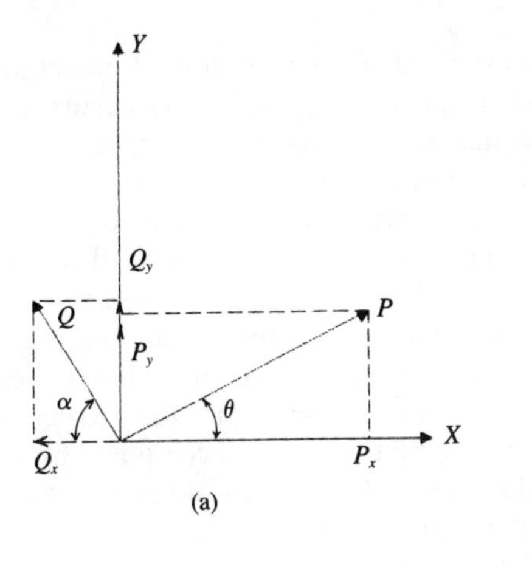

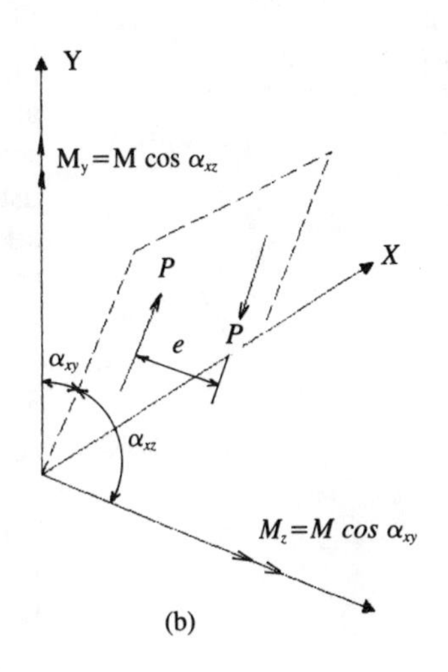

$$\left.\begin{aligned} P_x &= P\cos\theta \\ P_y &= P\cos(90^\circ-\theta) = P\sin\theta \\ Q_x &= Q\cos\alpha \\ Q_y &= Q\cos(90^\circ-\alpha) = Q\sin\alpha \end{aligned}\right\} \quad [1.1]$$

When resolving forces, we always use the cosine of the angle between the force and the direction of the resolved component. We also note that the combined effect of P and Q is $P_x - Q_x$ in the X direction and $P_y + Q_y$ in the Y direction.

If P and Q are orthogonal, as would be the case when considering axial force and shear force in a beam cross-section, then $\alpha = 90^\circ - \theta$; the resultant of P and Q in the X direction would then be $P\cos\theta - Q\sin\theta$ and the resultant in the Y direction would be $P\sin\theta + Q\cos\theta$. Moments may also be resolved about orthogonal axes as in Fig. 1.1(b). The forces P at distance e apart in a plane at angle α_{xy} to plane X–Y constitute a couple $M(=Pe)$ which resolves as follows:

$$M_z = Pe\cos\alpha_{xy} = M\cos\alpha_{xy}$$

$$M_y = M\cos\alpha_{xz}$$

The right-hand screw rule is used in the sign convention for moments, i.e. a right-hand thread is advanced positively by a positive moment looking in the positive direction of the axis. A double-headed arrow is used by convention to indicate a moment about the axis to which the arrows are applied.

1.2.2 Displacement

This is a general term representing any kind of finite deformation at a point in a structure. Transverse displacements of beams are usually called ‘deflections’; however, the term ‘displacement’ includes all linear and rotational movements. It is usual to focus attention on the displacements occurring at particular points in structures called 'nodes'. In framed structures the nodes will usually be the joints where members are connected together, but additional nodes may be identified at intermediate points within the length of a member if it is desired to obtain information specific to those points. In continuous structures without joints, the nodes will be defined at suitable points together constituting a pattern of structural elements called a 'mesh'. Axes of reference are chosen for the structure and the directions corresponding to these at each node are called 'coordinates'. It is usual for two different sets of coordinates to exist simultaneously. For the structure as a whole, a single set of coordinate axes will be defined and will apply to the complete structure; these are called 'global' coordinates. An individual element of the structure will have its own 'local' set of axes conveniently directed for that element. The transformation of forces and displacements from one set of coordinates to the other is an important operation in structural analysis. We shall look at it now.

1.2.3 Transformation of coordinate axes

Consider the simple truss member IJ shown in Fig. 1.2. We

Figure 1.2
Transformation of coordinate axes for simple truss element in X–Y plane

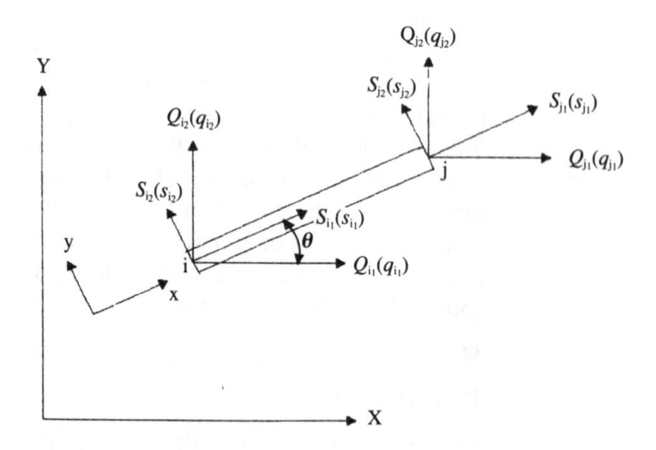

assume all actions to be in the X–Y plane. The local, natural, axes for the member are x, y and the global axes are X, Y. The x axis of the member is at the angle θ to the X axis. If the displacements under load of ends I and J in the local coordinate system are s_{i1}, s_{i2} and s_{j1}, s_{j2} respectively, and the corresponding displacements in the global coordinate system are q_{i1}, q_{i2}, q_{j1}, q_{j2}, then

$$\begin{bmatrix} s_{i1} \\ s_{i2} \\ s_{j1} \\ s_{j2} \end{bmatrix} = \begin{bmatrix} \cos\theta & \sin\theta & 0 & 0 \\ -\sin\theta & \cos\theta & 0 & 0 \\ 0 & 0 & \cos\theta & \sin\theta \\ 0 & 0 & -\sin\theta & \cos\theta \end{bmatrix} \begin{bmatrix} q_{i1} \\ q_{i2} \\ q_{j1} \\ q_{j2} \end{bmatrix} \qquad [1.2]$$

A more compact form is

$$\boldsymbol{S} = \boldsymbol{Aq} \qquad [1.3]$$

where

$$\boldsymbol{s} = \begin{bmatrix} s_{i1} \\ s_{i2} \\ s_{j1} \\ s_{j2} \end{bmatrix}; \qquad \boldsymbol{q} = \begin{bmatrix} q_{i1} \\ q_{i2} \\ q_{j1} \\ q_{j2} \end{bmatrix} \qquad [1.4]$$

and

$$\boldsymbol{A} = \begin{bmatrix} \cos\theta & \sin\theta & 0 & 0 \\ -\sin\theta & \cos\theta & 0 & 0 \\ 0 & 0 & \cos\theta & \sin\theta \\ 0 & 0 & -\sin\theta & \cos\theta \end{bmatrix} \qquad [1.5]$$

$\boldsymbol{A}$ is called the displacement transformation matrix.

The axial force in the member IJ can be found by considering the displacements s_{i1} and s_{j1} only. So we can omit the second and fourth rows of each of the matrices $\boldsymbol{s}$ and $\boldsymbol{A}$ and can therefore simplify the relationship to

$$\begin{bmatrix} s_{i1} \\ s_{j1} \end{bmatrix} = \begin{bmatrix} \cos\theta & \sin\theta & 0 & 0 \\ 0 & 0 & \cos\theta & \sin\theta \end{bmatrix} \begin{bmatrix} q_{i1} \\ q_{i2} \\ q_{j1} \\ q_{j2} \end{bmatrix} \qquad [1.6]$$

We can use a similar procedure to transform the forces from one set of coordinates to the other. Referring again to Fig. 1.2, and this time making the transformation in the direction 'local' to 'global', we obtain by simple resolution of forces,

$$\begin{bmatrix} Q_{i1} \\ Q_{i2} \\ Q_{j1} \\ Q_{j2} \end{bmatrix} = \begin{bmatrix} \cos\theta & 0 \\ \sin\theta & 0 \\ 0 & \cos\theta \\ 0 & \sin\theta \end{bmatrix} \begin{bmatrix} S_{i1} \\ S_{j1} \end{bmatrix} \qquad [1.7]$$

or

$$\boldsymbol{Q} = \boldsymbol{TS} \qquad [1.8]$$

where $\boldsymbol{T}$ is the force transformation matrix. We see that the displacement and force transformation matrices $\boldsymbol{A}$ and $\boldsymbol{T}$ are related; in fact the relationship is

$$\boldsymbol{T} = \boldsymbol{A}^{\mathrm{T}}$$

and we shall meet this important result again in later chapters.

1.2.4 Support constraints

Always an important aspect of a structure is the way it is supported and we shall need to be able to define support constraints in a very precise way. We can assume that there will always be a node at a point of support and so we can use the idea of coordinates to specify the nature of that support. Suppose a point of support of a structure has three linear coordinates defined as u, v and w, and three rotational coordinates θ_u, θ_v, θ_w, as in Fig. 1.3(a). Now if we equip the structure with a set of rollers as in Fig. 1.3(b) and if these rollers provide freedom of movement in the u direction only, then we say that the support 'constrains' the structure in the directions v and w. Similarly, if the structure is supported as in Fig. 1.3(c), then the constraints are in the u and w directions. If we now generalize this to include rotational constraints about three orthogonal axes, then we have a total of six possible constraints. Any or all of these may exist at a support. In a plane structure, subjected only to forces in its plane, we can reduce the number of coordinates, and hence possible constraints, to three. For example in Fig. 1.3, if we imagine a structure to lie in the u–v plane, then linear constraints can exist in the u and v directions and a rotational constraint can exist about the w axis.

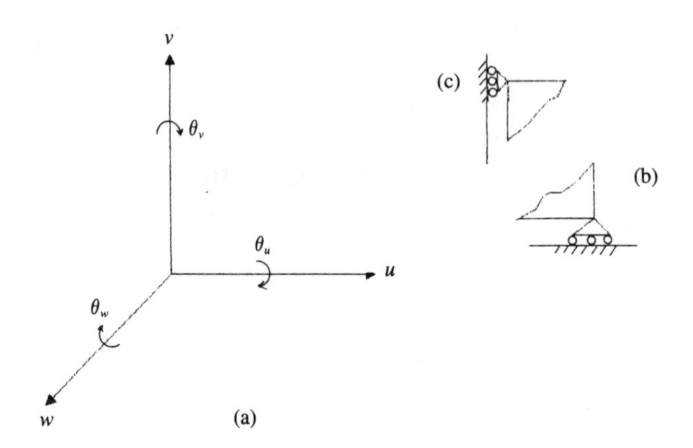

Figure 1.3 Coordinate axes and constraints

1.2.5 *Strength*

This is perhaps the most obvious property required by a structure. Clearly any structure must be strong enough to sustain the applied loads without failing or becoming unsatisfactory. We are now introducing terms which require engineering definitions; 'strong enough', 'failing' and 'applied loads' all require further explanation with a view to accurate interpretation in the context of civil engineering structures. For the moment it will be sufficient to observe that the assessment of the strength of a structure requires a detailed study of the way in which it supports the applied loads and the resulting distributions of stress within the elements of the structure. This requires a knowledge of the strength of materials and also the theory and analysis of structures.

1.2.6 *Stiffness*

The application of loads to a structure produces strains in the material. Under elastic conditions these strains are related to the elastic modulus of the material and a material with a high value of elastic modulus will suffer less strain than one with a low modulus. Two elastic moduli are of importance for us; Young's modulus E and the modulus of rigidity (shear modulus) G. In both cases the modulus is the constant value of the ratio of stress to strain in the material.

The effect of straining an engineering material is to produce distortions. If these distortions are large they may be unacceptable and the structure may need to be stiffened. This could of course be done, as we have just seen, by changing the material and using one with a higher modulus. However, it can usually be achieved much more effectively by changing the cross-sectional shape or dimensions of the members or by changing the geometry of the structure. We soon learn that the most effective way to make a beam stiffer is to make it deeper. This may however lead to other problems if we make the beam too deep. The most important factors affecting stiffness are

1. Type of structure
2. Geometry of structure
3. Cross-sections of members
4. Material used.

The designer therefore has considerable scope to control the stiffness of his structure.

1.2.7 *Structural joints*

The elements of a structure need to be connected together in a satisfactory way, i.e. the continuity actually provided by the joint must be what the designer intended. In steel structures, joints are generally made by welding or bolting, or a combination of the two. As a result joints vary from 'rigid', in which there is no relative movement of the parts joined, to 'pinned', in which unrestrained relative angular movement is possible. In concrete structures, joints are generally rigid owing to the high degree of continuity provided by the concrete and the steel reinforcement. In all structures design may deliberately provide for some relative movement between parts of the structure and one or more supports.

We have introduced the term 'node' and this is often used in structures to denote a particular point at which attention is to be focused. In jointed structures, it is usual for the terms 'node' and 'joint' to be synonymous. However, we should remember that a structural joint has finite dimensions whereas a 'node' is a convenient concept for a point in space at which we define coordinate axes. The structural design of the joint will certainly require that attention is given to its dimensions.

1.3 Statical and kinematical indeterminacies

1.3.1 Equilibrium and compatibility

These are important physical principles in structural analysis. The whole structure and every part of it must be in equilibrium under the actions of the force system both external to the structure and internally within the structural elements. The external force system is fairly easily visualized; the internal system can be more difficult to see since it may require extensive analysis to quantify it. The internal forces in the structure arise from stress distributions and, as we shall see in Chapter 3, the internal stresses can be integrated to produce 'stress resultants'. In a general three-dimensional situation, six stress resultants can exist at a section in a structural member: two bending moments, two shear forces, a twisting moment and a thrust. It is of course possible to work the other way round and interpret stress resultants in the form of stress distributions.

Just as the state of stress in a structure can be defined in terms of stress resultants at nodes, the state of deformation can be described by the displacements occurring at the nodes. It is necessary to know the relationships between nodal forces (stress resultants) and nodal displacements for the elements of the structure and this information is part of the data needed to carry out a structural analysis.

At certain points in a structure, continuity between one element and another may be interrupted by a 'release'. This is a device which imposes a zero value on one of the stress resultants at the point. For example a hinged joint between two members ensures that no bending moment exists at the point of connection in the coordinate corresponding to the release. This is a 'bending moment release'. Releases can exist consistent with zero axial force, shear force, thrust and so on. Each stress resultant requires one release to make it zero, so if we make a complete cut through a member in a three-dimensional structure we need six releases. Releases can be imagined to be mechanical devices either real, as they are in some structures, or imaginary for the purposes of analysis.

We have seen that each release introduces a discontinuity in the structure; however, the elements of the structure must fit together in a physically compatible way. This requirement of 'compatibility' is an important fundamental principle in structural analysis.

1.3.2 Statical indeterminacy

A structure is statically determinate if it is possible to obtain the forces in the structure from equilibrium conditions without reference to compatibility conditions. Consider the plane pin-jointed structure shown at (a) in Fig. 1.4. By resolving forces as necessary, we can write down two equations of equilibrium at

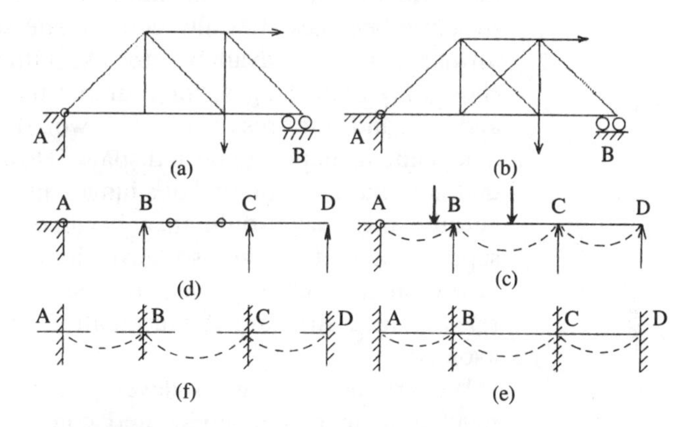

Figure 1.4 Statical and kinematical indeterminacy

each of the six joints and can therefore determine twelve unknown forces. The support conditions are such as to require, in general, horizontal and vertical components of reaction at A and a vertical reaction at B, i.e. three unknowns. There are nine members so the total number of unknowns is twelve, which just equals the number of equilibrium equations available to us. Structure (a) is therefore statically determinate. We cannot now remove any member from structure (a) or it would become 'deficient'. If we add another member to (a) as in (b) then we have thirteen unknowns but still only twelve equations. Structure (b) is statically indeterminate, sometimes called 'redundant', to the first degree.

In assessing the degree of statical indeterminacy of pin-connected structures, the total number N of unknowns should include all the reactive components, even when some of these are zero under certain loading conditions, and all the members. The degree of statical indeterminacy is then

$$n_s = N - \frac{2}{3}J \qquad [1.9]$$

where J is the number of joints, the coefficient 2 being used for plane frames and the coefficient 3 for space frames.

If we now examine the structure shown at (c) in Fig. 1.4, we see a continuous beam over four supports. This structure is statically indeterminate owing to the nature of its support. Let us apply the three conditions of equilibrium available for the complete beam. In the horizontal direction we have no applied forces and no horizontal reactions, so effectively only two equations of equilibrium are available to us. These will express vertical equilibrium and rotational equilibrium for the whole beam. We have however four vertical reactions to find and this cannot be done with only two equations. The structure is therefore statically indeterminate to degree 2. If two releases in the form of hinges are introduced in span BC, as at (d), the structure becomes statically determinate since we now have two further equations available to us. Other positions could be chosen for these hinges; for example they could be placed at B and C, and the result of this would be to produce three independent simply-supported spans. However, it should be seen that we could not place *both* hinges in either of the end spans because this span would then become a mechanism unable to support load. It would also leave the rest of the beam statically indeterminate. Clearly care is needed in the distribution of releases in a structure if the creation of a mechanism is to be avoided.

We are now going to develop a relationship between the number of members, nodes and constraints and the degree of statical indeterminacy. In the structure (c) of Fig. 1.4 the four supports A, B, C and D are all connected rigidly through the ground and there is no relative moment. This condition is represented by the dashed lines joining A, B, C and D. We now put the supports into a 'standard' condition by clamping them thus removing all releases. This is shown at (e) and we see that the structure now comprises three closed rings. We can easily make this structure statically determinate by cutting through each ring, as in diagram (f), and the resulting structure is simply a series of cantilevers each of which is statically determinate. At each cut three stress resultants are released and thus the degree of statical indeterminacy in condition (e) is three times the number of rings. A general relationship can be deduced between the number of members m, the number of nodes n, and the number of rings:

$$\text{number of rings} = m - n + 1$$
$$= 6 - 4 + 1 \quad \text{in diagram (e)}$$
$$= 3$$

We now restore the structure to its correct condition by inserting the required number of releases r. To revert from diagram (e) to diagram (c) requires the insertion of four moment releases, one each at A, B, C and D, and three axial force releases at B, C and D. Thus

$$n_s = 3(6 - 4 + 1) - 7$$
$$= 2$$

The general relationship for degree of statical indeterminacy is

$$n_s = \frac{3}{6}(m - n + 1) - r \qquad [1.10]$$

The coefficient 3 is used for plane frames and the 6 for space frames.

1.3.3 Kinematical indeterminacy

A structure is kinematically determinate if it is possible to obtain the nodal displacements from compatibility conditions without reference to equilibrium conditions. A fixed-end beam is kinematically determinate since the nodal (end) displacements (zero) are known from the support conditions. Consider again the structure shown at (e) in Fig. 1.4. This is the structure as at (c) but with standardized (clamped) supports. The structure has four nodes; the equivalent of one node must be considered as fixed in space so that, theoretically, the non-zero nodal displacements would be $3(n - 1)$. However, the support conditions at (e) impose constraints c on the development of nodal displacements. A 'constraint' is best defined as a device which ensures that one displacement is the same as another, usually zero. In diagram (e) we can see that the displacements at B, C, and D are all constrained, so $c = 9$. We now insert releases into the structure to restore it to condition (c), and this requires moment releases at A, B, C and D and axial force releases at B, C and D. Thus

$$n_k = 3(4 - 1) - 9 + 7$$
$$= 7$$

Since the horizontal displacements at B, C and D do not affect the bending moments and shears developing in the beam, we would not normally calculate these and would take $n_k = 4$ for structure (c). These displacements are of course the rotations of the axis of the beam at each of the four supports. They are called the 'degrees of freedom' of the structure since these displacements can be varied independently of all other nodal displacements. A general expression for kinematical indeterminacy is

$$n_k = \frac{3}{6}(n-1) - c + r \quad [1.11]$$

where n is the number of nodes, c the number of constrained displacements and r the number of releases.

Modern developments in structural analysis have focused almost entirely on the stiffness method, in which the relevant indeterminacy is the kinematical one n_k.

1.4 The role of the computer

The availability of the computer to solve large sets of simultaneous equations marked the beginning of the development of modern structural analysis. The stiffness method became increasingly attractive as it became apparent that the computer could accept more and more of the actual work of analysis. The development of an 'interactive' style of computing, made possible with the advent of the microcomputer, was a further step forward in structural analysis and, even more so, in structural design. Although the computer is a very powerful and sophisticated tool, it is imperative that the engineer remains firmly in control. This means that we must serve an 'apprenticeship' in learning how to use the tool. Would any of us want to drive across a bridge designed by a computer outside human control?

If we examine the facilities provided by some structural analysis packages, we find that they can

1. Advise on data input and carry out feasibility and consistency checks on data as it is typed in
2. Display drawings of the structure using high-resolution graphics screens
3. Allow screen editing of the data at any stage
4. Carry out the structural analysis
5. Output all results, displacements, reactions, internal force distributions (graphically) under the control of the user
6. Provide a 'help' facility at any stage.

Much microcomputer software is set in a WIMP environment – that is, using windows, icons, menus and pointers. The task of structural analysis is greatly simplified as the programs become more 'user friendly'. This does not however relieve the engineer of the responsibility for what he is doing. We can never blame the computer if something goes wrong. It is essential that engineers have sufficient knowledge of structures to be able to use the computer as a tool whilst retaining control.

The purpose of the chapters which now follow is to try to provide an educational basis for the development of this understanding, using the computer to assist us in doing it.

Exercises

1.1 In a plane truss, two members with forces P_1 and P_2 meet at a support as shown in Fig. 1.5. If the horizontal and vertical reactions are respectively H and V, show that

Figure 1.5 Exercise 1.1

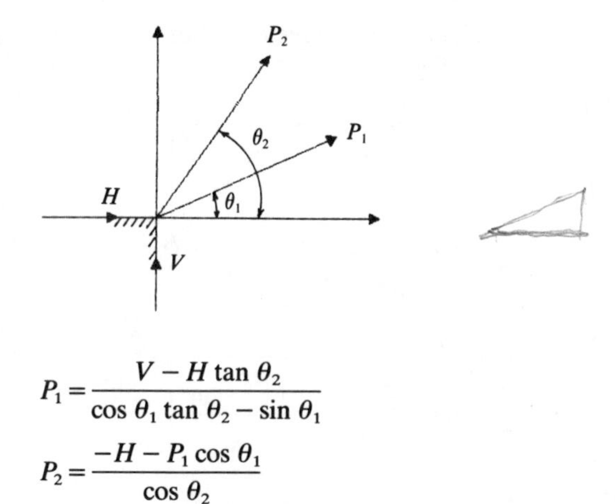

$$P_1 = \frac{V - H \tan \theta_2}{\cos \theta_1 \tan \theta_2 - \sin \theta_1}$$

$$P_2 = \frac{-H - P_1 \cos \theta_1}{\cos \theta_2}$$

1.2 By inserting numerical values for the variables, show that the equations for P_1 and P_2 in exercise 1.1 respond correctly for angles θ_1 and θ_2 from 0° to 360°.

1.3 An inclined beam is supported by reactions H and V as shown in Fig. 1.6. Show that the normal force N and shear force S are given by

Figure 1.6 Exercise 1.3

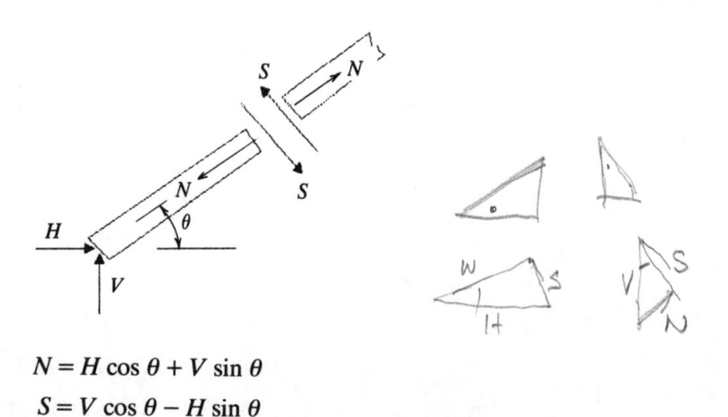

$$N = H \cos \theta + V \sin \theta$$

$$S = V \cos \theta - H \sin \theta$$

1.4 The structure shown in Fig. 1.7 has hinged supports at A and E, a hinged joint at B and a rigid joint at D. Calculate
(a) the reaction at A
(b) the reaction at E
(c) the bending moment at C
(d) the bending moment in the beam at D.

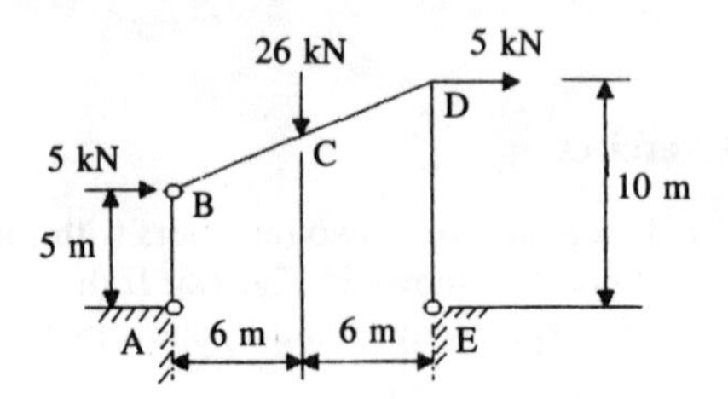

Figure 1.7 Exercise 1.4

1.5 Find the degree of statical indeterminacy of the structures shown in Fig. 1.8.

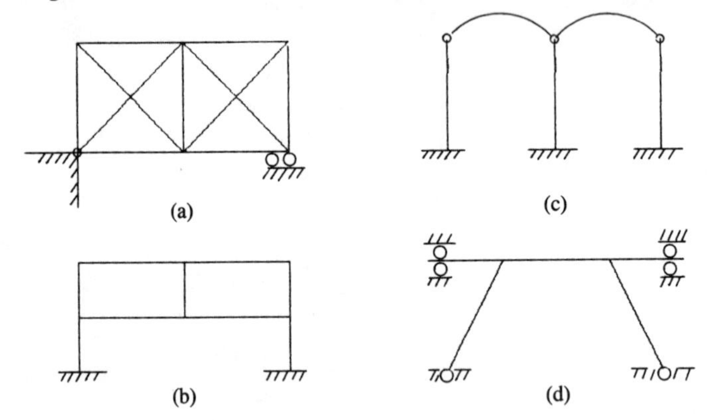

Figure 1.8 Exercises 1.5 and 1.6

1.6 Find the degree of kinematical indeterminacy of the structures shown in Fig. 1.8.

CHAPTER

2

Structural forms and materials

2.1 Introduction

Since the choice of structural form and material are closely related, it is appropriate to consider both at the same time. The principal materials used in civil engineering structures are steel and concrete; timber, aluminium and other materials such as glass reinforced plastic (GRP) are used to a lesser extent. Although an office block built in reinforced concrete may not appear to be very different from a similar building in structural steel, there will be important differences in the way the materials are used. Form and material should be complementary, so that the materials chosen suit the structural form desired and the structural form makes good use of the materials.

The environment is another important factor in choosing structural form and material. Reinforced concrete would be a very suitable choice for a retaining wall since its weight would assist its structural action and the material would be resistant to attack from the environment. Structural steel would be a good choice for a long-span, light-weight roof structure, where self-weight and snow and wind loading might be critical. If concrete were used in these circumstances then there would be a severe self-weight penalty. However, it may still be possible to use concrete by suitable changes in structural form. For example structural concrete can be very effective in thin shell form and when prestressing is used. By using suitable aggregate, concrete can also be produced in 'light-weight' form. Whatever material is used, care is needed in assessing the environmental aspects and the resistance which will be required from the material.

Other factors which are important in choosing structural form and material include the expected life of the structure and its maintenance. Ideally, the designer would like a maintenance-free structure. Although this is not possible in reality, maintenance requirements should be minimized or maintenance should be simple and cheap. An economical balance between initial cost and maintenance costs is possible; a low initial cost may incur higher maintenance costs and vice versa. The expected life of a structure may be an important consideration in design. Should a building be designed for a 50 year life or a 100 year life? Should a suspension bridge be designed for 100 years or 200 years? Structural steel, properly maintained, has a very long lifespan. Concrete also has a long-term reliability but is rather more susceptible than structural steel to sometimes unpredictable time-dependent effects and to the effects of the environment.

The choice of structural form and material is frequently influenced by the particular site for the construction. Severely limited site access will influence the structural design and the choice of material. Structural steel is generally fabricated in workshops and transported to site for erection. Site access and craneage will determine the size of the largest piece of structural steel which can be handled. Concrete can be used in a number of ways: it can be mixed on site, delivered as 'ready mixed' or used in precast form. The technology of concrete placing has developed very significantly in recent years and it is now fairly common to place concrete in situ by pumping, sometimes to a considerable height above ground level.

The last factor which will be mentioned in choosing structural form and material is that of appearance or 'aesthetics'. A structural engineer will generally work in a professional team in which the architect will take responsibility for the aesthetic aspects of the design. It is important however for the structural engineer to have some awareness of aesthetic factors and to give advice on what is possible structurally, in terms of form and material. The best design will be one which is structurally sound, aesthetically pleasing, practicable and at an economical cost. It is said that the best structural form is one where the path of the loads to the foundations is clearly evident in the form. A suspension bridge is a good example of this.

2.2 Structural elements

An engineering structure can be regarded as an assemblage of elements. This is a convenient definition from an analytical point of view and also from a practicable viewpoint. At a later stage we shall develop the structural properties of some of the more usual elements so for now we shall confine our attention to the basic functions of the elements.

2.2.1 Tie

This is perhaps the simplest of all elements. It is a linear structural member subjected principally to an axial tension as in Fig. 2.1(a)(i). It is conventional to represent the tensile force as applied by the member to the joints at the ends, so that we would represent the tie element as a line with the arrows pointing towards one another as in (a)(ii). Sometimes the application of load to the tie is not precisely axial and such 'eccentricity' of loading will induce bending in the element in addition to the axial tension. This effect is easily allowed for in design providing the axial force predominates.

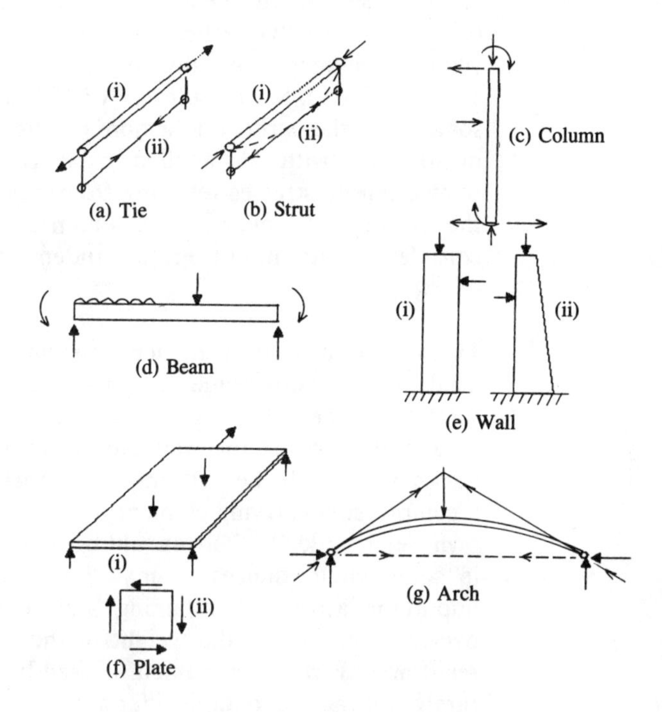

Figure 2.1 Structural elements

2.2.2 *Strut*

This element is subjected to an axial compression as in Fig. 2.1(b)(i). Again our structural representation is conveniently a single line coincident with the axis of the member, but this time the arrows point towards the joints representing the forces exerted on the joints by the element as in (b)(ii). The strut is predominantly a compression member and as such is subject to instability at certain 'critical' values of applied load. If the strut becomes unstable it bends sideways as shown by the dashed line in (b)(i). The applied load may have an eccentricity producing bending stresses in addition to the axial effect.

2.2.3 *Column*

A column is usually a vertical element in a structure carrying axial load and frequently bending caused either by eccentricity or by applied bending moments from other members as in Fig. 2.1(c). In general the column carries bending and axial compression and the bending can be as important as, or more important than, the axial load. Sometimes this element is called a 'beam-column', especially when it carries transverse loading. As with struts, instability is an important factor in column behaviour.

2.2.4 *Beam*

The beam is usually a horizontal element carrying transverse vertical loading and frequently end bending moments as in Fig. 2.1(d). The beam may also carry axial tension or compression and is thus similar to the column. The main distinction is that the column is usually vertical and carries significant axial load whereas the beam is usually horizontal and carries comparatively small axial load. Shearing forces will be important in a beam, and sometimes the structural action of the beam is improved by prestressing with an applied axial compression intended to reduce tensile stresses resulting from bending. The behaviour of the beam and column is sometimes complex and deserves considerable attention from the student of structural theory.

2.2.5 *Wall*

The wall is a vertical element carrying predominantly vertical axial loading with frequently 'small' eccentricity. There may be some transverse loading as in Fig. 2.1(e)(i), and if so the structural action of the wall can be improved by increasing the thickness towards the bottom as in (e)(ii). The wall is essentially a compression carrying element and its action is enhanced by its own weight. Ideally there should be no tension in a wall although in some circumstances a small tension can be accepted. An important aspect of retaining wall design, as in (e)(ii), is overturning. Clearly the weight of the wall must be such as to resist any overturning moment caused by the applied loads since rarely will the foundations offer any tensile resistance to the base of the wall.

2.2.6 *Plate*

The structural plate can be used in two principal ways: first as a bending element as in Fig. 2.1(f)(i), and secondly as a shear carrying element as in (f)(ii). In both cases there can be some tension or compression carried in the plane of the plate. A steel plate is a 'thin' structure and frequently stiffeners are used to enhance the structural performance. If a horizontal plate is made in concrete it is called a 'slab'.

2.2.7 *Arch*

The intrinsic structural value of the arch was of course recognized very early in the development of structural forms. The arch is able to support load in a very effective way because the ends are prevented from outward movement by abutments or by tying the ends together, as shown by the dashed line in Fig. 2.1(g). Ideally the arch is wholly in compression with no bending and therefore no tension is developed. Theoretically it is possible to achieve this but circumstances may dictate that some tension due to bending will occur under certain loading configurations. Arches may be 'two-hinged', as shown in Fig. 2.1(g); three-hinged, with the third hinge usually located at the crown; or 'fixed', where the supports provide rotational as well as translational restraint.

2.2.8 *Other elements*

In studying the various structural elements it should be understood that structures will generally contain many different types of element and it is important to know how these elements will interact in a totally unified way when the structure is functioning. Sometimes the presence of structural components may be ignored in order to simplify analysis and design; for example the stiffening effects of cladding to a building or infill panels in a structural frame may be ignored, although in certain circumstances these can have a significant effect on the structural behaviour.

Foundations also require full consideration in the analysis and design of a structure and must be regarded as an essential part of the whole for very obvious reasons. Just as with the superstructure, the analysis and design of the foundations or 'substructure' usually require some simplifying idealization in the treatment of stability, strength and stiffness.

2.3 Structural form

The investigation of possible structural forms and the final choice represent the most exciting stages in structural design. It is here that considerable scope is available to the engineer to match form and materials to the particular structural requirement (Gordon 1978), be it a building or a bridge. We shall look first of all at some of the simpler structural forms and then progress to more ambitious assemblages of structural elements.

In Fig. 2.2(a) we see the simplest possible form of a pin-jointed frame (truss) to support a single load. A triangle of three pin-connected members ABC with vertical supports at A and B will support the vertical load W at C. If the load W is not vertical it will have a horizontal component and this will need to be resisted at one or both supports. If horizontal support is provided at A and B, then the member AB could be removed. The senses of the forces in the members should be studied to identify ties and struts.

Structure (b) is a logical extension of (a) to longer spans and more loads. Note the 'triangulation' of the structure and identify the ties and struts. The arrowheads have been shown on one member to start the exercise. Structure (b) is of course a fairly standard form of roof truss. The principal sloping members, the rafters, support purlins which in turn support the roof sheeting and any imposed loads.

Figure 2.2 Examples of structural forms

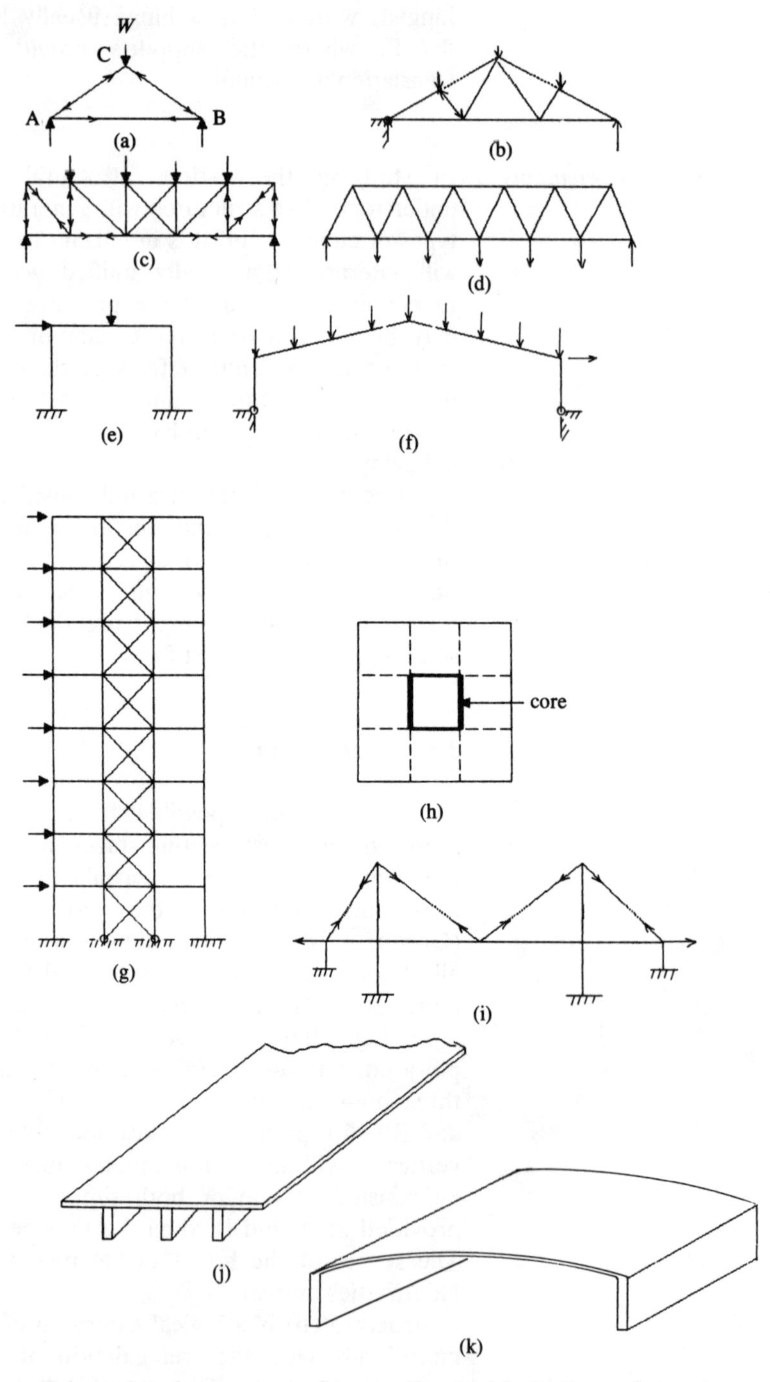

Triangulated trusses are frequently used in both roof structures and bridges. Structure (c) is a 'Pratt' truss and (d) is a Warren girder. Note the senses of the forces in the members of (c) and note the double diagonal bracing in the central panel. Under symmetrical loading there is no force in these members. How could this be proved? It is common practice to doubly brace panels in trusses where loading can produce changes in force from tension to compression in members. If (c) is a bridge girder then these conditions would prevail in the central panel under certain loading and double bracing would be required. Study also the senses of forces in the members of (d). Trusses can also be three-dimensional as 'space' frames. The basic principles are the same. The members are in tension or compression without bending (generally), and equilibrium conditions can be applied at the nodes to determine the forces providing the structure is statically determinate. Equation [1.9] can be used to assess statical determinacy. A facility with pin-jointed trusses and experience in analysing the forces in the members is important for all structural engineers.

We turn now to structures in which the members are predominantly in bending. Two examples of such structures are given in (e) and (f) of Fig. 2.2. Note that these structures are not triangulated. Their stiffness comes from rigidity in the joints and the ability of the members to resist bending actions. These frames are used in buildings where the internal space must be kept free and so diagonal bracing cannot be used. Structure (e) is a rectangular portal frame, and structure (f) is a pitched roof frame much used in industrial buildings. These frames are usually statically indeterminate and this can be assessed using eqn [1.10]. It is not difficult to design these frames to have adequate stiffness to support vertical and horizontal loads but the joint design needs careful attention to ensure that there is sufficient strength and stiffness.

A logical extension of the rectangular frame is the multi-storey frame of Fig. 2.2(g). It is most unlikely that the frame will have enough lateral stiffness to resist wind loading and so some other form of bracing must be introduced. This can be in the form of vertical triangulated panels as shown in (g). This vertical truss can form one wall of a lift shaft with three other similar walls forming a box. One of these will need special design to provide access. In a concrete building a convenient form of bracing against horizontal forces can be provided by a concrete box called a 'core'. If this is located at the centre of the building as shown in plan in diagram (h) the floors are then supported on beams between the core walls and the external columns. This has the advantage that the floor loadings are transferred partially to the concrete core, thus 'prestressing' it to assist in carrying the bending produced by the wind forces. Tall buildings of up to 30 or more storeys can be constructed in this way and careful design can produce a 'no-tension' condition in the core. It is then said that the wind loading is carried 'free'.

Under suitable circumstances, cables can be used very effectively in structures. The suspension bridge is of course an obvious example and represents the most effective structural form for really long spans. Cables are also used in structures like long-span roofs and 'cable-stayed' bridges such as the one shown in Fig. 2.2(i). This is not a suspension bridge in the usual sense because the main cables are attached directly to the bridge deck and not through suspenders. The back-stay cables may be anchored externally as they would be in a conventional suspension bridge, or they may be attached to the ends of the main girders. In this latter case, the bridge is a 'self-anchored' cable-stayed bridge. Note that the attachment of the cables to the ends of the bridge introduces a large compressive force in the main girders. This has a beneficial prestressing effect against bending but introduces problems of longitudinal instability.

A very useful class of structure is the one called 'thin-walled' (Zbirohowski-Koscia 1967). This can be a very elegant and effective structural form and can be in a number of materials. Both steel and concrete can be used in thin-walled form. A structure is thin-walled if the ratio of its principal cross-sectional dimension a to its thickness t is greater than about 10. The reinforced concrete core of example (h) in Fig. 2.2 would be deemed to be thin-walled because the ratio of the width of the box to the thickness of the wall will be high, probably of the order of 20. Other examples of thin-walled structures are shown in (j) a stiffened plate and (k) a thin concrete shell. The adoption of thin-walled behaviour allows certain simplifying assumptions to be made in the structural analysis, such as the assumption of uniform stress through the thickness of the material. Thin-walled structures need careful design from the instability point of view.

2.4 Structural materials

The two most important technical aspects of structural design, apart from cost, are strength and stiffness. In satisfying requirements under both these headings, the structural engineer combines structural form and geometry with the choice of material. In the previous section we considered some of the more important structural forms; both strength and stiffness will be influenced by the choice of structural form for a given set of loads to be supported. In this section we consider the more important materials available to the structural engineer and note the relative values of the material parameters which will influence the strength and stiffness of the resulting structure.

The strength of an engineering material is perhaps more easily visualized if we look at the 'ultimate' (breaking) strength in tension. Certainly, with a material like structural steel, the ultimate strength, quoted as say $400\,N/mm^2$, has an obvious relationship to what we might see in a structure. However, it is not sufficient to confine our attention to a single material parameter; for example, with concrete the ultimate strength in tension is rarely required or used since it is so low (1 to 3 N/mm^2) whereas the ultimate strength in compression is of crucial significance. The ultimate strength in tension will only be an 'active' parameter in design if failure will be by simple tension. An ultimate strength in compression, whilst crucial for a material like concrete, is unlikely to be an active parameter in design in structural steel or aluminium since instability is almost certain to occur long before that stress is reached. In these circumstances design will be controlled by buckling considerations and at a reduced value of compressive stress. The ultimate strength of a material is frequently a design variable in that the grade of steel and concrete to be used can be chosen by the designer to suit the circumstances. In composite materials such as GRP the material itself can be designed to a certain extent by controlling the proportions and properties of the elements of the composite. The most important material parameter influencing structural stiffness is the value of the elastic modulus E (kN/mm^2). Deflections in structures are generally inversely proportional to the value of E of the material, so that if we replace a material having a certain value of E with one having twice the value then the deflections will be halved. The control of stiffness by choice of material is not usual since there are more effective ways to do this. Structural geometry and cross-sectional geometrical properties of members are the most effective controls on structural stiffness.

For illustrative purposes the ultimate strengths and elastic modulus values of some of the more important structural materials are given in Table 2.1. The values quoted in Table 2.1

Table 2.1 Ultimate strength and elastic modulus of some structural materials

Material	Ultimate strength (N/mm^2)	Elastic modulus E (kN/mm^2)
Mild steel	430	205
High-tensile steel	540	205
Aluminium alloy	150–430	70
Reinforced concrete:		
tension	1.5–3.0	25–35
compression	20–50	
Glass reinforced composite	1600	60
Timber	50	7

are representative only and the reader is referred to British Standard specifications for the actual values to be used for a particular material. The *Civil Engineer's Reference Book* is a useful general source of information on material properties.

A comprehensive study of material properties is essential for all structural designers and should include the time-dependent properties of materials where these are significant, as they are with concrete for example, and the weathering and response to environmental attack. Other aspects of material behaviour which are often important are ductility in metal structures where plastic or 'inelastic' behaviour is likely, and behaviour under elevated or reduced temperatures. The degree of damping provided by a structural material in dynamic behaviour is also an important property of the material.

CHAPTER

3

Stress, strain and stress resultants

3.1 Introduction

When external forces are applied to a structure, internal distributions of force are set up. The intensities of these internal forces are called 'stresses' and the complete stress distribution at any point in the structure is described in terms of 'direct' and 'shear' stresses, or the 'components' of stress. Distortion also occurs and this is described in terms of 'strain'. Just as 'stress' is the fundamental description of force intensity at a point in the structure, 'strain' is the fundamental description of distortion. We shall see that we can progress from stress to force by defining 'stress resultants' and evaluating these from a knowledge of stress distributions, and similarly we can use strain distributions in our structure to quantify displacements such as deflections in beams.

In this chapter we look at the fundamentals of strain and stress at a point and develop some important relationships. We then go on to define principal stresses and principal planes. We then look at stress–strain relationships in two and three dimensions, stress transformations and the evaluation of stress resultants.

3.2 Strain at a point

In Fig. 3.1(a), A is a typical point in a structural material in the

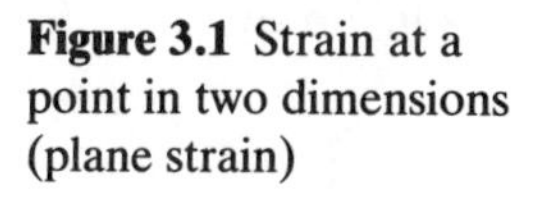
Figure 3.1 Strain at a point in two dimensions (plane strain)

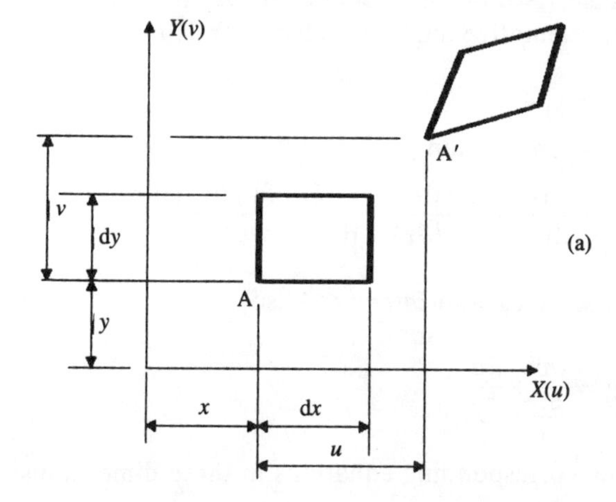

X–Y plane. Under load, point A moves to point A′. We can express the components of strain using purely geometrical considerations. We consider an elemental piece of material $dx \times dy$, for which we can identify three distinct deformations. First, in the X direction the change in length of the element is $(\partial u/\partial x)\,dx$. The strain ε_x in the X direction is defined as the change in length divided by the original length: thus

Figure 3.1 (*continued*)

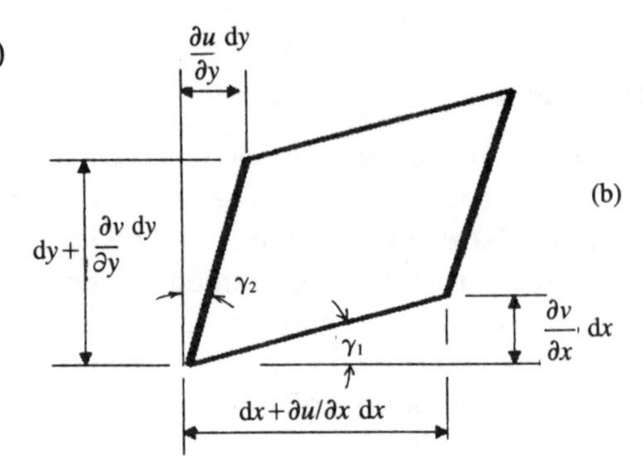

$$\varepsilon_x = \frac{\partial u}{\partial x} \qquad [3.1]$$

Similarly,

$$\varepsilon_y = \frac{\partial v}{\partial y} \qquad [3.2]$$

Now the third type of distortion is the purely angular one shown enlarged in Fig. 3.1(b). We describe this type of distortion as 'shear' strain and define it as the change in value of a representative angle in radians. Thus the shear strain is given by

$$\begin{aligned}\gamma_{xy} &= \gamma_1 + \gamma_2 \\ &\simeq \tan\gamma_1 + \tan\gamma_2 \\ &= \frac{(\partial v/\partial x)\,dx}{dx(1+\partial u/\partial x)} + \frac{(\partial u/\partial y)\,dy}{dy(1+\partial v/\partial y)}\end{aligned}$$

Now $\partial u/\partial x$ and $\partial v/\partial y \ll 1$, so

$$\gamma_{xy} = \frac{\partial v}{\partial x} + \frac{\partial u}{\partial y} \qquad [3.3]$$

The corresponding equations in three dimensions are

$$\left.\begin{aligned}
\varepsilon_x &= \frac{\partial u}{\partial x} \\
\varepsilon_y &= \frac{\partial v}{\partial y} \\
\varepsilon_z &= \frac{\partial w}{\partial z} \\
\gamma_{xy} &= \frac{\partial v}{\partial x} + \frac{\partial u}{\partial y} \\
\gamma_{yz} &= \frac{\partial w}{\partial y} + \frac{\partial v}{\partial z} \\
\gamma_{zx} &= \frac{\partial u}{\partial z} + \frac{\partial w}{\partial x}
\end{aligned}\right\} \qquad [3.4]$$

The six eqns [3.4] express the components of strain at a point as functions of the three linear displacements u, v and w. If these displacements are eliminated from eqns [3.4] by suitable operations of differentiation, we are left with three equations relating derivatives of the strains. These are the 'compatibility' conditions of strain. The actual compatibility equations obtained depend on how the elimination is carried out, but one frequently used set is

$$\left.\begin{aligned}
\frac{\partial^2 \varepsilon_x}{\partial y^2} + \frac{\partial^2 \varepsilon_y}{\partial x^2} - \frac{\partial^2 \gamma_{xy}}{\partial x\, \partial y} &= 0 \\
\frac{\partial^2 \varepsilon_y}{\partial z^2} + \frac{\partial^2 \varepsilon_z}{\partial y^2} - \frac{\partial^2 \gamma_{yz}}{\partial y\, \partial z} &= 0 \\
\frac{\partial^2 \varepsilon_z}{\partial x^2} + \frac{\partial^2 \varepsilon_x}{\partial z^2} - \frac{\partial^2 \gamma_{zx}}{\partial z\, \partial x} &= 0
\end{aligned}\right\} \qquad [3.5]$$

Displacements u, v and w must be such as to produce strains from eqns [3.4] which satisfy eqns [3.5].

3.3 Stress at a point

The stress at a point in a structural material is described by direct and shear stress components σ and τ respectively. Consider an element of material $dx \times dy \times dz$ in the X–Y plane (Fig. 3.2). The normal (direct) stresses are σ_x and σ_y and the shear stresses

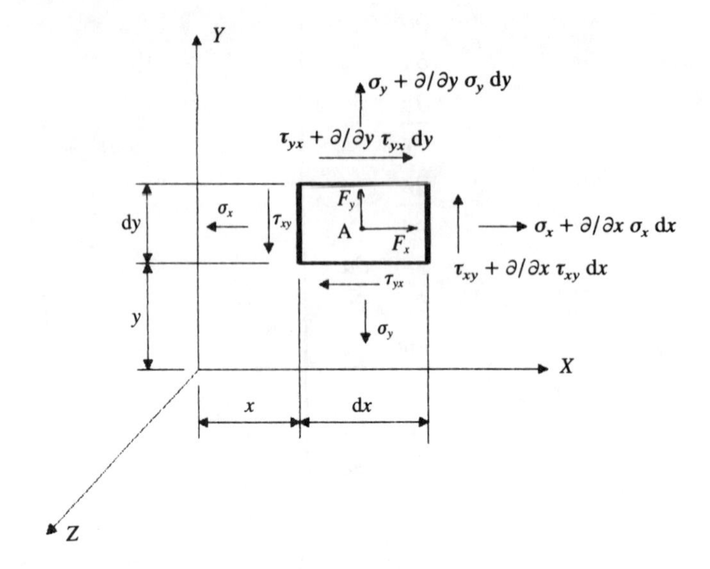

Figure 3.2 Stress at a point

are τ_{xy} and τ_{yx}. We introduce a word of explanation regarding the double subscript convention for shear stress. The shear stress τ_{xy} acts on a surface for which x (first subscript) is constant and in the Y direction (second subscript y). Similarly τ_{yx} acts on a surface for which y is constant in the X direction. F_x and F_y are the 'body' forces (forces per unit volume). For equilibrium in the X direction,

$$\left(\sigma_x + \frac{\partial \sigma_x}{\partial x}\mathrm{d}x\right)\mathrm{d}y\,\mathrm{d}z - \sigma_x\,\mathrm{d}y\,\mathrm{d}z + \left(\tau_{yx} + \frac{\partial \tau_{yx}}{\partial y}\mathrm{d}y\right)\mathrm{d}x\,\mathrm{d}z$$
$$- \tau_{yx}\,\mathrm{d}x\,\mathrm{d}z + F_x\,\mathrm{d}x\,\mathrm{d}y\,\mathrm{d}z = 0$$

That is,

$$\frac{\partial \sigma_x}{\partial x} + \frac{\partial \tau_{yx}}{\partial y} + F_x = 0 \qquad [3.6]$$

Similarly, expressing equilibrium in the Y direction, we obtain

$$\frac{\partial \sigma_y}{\partial y} + \frac{\partial \tau_{xy}}{\partial x} + F_y = 0 \qquad [3.7]$$

Equations [3.6] and [3.7] are the equations of linear equilibrium in two dimensions. If we now take moments about A,

$$\tau_{yx}\,\mathrm{d}x\,\mathrm{d}z\frac{\mathrm{d}y}{2} + \left(\tau_{yx} + \frac{\partial \tau_{yx}}{\partial y}\mathrm{d}y\right)\mathrm{d}x\,\mathrm{d}z\frac{\mathrm{d}y}{2}$$
$$= \tau_{xy}\,\mathrm{d}y\,\mathrm{d}z\frac{\mathrm{d}x}{2} + \left(\tau_{xy} + \frac{\partial \tau_{xy}}{\partial x}\mathrm{d}x\right)\mathrm{d}y\,\mathrm{d}z\frac{\mathrm{d}x}{2}$$

which, in the limit as $dx \to 0$ and $dy \to 0$, becomes

$$\tau_{yx} = \tau_{xy} \tag{3.8}$$

This is the condition for rotational equilibrium in plane stress. The two equal shear stresses τ_{yx} and τ_{xy} are called 'complementary' shear stresses.

Equations [3.6], [3.7] and [3.8] are the stress equilibrium equations in plane stress and are independent of z. In three-dimensional stress the corresponding equations are

$$\left.\begin{aligned}
&\frac{\partial \sigma_x}{\partial x} + \frac{\partial \tau_{yx}}{\partial y} + \frac{\partial \tau_{zx}}{\partial z} + F_x = 0 \\
&\frac{\partial \sigma_y}{\partial y} + \frac{\partial \tau_{zy}}{\partial z} + \frac{\partial \tau_{xy}}{\partial x} + F_y = 0 \\
&\frac{\partial \sigma_z}{\partial z} + \frac{\partial \tau_{xz}}{\partial x} + \frac{\partial \tau_{yz}}{\partial y} + F_z = 0 \\
&\tau_{xy} = \tau_{yx} \\
&\tau_{yz} = \tau_{zy} \\
&\tau_{zx} = \tau_{xz}
\end{aligned}\right\} \tag{3.9}$$

3.4 Transformation of stress: principal stresses

Figure 3.3 shows an elementary block of material of thickness t

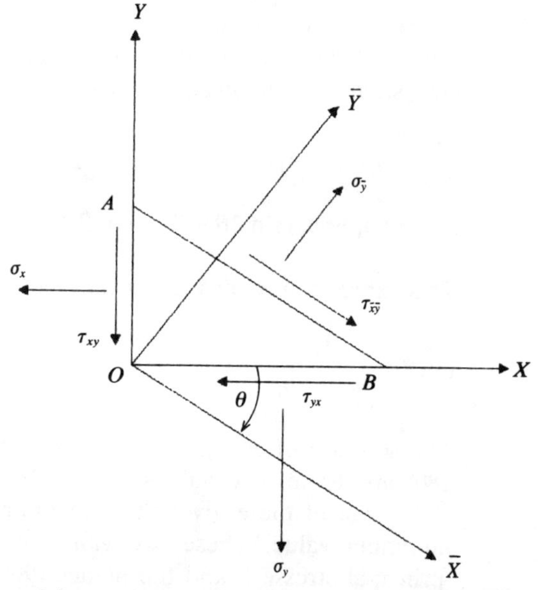

Figure 3.3
Transformation of stress in two dimensions

subject to normal stresses σ_x and σ_y and shear stresses τ_{xy} and τ_{yx}. If the X axis is rotated to $\bar{X}$ we can resolve forces to find the normal and tangential stresses on the plane $O\bar{X}$. Resolving forces in the direction of $\sigma_{\bar{y}}$, we obtain

$$\sigma_{\bar{y}}(AB)t = \sigma_x(AO)t \sin\theta + \sigma_y(BO)t \cos\theta + \tau_{xy}(AO)t \cos\theta + \tau_{yx}(BO)t \sin\theta$$

Substituting $AO = AB \sin\theta$ and $BO = AB \cos\theta$, then

$$\sigma_{\bar{y}} = \sigma_x \sin^2\theta + \sigma_y \cos^2\theta + \tau_{xy} \sin\theta \cos\theta + \tau_{yx} \sin\theta \cos\theta$$

and since $\tau_{xy} = \tau_{yx}$,

$$\sigma_{\bar{y}} = \sigma_x \sin^2\theta + \sigma_y \cos^2\theta + \tau_{xy} \sin 2\theta \qquad [3.10]$$

Now resolving forces parallel to AB,

$$\tau_{\bar{y}\bar{x}}(AB)t = \sigma_x(AO)t \cos\theta - \sigma_y(BO)t \sin\theta - \tau_{xy}(AO)t \sin\theta + \tau_{yx}(BO)t \cos\theta$$

$$\tau_{\bar{y}\bar{x}} = \frac{\sigma_x - \sigma_y}{2} \sin 2\theta + \tau_{xy} \cos 2\theta \qquad [3.11]$$

Similarly we could resolve forces on a plane at right angles to AB to obtain the corresponding expressions for $\sigma_{\bar{x}}$ and $\tau_{\bar{x}\bar{y}}$. However, we do not need to do this since the results can be deduced from eqns [3.10] and [3.11].

For given values of σ_x, σ_y and τ_{xy}, the stress $\sigma_{\bar{y}}$ is a function only of θ. Thus we can differentiate eqn [3.10] to find the angle θ consistent with the maximum value of $\sigma_{\bar{y}}$:

$$\frac{d\sigma_{\bar{y}}}{d\theta} = \sigma_x(\sin 2\theta) + \sigma_y(-\sin 2\theta) + 2\tau_{xy} \cos 2\theta$$

$$= (\sigma_x - \sigma_y) \sin 2\theta + 2\tau_{xy} \cos 2\theta$$

This expression is zero for turning points in $\sigma_{\bar{y}}$, and hence

$$\tan 2\theta = \frac{2\tau_{xy}}{\sigma_y - \sigma_x} \qquad [3.12]$$

The solution of eqn [3.12] consists of two values of angle 2θ at 180° apart, i.e. two values of θ at 90° apart in the range 0° to 180°. One of these gives the maximum value of $\sigma_{\bar{y}}$ and one the minimum value. These two values of normal stress are called 'principal stresses', and the planes (90° apart) on which they act are called 'principal planes'.

Substituting for 2θ from eqn [3.12] in eqn [3.11] we obtain

$$\frac{\tau_{\bar{y}\bar{x}}}{\cos 2\theta} = \frac{\sigma_x - \sigma_y}{2} \tan 2\theta + \tau_{xy}$$

$$= \frac{\sigma_x - \sigma_y}{2} \frac{2\tau_{xy}}{\sigma_y - \sigma_x} + \tau_{xy}$$

$$= 0$$

We thus obtain the important result that the shear stress is zero on principal planes. If we substitute the value of 2θ corresponding to principal planes in eqn [3.10] we can obtain expressions for the principal stresses:

$$\sigma_P = \sigma_x \frac{1}{2}(1 - \cos 2\theta) + \sigma_y \frac{1}{2}(1 + \cos 2\theta) + \tau_{xy} \sin 2\theta$$

$$\left(\sigma_P - \frac{\sigma_x + \sigma_y}{2}\right)^2 = \left(\frac{\sigma_x - \sigma_y}{2}\right)^2 + \tau_{xy}^2$$

or

$$\sigma_P = \frac{\sigma_x + \sigma_y}{2} \pm \sqrt{\left[\left(\frac{\sigma_x - \sigma_y}{2}\right)^2 + \tau_{xy}^2\right]} \qquad [3.13]$$

The two values of σ_P resulting from eqn [3.13] are the principal stresses:

$$\sigma_{P1} = \frac{\sigma_x + \sigma_y}{2} + \sqrt{\left[\left(\frac{\sigma_x - \sigma_y}{2}\right)^2 + \tau_{xy}^2\right]}$$

$$\sigma_{P2} = \frac{\sigma_x + \sigma_y}{2} - \sqrt{\left[\left(\frac{\sigma_x - \sigma_y}{2}\right)^2 + \tau_{xy}^2\right]} \qquad [3.14]$$

If we require values for the two principal stresses we can either use eqn [3.14] directly or we can determine the principal planes using eqn [3.12] and then substitute two values of θ (90° apart) in eqn [3.10] to obtain the principal stresses. The latter method is preferred since it enables the principal stresses to be related to corresponding principal planes.

We can rewrite eqn [3.10] in the forms

$$\sigma_{P(\theta)} = \sigma_x \sin^2 \theta + \sigma_y \cos^2 \theta + \tau_{xy} \sin 2\theta \qquad [3.15]$$

and

$$\sigma_{P(\theta+90°)} = \sigma_x \sin^2 (\theta + 90°) + \sigma_y \cos^2 (\theta + 90°) + \tau_{xy} \sin 2(\theta + 90°) \qquad [3.16]$$

Then eqn [3.15] gives the principal stress on $O\bar{X}$ obtained by rotating OX to $O\bar{X}$ through angle θ (Fig. 3.3), and eqn [3.16] gives the principal stress on a plane obtained by rotating OX through $(\theta + 90°)$.

The equivalent transformations in three dimensions can be obtained in a similar way by considering equilibrium of a block of material in the form of a tetrahedron. The resulting expressions are more complex.

Example 3.1

At a point in a block of material, given that $\sigma_x = 0$, $\sigma_y = 35\,\text{N/mm}^2$ and $\tau_{xy} = 30\,\text{N/mm}^2$, we wish to find the principal planes and principal stresses. From eqn [3.12],

$$\tan 2\theta = \frac{2 \times 30}{35 - 0} = 1.714$$

$$2\theta = 59.74°$$

$$\theta = 29.87°$$

$$\sigma_{P(\theta)} = 35\cos^2(29.87°) + 30\sin(59.74°)$$

$$= 52.23\,\text{N/mm}^2$$

$$\sigma_{P(\theta+90°)} = 35\cos^2(119.87°) + 30\sin(239.74°)$$

$$= -17.23\,\text{N/mm}^2$$

The data and the results are shown in Fig. 3.4.

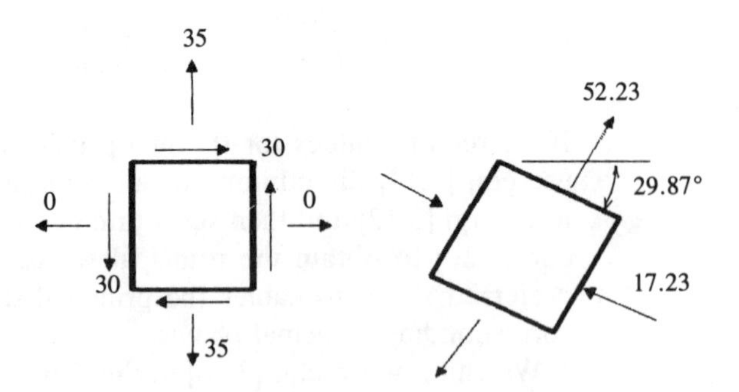

Figure 3.4 Example 3.1

3.5 Stress–strain relationships

The relationships between stresses and strains are properties of the material and reflect the assumptions made regarding the type of material. It is usual to assume that the material is homogeneous, isotropic and linearly elastic for the purposes of setting up the fundamental relationships. In homogeneous materials, properties do not vary from point to point. In isotropic materials, properties do not vary with direction within the material. Elastic materials are capable of full recovery to the unstrained state on removal of load. It is further assumed that the deformations are sufficiently small to leave the geometry of the structure unchanged. It is immediately apparent that many engineering materials do not satisfy all these conditions. In these circumstances the stress–strain relationships will be changed from what follows. For a full discussion of this topic the reader is referred to books on strength of materials such as that by Budynas (1977).

Consider a simple one-dimensional bar to which we apply a force N in the axial direction. The stress in the bar will be

$$\sigma_x = N/A \qquad [3.17]$$

where A is the cross-sectional area of the bar. If the bar has a length L and if the force N produces an extension δ, then the strain in the bar is

$$\varepsilon_x = \delta/L$$

Now this is the strain in the axial direction X, but the bar will exhibit lateral strains (Y and Z directions) since there will be a transverse contraction of the material. The lateral strains will be

$$\varepsilon_y = \varepsilon_z = -\nu\varepsilon_x \qquad [3.18]$$

where ν is Poisson's ratio, which is assumed to be the same in tension and compression.

Now the ratio of stress to strain is E, the modulus of elasticity (Young's modulus). This is constant for the material and is sensibly the same in tension and compression. The units are force per unit area. Thus due to σ_x alone,

$$\varepsilon_x = \sigma_x/E \qquad [3.19]$$

If we now imagine that we have an elemental block of material (Fig. 3.5) which is subjected to stresses of σ_x, σ_y and σ_z in the direction of the axes X, Y and Z, then each stress will produce extension along its own axis and pure contraction in the direction of the other two axes. Thus no shear strains will be developed by σ_x, σ_y or σ_z or, consequently, by all three together. We can make a similar observation about the shear stresses and conclude that they will produce only shear strains. The direct and shear strains are said to be 'uncoupled', a result of the assumed isotropy of the material.

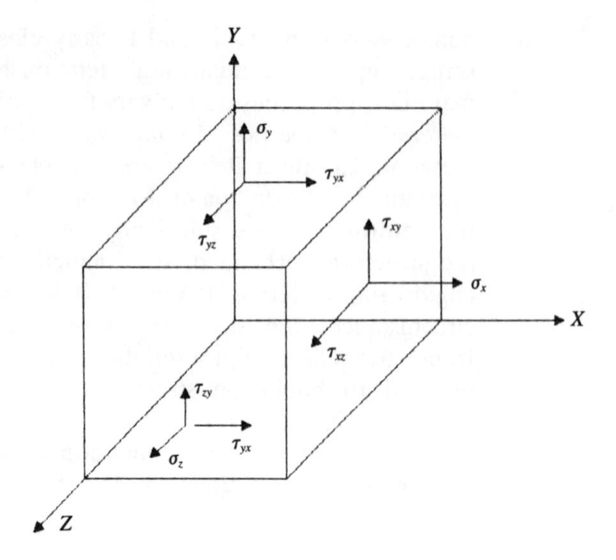

Figure 3.5 Three-dimensional stresses

Now the total strain in the X direction is, from eqns [3.18] and [3.19],

$$\left.\begin{aligned} \varepsilon_x &= \frac{1}{E}[\sigma_x - \nu(\sigma_y + \sigma_z)] \\ \varepsilon_y &= \frac{1}{E}[\sigma_y - \nu(\sigma_z + \sigma_x)] \\ \varepsilon_z &= \frac{1}{E}[\sigma_z - \nu(\sigma_x + \sigma_y)] \end{aligned}\right\} \quad [3.20]$$

If we now consider the shear strains produced by the shear stresses, we have

$$\left.\begin{aligned} \gamma_{xy} &= \tau_{xy}/G \\ \gamma_{yz} &= \tau_{yz}/G \\ \gamma_{zx} &= \tau_{zx}/G \end{aligned}\right\} \quad [3.21]$$

where G is the 'shear modulus' or 'modulus of rigidity' and is related to E and ν by

$$G = \frac{E}{2(1+\nu)} \quad [3.22]$$

Then we can collect the results for all six strains and corresponding stresses using eqns [3.20], [3.21] and [3.22]:

$$\begin{bmatrix} \varepsilon_x \\ \varepsilon_y \\ \varepsilon_z \\ \hline \gamma_{xy} \\ \gamma_{yz} \\ \gamma_{zx} \end{bmatrix} = \frac{1}{E} \left[\begin{array}{ccc|ccc} 1 & -\nu & -\nu & & & \\ -\nu & 1 & -\nu & & 0 & \\ -\nu & -\nu & 1 & & & \\ \hline & & & 2(1+\nu) & 0 & 0 \\ & 0 & & 0 & 2(1+\nu) & 0 \\ & & & 0 & 0 & 2(1+\nu) \end{array}\right] \begin{bmatrix} \sigma_x \\ \sigma_y \\ \sigma_z \\ \hline \tau_{xy} \\ \tau_{yz} \\ \tau_{zx} \end{bmatrix} \qquad [3.23]$$

It will be found more useful in the future to express eqn [3.23] in the inverse form, so solving for the stresses we obtain

$$\begin{bmatrix} \sigma_x \\ \sigma_y \\ \sigma_z \\ \hline \tau_{xy} \\ \tau_{yz} \\ \tau_{zx} \end{bmatrix} = \frac{E}{(1+\nu)(1-2\nu)} \left[\begin{array}{ccc|ccc} (1-\nu) & \nu & \nu & & & \\ \nu & (1-\nu) & \nu & & 0 & \\ \nu & \nu & (1-\nu) & & & \\ \hline & & & \frac{(1-2\nu)}{2} & 0 & 0 \\ & 0 & & 0 & \frac{(1-2\nu)}{2} & 0 \\ & & & 0 & 0 & \frac{(1-2\nu)}{2} \end{array}\right] \begin{bmatrix} \varepsilon_x \\ \varepsilon_y \\ \varepsilon_z \\ \hline \gamma_{xy} \\ \gamma_{yz} \\ \gamma_{zx} \end{bmatrix} \qquad [3.24]$$

Equation [3.24] is the general three-dimensional stress–strain relationship for a homogeneous, isotropic, elastic material. The condensed form of the stress–strain relationship in the X–Y plane is

$$\begin{bmatrix} \sigma_x \\ \sigma_y \\ \tau_{xy} \end{bmatrix} = \frac{E}{(1-\nu^2)} \begin{bmatrix} 1 & \nu & 0 \\ \nu & 1 & 0 \\ 0 & 0 & \frac{1}{2}(1-\nu) \end{bmatrix} \begin{bmatrix} \varepsilon_x \\ \varepsilon_y \\ \gamma_{xy} \end{bmatrix} \qquad [3.25]$$

3.6 Stress resultants

The stress distribution over a cross-section of a structural member can be described in terms of the equivalent generalized forces or 'stress resultants'. In a general three-dimensional situation (Fig. 3.6) we can define a total of six such forces at a typical point. With X, Y and Z axes directed as shown in diagram (a), we can identify three forces in the directions of the axes and three moments about the axes. These are labelled N, S_y, and S_z in the X, Y and Z directions respectively, and M_x, M_y and M_z about the axes X, Y and Z respectively. We can collect these forces into a vector

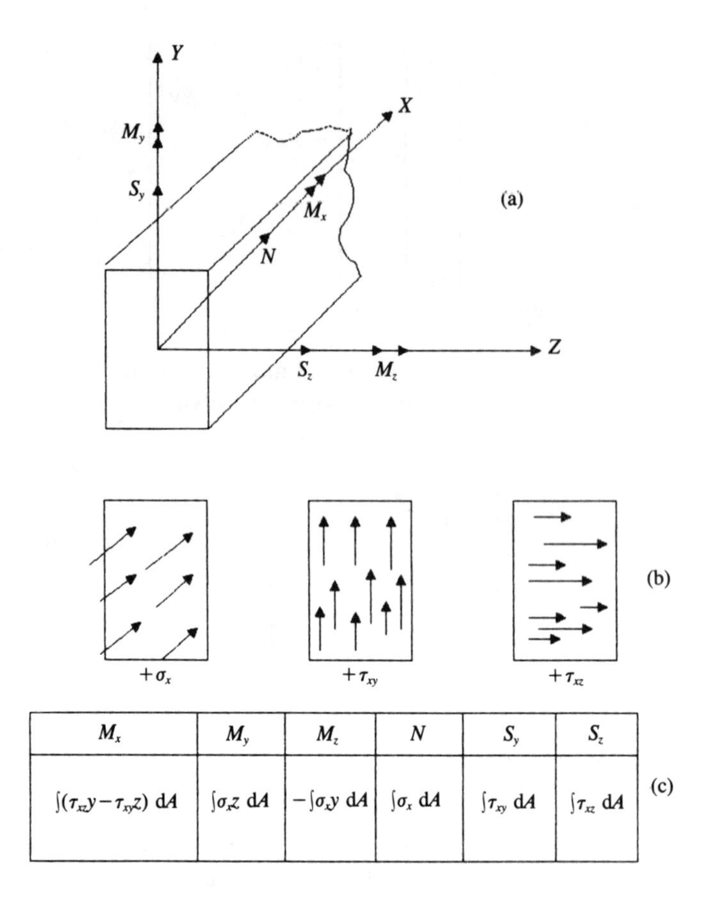

M_x	M_y	M_z	N	S_y	S_z
$\int(\tau_{xz}y-\tau_{xy}z)\,dA$	$\int\sigma_x z\,dA$	$-\int\sigma_x y\,dA$	$\int\sigma_x\,dA$	$\int\tau_{xy}\,dA$	$\int\tau_{xz}\,dA$

Figure 3.6 Coordinate axes, sign conventions and stress resultants

$$\boldsymbol{S}=\begin{bmatrix} M_x \\ M_y \\ M_z \\ N \\ S_y \\ S_z \end{bmatrix} \qquad [3.26]$$

Some of these stress resultants are easily recognized: N is the thrust and M_x the twisting moment; S_y and S_z are the shear forces in the plane of the cross-section (Y–Z plane); and M_y and M_z are the bending moments about axes Y and Z respectively. It is conventional, as we have already observed, to adopt the X axis as coincident with the longitudinal axis of the member. The sign convention needs careful interpretation; N, S_y and S_z are positive if acting in the *positive* directions of the corresponding axes. The moments M_x, M_y and M_z are positive if acting clockwise when viewed in the *positive* direction of the respective axis. The sign convention for the moments needs care when the relevant axis is directed towards the viewer.

We now consider the stresses on the typical cross-section. There are two types: normal stress σ and shear stress τ. Whilst there is usually only one normal stress σ_x in a linear structural member, there are two shear stresses, τ_{xy} in the Y direction and τ_{xz} in the Z direction. The positive directions of these three stresses are shown in Fig. 3.6(b).

It should now be a straightforward matter to convert the stress distribution into stress resultants by carrying out area integrations as set out in Fig. 3.6(c).

The situation can be simplified for a member of a plane frame. For a structural member lying in the X–Y plane subjected to loads in that plane only, we can take $M_y = 0$, $S_z = 0$ and $M_x = 0$. The stress resultant vector will be

$$\boldsymbol{S} = \begin{bmatrix} M_z \\ N \\ S_y \end{bmatrix} \qquad [3.27]$$

Example 3.2

A structural member has the cross-section shown in Fig. 3.7(a). It is found that the normal stress on the cross-section varies uniformly from 26.35 N/mm² compression, constant along the edge AB, to 4.50 N/mm² tension, constant along the edge CD. Determine the stress resultants N, M_z and M_y.

Figure 3.7 Example 3.2

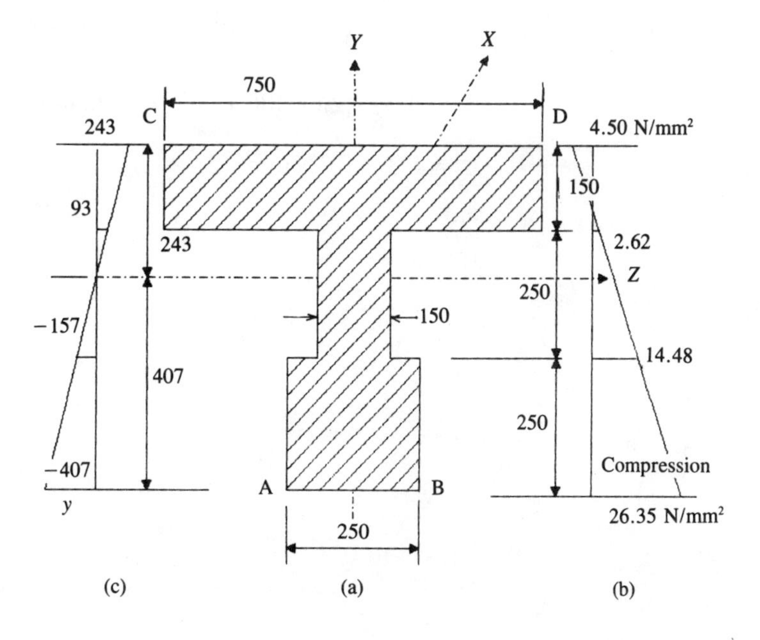

$$\sigma_x \text{ (at centroid)} = (26.35 + 4.50)\frac{243}{650} - 4.50$$
$$= 7.033 \text{ N/mm}^2$$

Therefore

$$\sigma_x = f(y) = 7.033 - \frac{(26.35 + 4.50)y}{650}$$
$$= 7.033 - 0.04\,746y$$

Referring to Fig. 3.6, the normal stress resultant is,

$$N = \int \sigma_x \, dA = \int (7.033 - 0.04\,746y)\, dA$$

This integral is easily evaluated using the table of product integrals in Appendix 4. Hence

$$N = \frac{250 \times 250}{2}(26.35 + 14.48) + \frac{250 \times 150}{2}(14.48 + 2.62)$$
$$+ \frac{750 \times 150}{2}(2.62 - 4.50)$$
$$= 1500 \text{ kN}$$

Now

$$M_z = -\int \sigma_x y \, dA = -\int \sigma_x yb \, dy$$
$$= \left\{\frac{250}{6}[26.35(-814 - 157) + 14.48(-314 - 407)]250\right.$$
$$+ \frac{250}{6}[14.48(-314 + 93) + 2.62(186 - 157)]150$$
$$\left. + \frac{150}{6}[2.62(186 + 243) - 4.50(486 + 93)]750\right\}$$
$$= 423 \times 10^6 \text{ N mm}$$

and

$$M_y = 0$$

CHAPTER

4

Basic concepts and structural theorems

4.1 Introduction

The purpose of structural analysis is to determine the distribution of internal forces and displacements in structures. Analysis may be necessary in order to verify the adequacy of a structural design and to assess the behaviour of the structure under load, or it may be more closely integrated with the design process. It is necessary to analyse structures in order to obtain a general understanding of their behaviour in practice. Precise analysis is seldom required; indeed, many structures are of a complexity which does not permit precise analysis. Generally speaking an 'engineering' approach is required in which reasonable assumptions and approximations are made. Consequently a thorough understanding of fundamental principles is essential so that the validity of a particular analysis can be assessed.

It is convenient to regard structures as assemblages of finite or discrete parts called elements. In framed structures the elements are usually the individual members, although it is sometimes useful to treat parts of members as discrete elements. The elements are connected at the nodes or joints. In the case of a framed structure the nodes are 'real', but in the case of a plate which has been internally subdivided into (imaginary) elements the nodes are imaginary. The fundamental basis of the stiffness method is the equilibrium of the structure including the equilibrium of each element of which the structure is composed.

There is a need to develop compact methods of describing the structural properties of the elements since some structures will contain large numbers of elements (Desai 1979; Cheung, Yeo 1979; Elias 1986). For this reason, matrix methods have been extensively adopted in structural analysis. Furthermore it is found that the various manipulations of the quantities are frequently standard matrix operations. The ability to follow matrix expressions and handle matrix operations is essential to the structural analyst.

The purpose of this chapter is to state and illustrate the fundamental principles of equilibrium of forces and compatibility of displacements and to state the basic structural theorems used to establish a theoretical numerical model of a structure. Once this has been done, the work of analysis is completed by solving a set of simultaneous equations, and we can now assume that this will always be done by a computer. Indeed we shall see that virtually the whole process of analysis can be carried out on the computer.

We shall need to become familiar with the handling of systems of displacements and systems of forces in structures. We shall need to take care with the ordering of the individual displacements and forces so that they are mutually consistent. This is necessary because much of our time will be spent manipulating ordered arrays of numbers. It is very probable that we shall meet difficulties with sign conventions and their interpretation. These difficulties will only be overcome with diligence and practice. More of this later!

We start with the basic concepts of flexibility and stiffness. The sooner we become thoroughly familiar with these words the better.

4.2 Flexibility and stiffness of a structure

Before attempting a more rigorous and general treatment of these concepts we shall look at a very simple structure in order to introduce the ideas. Consider the two-bar structure ABC shown in Fig. 4.1(a). If we only wanted to know the forces in the

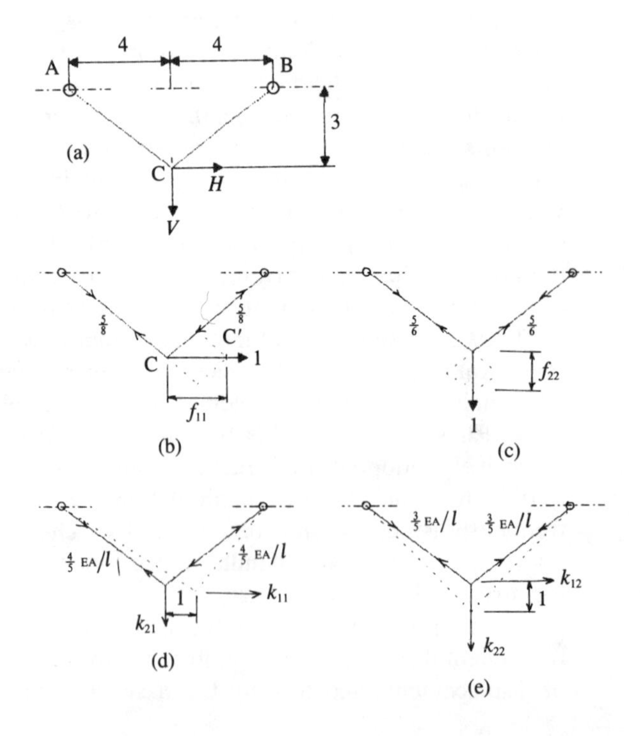

Figure 4.1 Flexibilities and stiffnesses of simple structure

members due to the applied loads H and V we could resolve forces at C both horizontally and vertically and obtain

$$P_{AC} = \frac{20V + 15H}{24} \quad \text{and} \quad P_{BC} = \frac{20V - 15H}{24}$$

If, in addition, we wish to know the horizontal and vertical components of the displacement at C, then this would involve a little more work but would not be difficult to do from first principles. However, the purpose of this exercise is to introduce a more general approach yielding both member forces and nodal displacements using the concepts of flexibility and stiffness. We first of all remove the applied loads since these will not affect the mathematical description of the structure we are going to set up. Now, 'flexibility' is defined as a displacement at some point in a structure caused by a unit force at the same point or at some other point. At these specified points in the structure – the nodes – we define 'coordinates', being the directions in which we will study forces and displacements. If the coordinate at which the flexibility is defined is i and the coordinate at which the unit force is applied is j then the flexibility is f_{ij}. If we consider just the two coordinates i and j, then this gives rise to four flexibilities f_{ii}, f_{jj}, f_{ij} and f_{ji}. The first two are called 'direct' flexibilities, since the displacement and the unit force are at the same coordinate; whereas the second two are called 'cross' flexibilities, the displacement and force being at different coordinates. We shall show later that $f_{ij} = f_{ji}$. It is important to know that both coordinates i and j could be defined at the same node in the structure – one might be horizontal and the other vertical, for example – or they could be defined at different nodes.

Stiffness is defined in a similar way to flexibility, with 'force' and 'displacement' interchanged throughout. Thus a stiffness k_{ij} is the force at coordinate i corresponding to a unit displacement at coordinate j. Again we have, with two coordinates, four stiffnesses k_{ii}, k_{jj}, k_{ij} and k_{ji}, the first two being 'direct' stiffnesses and the second two 'cross' stiffnesses.

Let us now turn to the structure of Fig. 4.1(a) and consider first of all the flexibilities at joint C. We consider both horizontal and vertical coordinates. Quite arbitrarily we number the two coordinates in the order: horizontal 1 and vertical 2. We apply unit force at coordinate 1 as in Fig. 4.1(b). The member forces are obtained by resolving horizontally at C and noting that, for vertical equilibrium, the forces will be numerically equal and opposite in sign. Knowing the forces in the members we can evaluate the changes in length as $(5/8)(l/EA)$, and from geometrical considerations the horizontal displacement of C is

5/4 times the member extension. Thus $f_{11} = (25/32)(l/EA)$. With the member forces equal and opposite, it is clear that there is no vertical displacement of C under unit horizontal load; hence $f_{21} = 0$. A similar analysis is carried out for a unit vertical load, and we obtain $f_{22} = (25/18)(l/EA)$ and $f_{12} = 0$ (due to symmetry). In passing, we note that $f_{12} = f_{21}$. Thus we can gather these flexibilities into a matrix:

$$\boldsymbol{F} = \begin{bmatrix} f_{11} & f_{12} \\ f_{21} & f_{22} \end{bmatrix} = \frac{l}{EA} \begin{bmatrix} 25/32 & 0 \\ 0 & 25/18 \end{bmatrix} \qquad [4.1]$$

We pause now to explain how matrix $\boldsymbol{F}$ can be used in practice. The matrix is a statement of the flexibility of the structure, in this case focused on the only free node C. Since the flexibilities correspond to unit loads in the coordinate directions, if we multiply by the actual loads we obtain the displacements

$$\boldsymbol{F} \begin{bmatrix} H \\ V \end{bmatrix} = \begin{bmatrix} \Delta_H \\ \Delta_V \end{bmatrix} \qquad [4.2]$$

from which

$$\left.\begin{aligned} \Delta_H &= \frac{25}{32} \frac{l}{EA} H \\ \Delta_V &= \frac{25}{18} \frac{l}{EA} V \end{aligned}\right\} \qquad [4.3]$$

It is worth noting that Δ_H is independent of V and Δ_V is independent of H, owing to the diagonal nature of the matrix $\boldsymbol{F}$. This will not generally be so.

We shall now carry out a similar exercise to obtain the stiffness matrix. We impose unit displacements at C, horizontally in Fig. 4.1(d) and vertically in (e). The member forces are now obtained by using the geometry to evaluate the changes in length and then converting this to a force. For example the change in length of both members in (d) is $1 \times 4/5$ and therefore the forces are $(4/5)(EA/l)$, tension in AC and compression in BC. Now we seek the forces, horizontally and vertically, at C consistent with this imposed displacement; these are the stiffnesses

$$k_{11} = 2 \times \frac{4}{5} \frac{EA}{l} \times \frac{4}{5} = \frac{32}{25} \frac{EA}{l}$$

$$k_{21} = 0 \qquad \text{since } P_{AC} = -P_{BC}$$

Similarly, in (e) we impose a unit vertical displacement at C and obtain

$$k_{22} = \frac{18}{25}\frac{EA}{l}$$

$k_{12} = 0$ by symmetry

Thus

$$\boldsymbol{K} = \begin{bmatrix} k_{11} & k_{12} \\ k_{21} & k_{22} \end{bmatrix} = \frac{EA}{l}\begin{bmatrix} 32/25 & 0 \\ 0 & 18/25 \end{bmatrix} \qquad [4.4]$$

If we examine $\boldsymbol{F}$ and $\boldsymbol{K}$ we see that each is the inverse of the other, i.e.

$$\boldsymbol{K} = \boldsymbol{F}^{-1} \qquad [4.5]$$

This is not a surprising result if we trace through the operations we have performed to arrive at $\boldsymbol{F}$ and $\boldsymbol{K}$. However, the result has considerable significance for our future work since it gives us an indirect method of finding $\boldsymbol{K}$ if $\boldsymbol{F}$ is easier to obtain, as it sometimes is.

Before moving on it would be useful to review what we have just done in evaluating flexibilities (displacements) and stiffnesses (forces) by examining the geometry of the member extensions and the equilibrium of the forces at C respectively. In both cases we worked from first principles, but with more complex structures we shall use a more efficient approach in which the resolution of forces and displacements is carried out by the process of matrix transformation, which we introduced in section 1.2. It is also worth noting that since the structure is statically determinate we were able to deduce member forces in Fig. 4.1(b) and (c) directly from equilibrium considerations at C. With statically indeterminate structures this will not be possible and we shall require some further analysis involving the introduction of principles of compatibility of displacements.

4.3 Virtual work and the unit load and displacement theorems

The principle of virtual work states that an elastic structure is in equilibrium under a given set of n forces $\boldsymbol{P}$ if the total work done by the forces is zero for any geometrically compatible set of virtual displacements $\bar{\boldsymbol{\Delta}}$. A convenient mathematical expression of this principle is

$$\sum_{1}^{n} P_i\bar{\Delta}_i = \int \boldsymbol{\sigma}^{\mathrm{T}}\bar{\boldsymbol{\varepsilon}}\, \mathrm{d}vol \qquad [4.6]$$

where

P_i is the ith force of a set of actual forces
$\bar{\Delta}_i$ is the ith displacement of a set of virtual displacements, the n coordinates of the set corresponding to those of $\boldsymbol{P}$; however, the choice of $\bar{\boldsymbol{\Delta}}$ is arbitrary providing the displacements form a geometrically compatible set
$\boldsymbol{\sigma}$ is a column matrix of actual stresses corresponding to $\boldsymbol{P}$
$\bar{\boldsymbol{\varepsilon}}$ is a column matrix of virtual strains corresponding to $\bar{\boldsymbol{\Delta}}$

If $\bar{\Delta}_i$ is successively chosen so that there is a single non-zero displacement of unit value at each of the coordinates i in turn, then eqn [4.6] becomes

$$P_i = \int \boldsymbol{\sigma}^{\mathrm{T}} \bar{\boldsymbol{\varepsilon}}_i \, \mathrm{d}vol \qquad [4.7]$$

where $\bar{\boldsymbol{\varepsilon}}_i$ is a matrix of strains corresponding to each virtual unit displacement. This is the unit displacement theorem, which is used in the calculation of stiffnesses of structural elements.

Alternatively we can express the principle of virtual work in a different way and interpret $\boldsymbol{P}$ as a set of virtual forces $\bar{\boldsymbol{P}}$ with $\bar{\boldsymbol{\sigma}}$ the corresponding virtual stresses, and then $\boldsymbol{\Delta}$ is interpreted as a set of actual displacements with $\boldsymbol{\varepsilon}$ the corresponding actual strains. In this case the forces $\bar{\boldsymbol{P}}$ are virtual and arbitrary providing they form an equilibrium set. In particular we may choose a unit force at each coordinate in turn, the other forces being zero. We then obtain, by substitution in eqn [4.6],

$$\Delta_i = \int \bar{\boldsymbol{\sigma}}_i^{\mathrm{T}} \boldsymbol{\varepsilon} \, \mathrm{d}vol \qquad [4.8]$$

where $\bar{\boldsymbol{\sigma}}_i$ is now a matrix of virtual stresses corresponding to the n applications of unit force in place of $\boldsymbol{P}$. This is now the unit load theorem.

4.4 The flexibility method

Having introduced the fundamental structural theorems needed, we shall now look at the two basic approaches to structural analysis: the flexibility and the stiffness methods. We shall look first at the flexibility method. We try to make the treatment as concise as possible, since we do not want to dwell too long on this method in view of the importance we must attach to the stiffness method.

4.4.1 *Application of the unit load theorem to obtain element flexibilities*

We define a set of member forces for a structural element as $\boldsymbol{S}$ and the corresponding displacements at coordinates as $\boldsymbol{s}$. Applying the unit load theorem, the typical ith displacement is

$$s_i = \int \bar{\boldsymbol{\sigma}}_i^{\mathrm{T}} \boldsymbol{\varepsilon} \, \mathrm{d}vol \qquad [4.9]$$

where $\bar{\boldsymbol{\sigma}}_i$ are the stresses due to unit load in place of S_i, and $\boldsymbol{\varepsilon}$ are the strains due to all forces $\boldsymbol{S}$. We can therefore express all the displacements in a matrix equation

$$\boldsymbol{s} = \int \bar{\boldsymbol{\sigma}}^{\mathrm{T}} \boldsymbol{\varepsilon} \, \mathrm{d}vol \qquad [4.10]$$

Now if the behaviour is assumed to be linear then the stresses are proportional to the forces:

$$\begin{aligned} \boldsymbol{\sigma} &= \boldsymbol{CS} \\ &= \boldsymbol{C} \qquad \text{for unit loads in place of } \boldsymbol{S} \end{aligned} \qquad [4.11]$$

Now

$$\boldsymbol{\sigma} = \boldsymbol{D\varepsilon} \qquad [4.12]$$

where $\boldsymbol{D}$ is the ‘elasticity’ matrix. Hence

$$\boldsymbol{\varepsilon} = \boldsymbol{D}^{-1}\boldsymbol{\sigma} = \boldsymbol{D}^{-1}\boldsymbol{CS}$$

and

$$\boldsymbol{s} = \int \bar{\boldsymbol{\sigma}}^{\mathrm{T}} \boldsymbol{D}^{-1} \boldsymbol{\sigma} \, \mathrm{d}vol \qquad [4.13]$$

$$= \int \boldsymbol{C}^{\mathrm{T}} \boldsymbol{D}^{-1} \boldsymbol{CS} \, \mathrm{d}vol \qquad [4.14]$$

$$= \left(\int \boldsymbol{C}^{\mathrm{T}} \boldsymbol{D}^{-1} \boldsymbol{C} \, \mathrm{d}vol \right) \boldsymbol{S} \qquad [4.15]$$

Now, by definition, displacement = flexibility × force. Hence

$$\boldsymbol{s} = \boldsymbol{fS} \qquad [4.16]$$

and therefore

$$\boldsymbol{f} = \int \boldsymbol{C}^{\mathrm{T}} \boldsymbol{D}^{-1} \boldsymbol{C} \, \mathrm{d}vol \qquad [4.17]$$

Example 4.1

For the structure shown in Fig. 4.2 we remove the rigid body axial displacement by putting either $s_1 = 0$ or $s_2 = 0$. Say we put $s_1 = 0$; then $s_2 = s$, $S_2 = S$ and $S_1 = -S$. Therefore $\sigma = CS = S/A = 1/A$ for unit S. Substituting in eqn [4.17],

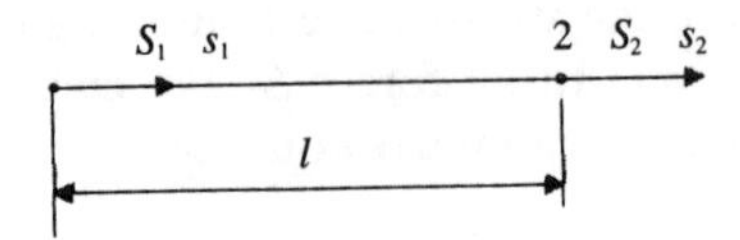

Figure 4.2 Example 4.1

$$f = \int 1/A \;\; 1/E \;\; 1/A \, dvol = l/EA$$

4.4.2 Flexibilities using Castigliano's second theorem

A more convenient method of obtaining flexibilities of individual elements of a structure is to use Castigliano's second theorem, which states that the partial derivative of the total strain energy of a structure with respect to an applied force P_i gives the displacement of that force:

$$\Delta_i = \frac{\partial U}{\partial P_i} \qquad [4.18]$$

We illustrate this approach by considering a beam element of length l with applied bending moments M_1 and M_2 at ends 1 and 2 respectively, as in Fig. 4.3(a). Now the strain energy due to bending in a beam can be shown to be

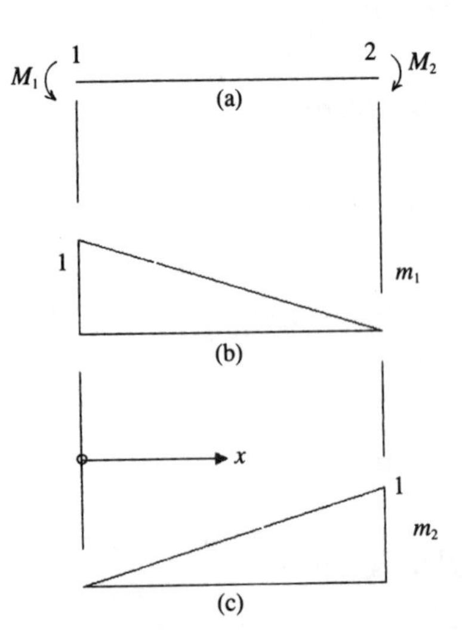

Figure 4.3 Flexibilities using Castigliano's second theorem

$$U = \int \frac{M^2}{2EI} \mathrm{d}x \qquad [4.19]$$

where M is the bending moment, a function of x along the beam and EI is the flexural rigidity. If we differentiate eqn [4.19] according to eqn [4.18] we obtain

$$\theta_i = \int \frac{M}{EI} \frac{\partial M}{\partial M_i} \mathrm{d}x \qquad [4.20]$$

In order to obtain rotational flexibilities at ends 1 and 2, we apply unit moment at end 1 and then, separately, at end 2. The resulting bending moment diagrams are shown in Fig. 4.3(b) and (c). Reminding ourselves of the definition of flexibility in section 4.2, we obtain

$$f_{11} = \int_0^l \frac{m_1 m_1}{EI} \mathrm{d}x = \int_0^l \frac{(1 - x/l)^2}{EI} \mathrm{d}x = \frac{l}{3EI}$$

$$f_{12} = \int_0^l \frac{m_1 m_2}{EI} \mathrm{d}x = \int_0^l \frac{(1 - x/l)(x/l)}{EI} \mathrm{d}x = \frac{l}{6EI}$$

Similarly,

$$f_{22} = \frac{l}{3EI}$$

$$f_{21} = \frac{l}{6EI} = f_{12}$$

Thus the flexibility of the member is

$$f = \frac{l}{6EI} \begin{bmatrix} 2 & 1 \\ 1 & 2 \end{bmatrix} \qquad [4.21]$$

The integrations can be avoided if tables of product integrals are used (Appendix 4).

This approach to obtaining element flexibilities can be extended quite easily from element to complete structure. We first make the structure statically determinate by inserting n_s suitable releases and then calculate the flexibility influence coefficients using

$$f_{ii} = \int \frac{m_i^2}{EI} \mathrm{d}x \qquad 1 \leqslant i \leqslant n_s$$

$$f_{ij} = f_{ji} = \int \frac{m_i m_j}{EI} \mathrm{d}x \qquad 1 \leqslant i \leqslant n_s, 1 \leqslant j \leqslant n_s \qquad [4.22]$$

The displacements at the releases due to the applied loading are calculated as

$$u_i = \int \frac{m_0 m_i}{EI} \mathrm{d}x \qquad 1 \leqslant i \leqslant n_s \qquad [4.23]$$

where m_0 is the bending moment distribution in the released structure due to the applied loading. The flexibility matrix of the released structure is then

$$\boldsymbol{F}_x = \begin{bmatrix} f_{11} & f_{12} & \dots & f_{1n_s} \\ f_{21} & f_{22} & \dots & f_{2n_s} \\ \vdots & & & \vdots \\ f_{n_s1} & f_{n_s2} & \dots & f_{n_sn_s} \end{bmatrix} \qquad [4.24]$$

The redundants $\boldsymbol{X}$ are then calculated by solving the compatibility equations

$$\boldsymbol{F}_x\boldsymbol{X} + \boldsymbol{U} = \boldsymbol{0} \qquad [4.25]$$

4.4.3 Matrix formulation of the flexibility method

We shall now outline a general matrix formulation of the flexibility method applied to statically indeterminate structures. If the degree of statical indeterminacy is n_s then it is necessary to find n_s stress resultants in the structure from conditions of compatibility of displacements. We shall need to set up n_s simultaneous equations expressing these conditions. The sequence of events is as follows:

1. Determine n_s and insert n_s releases into the structure, rendering it statically determinate. Care is needed in choosing the positions of the releases to ensure that the resulting structure is statically determinate. The released stress resultants $\boldsymbol{X}$ are what we want to find.
2. Apply unit value of each $\boldsymbol{X}$ in turn and evaluate displacements at each of the n_s releases. These are the flexibilities and collectively they constitute the flexibility matrix of the released structure $\boldsymbol{F}$.
3. Apply the actual loads to the released structure and calculate the resulting displacements u_i $(i = 1, \dots, n_s)$ at the n_s releases. These displacements constitute a matrix $\boldsymbol{U}$.
4. Form the compatibility equations at the releases as in eqn [4.25], i.e.

   $$\boldsymbol{F}_x\boldsymbol{X} + \boldsymbol{U} = \boldsymbol{O}$$

   and solve for the stress resultants $\boldsymbol{X}$.
5. Determine all stress resultants required using appropriate values from $\boldsymbol{X}$ and conditions of equilibrium.

6. Determine displacements in the actual structure due to the applied loads if desired.

We now develop matrix procedures which will follow the scheme outlined above. First we identify the elements of the structure, usually the members in a framed structure, and define their nodal forces in matrix $\boldsymbol{Y}$. If we now consider the two systems of forces $\boldsymbol{X}$ and $\boldsymbol{Y}$, we can relate them by a transformation

$$\boldsymbol{S} = \boldsymbol{T}\boldsymbol{X} \qquad [4.26]$$

We can obtain the transformation matrix $\boldsymbol{T}$ without knowing $\boldsymbol{X}$ or $\boldsymbol{Y}$ simply by giving the $\boldsymbol{X}$ forces unit value in turn and evaluating the corresponding member forces $\boldsymbol{S}$. If we now express the relationship between the corresponding sets of displacements as $\boldsymbol{U}_s = \boldsymbol{A}\boldsymbol{U}_x$ then it can be shown that $\boldsymbol{A} = (\boldsymbol{T}^T)^{-1}$. It is said that corresponding sets of forces and displacements transform 'contragrediently'. Thus,

$$\boldsymbol{U}_x = \boldsymbol{T}^T\boldsymbol{U}_s \qquad [4.27]$$

where displacements $\boldsymbol{U}_x$ correspond to forces $\boldsymbol{X}$ and displacements $\boldsymbol{U}_s$ correspond to forces $\boldsymbol{S}$.

Now, by definition, displacement = flexibility × force, so

$$\boldsymbol{U}_s = \boldsymbol{F}_s\boldsymbol{S} \qquad [4.28]$$

and

$$\boldsymbol{U}_x = \boldsymbol{F}_x\boldsymbol{X} \qquad [4.29]$$

Substituting for $\boldsymbol{S}$ and $\boldsymbol{U}_s$ from eqns [4.26] and [4.27] in eqn [4.28], we have

$$\boldsymbol{F}_s\boldsymbol{T}\boldsymbol{X} = (\boldsymbol{T}^T)^{-1}\boldsymbol{U}_x$$

or

$$(\boldsymbol{T}^T\boldsymbol{F}_s\boldsymbol{T})\boldsymbol{X} = \boldsymbol{U}_x$$

So

$$\boldsymbol{F}_x = \boldsymbol{T}^T\boldsymbol{F}_s\boldsymbol{T} \qquad [4.30]$$

Thus the flexibility matrix of the released structure can be obtained from the individual element flexibilities by the transformation indicated in eqn [4.30]. The transformation matrix $\boldsymbol{T}$ is obtained by finding member forces $\boldsymbol{S}$ corresponding successively to unit values of the forces $\boldsymbol{X}$ according to eqn [4.26].

A similar transformation is used to find the displacements at releases $\boldsymbol{U}_{x0}$ caused by the applied loads. We arrange the applied loads in a matrix $\boldsymbol{R}$ and relate these, in the released structure, to member forces $\boldsymbol{S}_0$ by

$$\boldsymbol{S}_0 = \boldsymbol{T}_0\boldsymbol{R} \quad [4.31]$$

The matrix $\boldsymbol{T}_0$ can be determined by making each load in $\boldsymbol{R}$ equal to unity successively. The member displacements $\boldsymbol{U}_{s0}$ corresponding to member forces $\boldsymbol{S}_0$ are given by

$$\boldsymbol{U}_{s0} = \boldsymbol{F}_s\boldsymbol{S}_0 \quad [4.32]$$

Applying the principle of contragredience again to eqn [4.26],

$$\boldsymbol{U}_{x0} = \boldsymbol{T}^{\mathrm{T}}\boldsymbol{U}_{s0} \quad [4.33]$$

Substituting for $\boldsymbol{S}_0$ and $\boldsymbol{U}_{s0}$ from eqns [4.31] and [4.33] in eqn [4.32],

$$(\boldsymbol{T}^{\mathrm{T}})^{-1}\boldsymbol{U}_{x0} = \boldsymbol{F}_s\boldsymbol{T}_0\boldsymbol{R}$$

Hence

$$(\boldsymbol{T}^{\mathrm{T}}\boldsymbol{F}_s\boldsymbol{T}_0)\boldsymbol{R} = \boldsymbol{U}_{x0} \quad [4.34]$$

On the understanding that $\boldsymbol{U}_{x0} = \boldsymbol{U}$, then we have the general form of the compatibility equations as anticipated in eqn [4.25]. Just as the matrix $\boldsymbol{T}$ transformed redundant forces, the matrix $\boldsymbol{T}_0$ transforms external loads.

The calculation of member forces applies the principle of superposition to the released structure in that the forces due to each redundant in turn are added to those due to the applied loads, and the total member forces are therefore

$$\boldsymbol{S} = \boldsymbol{T}\boldsymbol{X} + \boldsymbol{T}_0\boldsymbol{R} \quad [4.35]$$

If the applied loads $\boldsymbol{R}$ are nodal loads statically equivalent to a set of internal loads and fixed-end moments, then this latter system must be added to that obtained using eqn [4.35].

The matrix formulation of the flexibility method can be extended to the calculation of displacements in structures, but the manipulations needed to do this are rather more complex and less efficient than the use of the stiffness method. The latter can produce the flexibility matrix $\boldsymbol{F}$ of the complete structure from the relationship

$$\boldsymbol{F} = \boldsymbol{K}^{-1} \quad [4.36]$$

We shall now illustrate the flexibility method using the following example.

Example 4.2

We shall analyse the five-span continuous beam shown in Fig. 4.4(a). The intention is eventually to obtain influence lines (see Chapter 9) for the internal bending moments in the beam at the interior supports. One way to do this is to place a unit load successively at points along the beam and calculate ordinates to the influence lines at each load position. The structure has a statical indeterminacy $n_s = 4$, and we convert to a statically determinate state by introducing internal moment releases at the interior supports. The redundancies are then

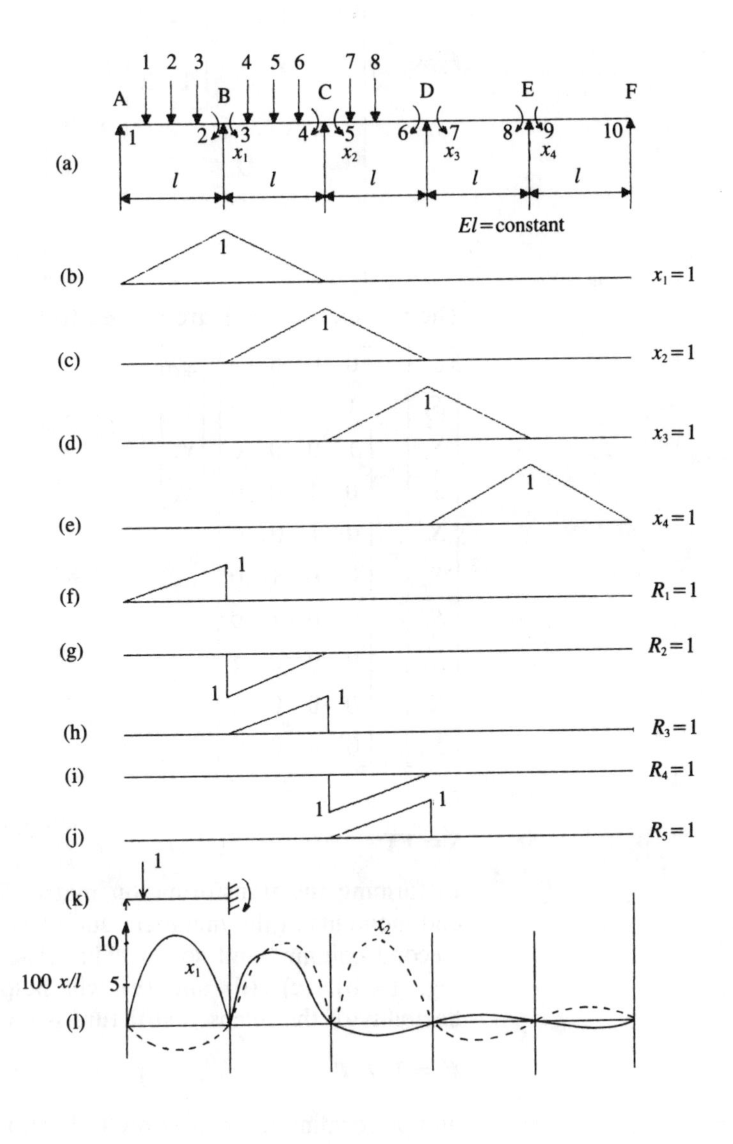

Figure 4.4 Example 4.2

$$X = \begin{bmatrix} x_1 \\ x_2 \\ x_3 \\ x_4 \end{bmatrix} \qquad [4.37]$$

The flexibility matrix $\boldsymbol{F}_s$ of the unassembled elements is

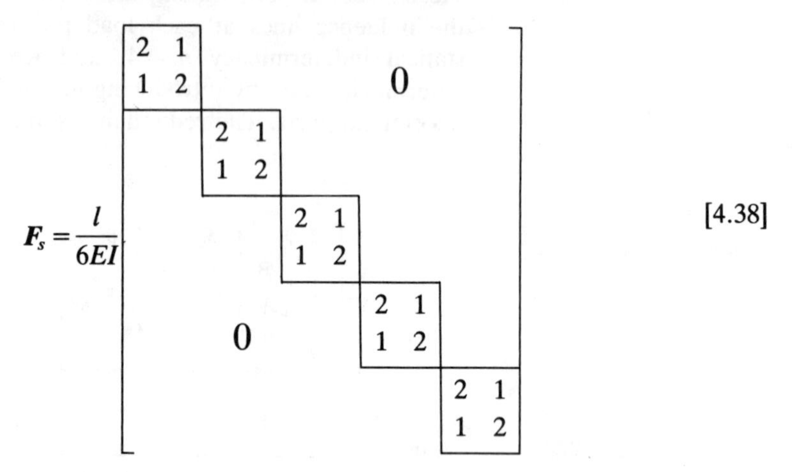

[4.38]

The member forces $\boldsymbol{S}$ are related to the redundants $\boldsymbol{X}$ by

$$\begin{bmatrix} S_1 \\ S_2 \\ S_3 \\ S_4 \\ S_5 \\ S_6 \\ S_7 \\ S_8 \\ S_9 \\ S_{10} \end{bmatrix} = \begin{bmatrix} 0 & 0 & 0 & 0 \\ 1 & 0 & 0 & 0 \\ 1 & 0 & 0 & 0 \\ 0 & 1 & 0 & 0 \\ 0 & 1 & 0 & 0 \\ 0 & 0 & 1 & 0 \\ 0 & 0 & 1 & 0 \\ 0 & 0 & 0 & 1 \\ 0 & 0 & 0 & 1 \\ 0 & 0 & 0 & 0 \end{bmatrix} \begin{bmatrix} x_1 \\ x_2 \\ x_3 \\ x_4 \end{bmatrix}$$

or

$$\boldsymbol{S} = \boldsymbol{TX} \qquad [4.39]$$

In forming the transformation matrix $\boldsymbol{T}$ we list in columns the end moments in the members due to $x_1 = 1$ (first column), $x_2 = 1$ (second column) and so on. The bending moment diagrams in Fig. 4.4(b), (c), (d) and (e) will help to obtain $\boldsymbol{T}$. Now the flexibility of the released structure is given by

$$\boldsymbol{F}_x = \boldsymbol{T}^{\mathrm{T}}\boldsymbol{F}_s\boldsymbol{T}$$

and performing the matrix multiplications we get

$$\boldsymbol{F}_x = \frac{l}{6EI}\begin{bmatrix} 4 & 1 & 0 & 0 \\ 1 & 4 & 1 & 0 \\ 0 & 1 & 4 & 1 \\ 0 & 0 & 1 & 4 \end{bmatrix} \qquad [4.40]$$

Since we are to analyse this structure for eight separate unit load positions it will be more efficient to invert the matrix $\boldsymbol{F}_x$ rather than to solve the eqn [4.25] eight times. The inverse of $\boldsymbol{F}_x$ in eqn [4.40] is

$$\boldsymbol{F}_x^{-1} = \frac{6EI}{l}\begin{bmatrix} 0{\cdot}2679 & -0{\cdot}07\,177 & 0{\cdot}01\,914 & -0{\cdot}004\,785 \\ & 0{\cdot}2871 & -0{\cdot}07\,656 & 0{\cdot}01\,914 \\ & & 0{\cdot}2871 & -0{\cdot}07\,177 \\ \text{symmetrical} & & & 0{\cdot}2679 \end{bmatrix}$$

We now proceed to compose the matrix $\boldsymbol{U}$ in the compatibility conditions of eqn [4.25] using the general form eqn [4.34], i.e.

$$\boldsymbol{U} = (\boldsymbol{T}^{\mathrm{T}}\boldsymbol{F}_s\boldsymbol{T}_0)\boldsymbol{R}$$

We now need the matrices $\boldsymbol{T}_0$ and $\boldsymbol{R}$. First we find T_0 from eqn [4.31], i.e.

$$\boldsymbol{S}_0 = \boldsymbol{T}_0\boldsymbol{R}$$

The applied loads are generalized as external moments R_i ($i = 1, \ldots, 5$) applied to the released structure, at the supports B, C and D respectively. Providing equilibrium is maintained, the applied moment $\boldsymbol{R}$ can be shared in any proportion between the two adjacent simply-supported spans; however, for simplicity we absorb the whole moment into a single span each time. Thus we obtain the bending moment diagrams of Fig. 4.4(f) to (j). (The reader might like to experiment with different diagrams for R_i ($i = 1, \ldots, 5$).) To obtain $\boldsymbol{T}_0$ from eqn [4.31] we list, in columns, the end moments in the spans due to $R_i = 1$ ($i = 1, \ldots, 5$). Using diagrams (f) to (j), we obtain

$$\boldsymbol{T}_0 = \begin{bmatrix} 0 & 0 & 0 & 0 & 0 \\ 1 & 0 & 0 & 0 & 0 \\ 0 & -1 & 0 & 0 & 0 \\ 0 & 0 & 1 & 0 & 0 \\ 0 & 0 & 0 & -1 & 0 \\ 0 & 0 & 0 & 0 & 1 \\ 0 & 0 & 0 & 0 & 0 \\ 0 & 0 & 0 & 0 & 0 \\ 0 & 0 & 0 & 0 & 0 \\ 0 & 0 & 0 & 0 & 0 \end{bmatrix}$$

Hence

$$(\boldsymbol{T}^{\mathrm{T}}\boldsymbol{F}_s\boldsymbol{T}_0)\boldsymbol{R} = \frac{l}{6EI}\begin{bmatrix} 2 & -2 & 1 & 0 & 0 \\ 0 & -1 & 2 & -2 & 1 \\ 0 & 0 & 0 & -1 & 2 \\ 0 & 0 & 0 & 0 & 0 \end{bmatrix}\boldsymbol{R}$$

Substituting in eqn [4.25] in the form

$$\boldsymbol{X} = -\boldsymbol{F}_x^{-1}\boldsymbol{U} \qquad [4.41]$$

we obtain

$$\boldsymbol{X} = \begin{bmatrix} -0.5358 & 0.46\,403 & -0.12\,436 & -0.1244 & 0.03\,349 \\ 0.14\,354 & 0.14\,356 & -0.50\,243 & 0.49\,764 & -0.13\,398 \\ -0.03\,828 & -0.03\,828 & 0.13\,398 & 0.13\,398 & -0.49\,764 \\ 0.00\,957 & 0.00\,957 & -0.03\,3495 & -0.03\,349 & 0.12\,44 \end{bmatrix}\begin{bmatrix} R_1 \\ R_2 \\ R_3 \\ R_4 \\ R_5 \end{bmatrix} \qquad [4.42]$$

To get the applied loads for the matrix $\boldsymbol{R}$ we first consider the beam to be fixed at the interior supports, as for example in Fig. 4.4(k). For position 1 of the unit load, the fixed-end moment at B (Appendix 3) is $15l/128$ and is clockwise. To remove this from the actual structure we impose an anticlockwise moment $R = -15l/128$. Repeating this for all eight load positions we obtain

$$\boldsymbol{R} = l\begin{bmatrix} -0.1172 & -0.1875 & -0.1641 & 0 & 0 & 0 & 0 & 0 \\ 0 & 0 & 0 & 0.1406 & 0.125 & 0.0469 & 0 & 0 \\ 0 & 0 & 0 & -0.0469 & -0.125 & -0.1406 & 0 & 0 \\ 0 & 0 & 0 & 0 & 0 & 0 & 0.1406 & 0.125 \\ 0 & 0 & 0 & 0 & 0 & 0 & -0.0469 & -0.125 \end{bmatrix} \qquad [4.43]$$

The columns of matrix $\boldsymbol{R}$ contain the values of R_1 to R_5 for each unit load position 1 to 8. To find the values of x_1 to x_4 we perform the matrix multiplications of eqn [4.42] and obtain

$$\begin{bmatrix} x_1 \\ x_2 \\ x_3 \\ x_4 \end{bmatrix} = \frac{l}{10^2}\begin{bmatrix} 6.28 & 10.05 & 8.79 & 7.11 & 7.36 & 3.93 & -1.91 & -1.97 \\ -1.68 & -2.69 & -2.36 & 4.38 & 8.08 & 7.74 & 7.63 & 7.90 \\ 0.45 & 0.72 & 0.63 & -1.17 & -2.15 & -2.06 & 4.22 & 7.90 \\ -0.112 & -0.179 & -0.157 & 0.292 & 0.538 & 0.516 & -1.054 & -1.97 \end{bmatrix} \qquad [4.44]$$

Although we have applied the unit load only to the left-hand half of the structure, we can of course use the symmetry to obtain the ordinates for the right-hand half. If we plot ordinates for say x_1 up to load point 8, then the ordinates for x_1 on the right-hand side are the same as those for x_4 on the left-hand side. We can relate x_2 and x_3 similarly and hence obtain complete influence lines for x_1 and x_2 over the whole structure. These are shown plotted in Fig. 4.4(l).

The reader will have noticed that considerable arithmetic is involved in the handling of the matrices. In addition, many of them contain large numbers of zeros; the term 'sparse' is used to describe matrices like this, and sparse matrices are a feature of structural analysis. The procedures just followed in example 4.2 could be improved by careful study of the patterns exhibited in the matrices. However, as we shall see later, the stiffness method is generally more powerful and economical in terms of programming effort and computer space needed. The preparation of data is also much simpler. In these circumstances there is little to be gained by further development of this general matrix formulation of the flexibility method.

4.5 The stiffness method

In this section we apply the basic structural theorems of section 4.3 to obtain the stiffnesses of simple structural elements. We then go on to develop a matrix formulation of the stiffness method comparable in style to that developed for the flexibility method in the previous section.

4.5.1 Application of the unit displacement theorem to obtain element stiffnesses

We recall eqn [4.7], this time in the form

$$S_i = \int \boldsymbol{\sigma}^{\mathrm{T}} \bar{\boldsymbol{\varepsilon}}_i \, \mathrm{d}vol$$

where $\bar{\boldsymbol{\varepsilon}}_i$ are the strains due to unit displacements in all S_i directions in turn, and $\boldsymbol{\sigma}$ is the stress matrix due to applied forces.

For all member forces $\boldsymbol{S}$ collectively,

$$\boldsymbol{S} = \int \boldsymbol{\sigma}^{\mathrm{T}} \bar{\boldsymbol{\varepsilon}} \, \mathrm{d}vol \qquad [4.45]$$

Now if the behaviour is linear, strain is proportional to displacement, and so we can put

$$\boldsymbol{\varepsilon} = \boldsymbol{B}\boldsymbol{s} \qquad [4.46]$$

Then, for unit displacements

$$\bar{\boldsymbol{\varepsilon}} = \boldsymbol{B}$$

i.e. $\boldsymbol{B}$ is a matrix of strains due to unit displacements at coordinates corresponding to $\boldsymbol{S}$.

Now

$$\boldsymbol{\sigma} = \boldsymbol{D}\boldsymbol{\varepsilon} \qquad [4.47]$$

where $\boldsymbol{D}$ is the 'elasticity' matrix. Hence in general

$$\boldsymbol{\sigma} = (\boldsymbol{DB})\boldsymbol{s} \qquad [4.48]$$

where $(\boldsymbol{DB})$ is sometimes called the 'stress matrix'.
Substituting eqns [4.48] and [4.47] in eqn [4.45],

$$\boldsymbol{S} = \int \boldsymbol{s}^{\mathrm{T}}\boldsymbol{B}^{\mathrm{T}}\boldsymbol{D}^{\mathrm{T}}\boldsymbol{B}\,\mathrm{d}vol$$

Noting that the stiffness matrix is symmetric, we can write

$$\boldsymbol{S} = \left(\int \boldsymbol{B}^{\mathrm{T}}\boldsymbol{DB}\,\mathrm{d}vol\right)\boldsymbol{s}$$

$$= \boldsymbol{ks}$$

where $\boldsymbol{k}$ is the stiffness matrix, given by

$$\boldsymbol{k} = \int \boldsymbol{B}^{\mathrm{T}}\boldsymbol{DB}\,\mathrm{d}vol \qquad [4.49]$$

Example 4.3

Consider the pin-ended member of Fig. 4.2. We have

$$s = s_1 + (s_2 - s_1)x/l$$

$$\boldsymbol{\varepsilon} = \frac{\mathrm{d}s}{\mathrm{d}x} = \frac{1}{l}(s_2 - s_1) = \frac{1}{l}[-1 \quad 1]\begin{bmatrix} s_1 \\ s_2 \end{bmatrix}$$

Hence $\boldsymbol{B} = (1/l)[-1 \quad 1]$ and $\boldsymbol{D} = E$. Hence

$$\boldsymbol{k} = \int_0^l \frac{1}{l}\begin{bmatrix} -1 \\ 1 \end{bmatrix}\frac{E}{l}[-1 \quad 1]A\,\mathrm{d}x$$

$$= \frac{EA}{l}\begin{bmatrix} 1 & -1 \\ -1 & 1 \end{bmatrix}$$

4.5.2 *Stiffnesses using Castigliano's first theorem*

This theorem (Przemieniecki 1968) states that if forces S_i ($i = 1, \ldots, n$) act on a structure and produce displacements at corresponding coordinates s_i ($i = 1, \ldots, n$), and if the strain energy is expressed in terms of displacements $\boldsymbol{s}$, then n equilibrium relationships can be expressed as

$$\frac{\partial U}{\partial s_i} = S_i \qquad i = 1, \ldots, n \qquad [4.50]$$

Now

$$U = \frac{1}{2}\int \boldsymbol{\sigma}^{\mathrm{T}}\boldsymbol{\varepsilon}\,\mathrm{d}vol$$

and for linear elastic behaviour $\sigma \propto \varepsilon$. In fact, $\boldsymbol{\sigma} = \boldsymbol{D\varepsilon}$ and $\boldsymbol{\varepsilon} = \boldsymbol{Bs}$ from eqns [4.47] and [4.46]. Hence

$$U = \frac{1}{2}\int \boldsymbol{s}^{\mathrm{T}}\boldsymbol{B}^{\mathrm{T}}\boldsymbol{D}^{\mathrm{T}}\boldsymbol{Bs}\,\mathrm{d}vol \qquad [4.51]$$

Typically, for $s = s_i$,

$$\frac{\partial U}{\partial s_i} = S_i = \left(\int \boldsymbol{B}^{\mathrm{T}}\boldsymbol{D}^{\mathrm{T}}\boldsymbol{B}\,\mathrm{d}vol\right)s_i$$

Hence for all s,

$$\boldsymbol{S} = \left(\int \boldsymbol{B}^{\mathrm{T}}\boldsymbol{D}^{\mathrm{T}}\boldsymbol{B}\,\mathrm{d}vol\right)\boldsymbol{s}$$

or

$$\boldsymbol{k} = \int \boldsymbol{B}^{\mathrm{T}}\boldsymbol{DB}\,\mathrm{d}vol$$

which is the same result as eqn [4.49].

4.5.3 Matrix formulation of the stiffness method

Stiffnesses of structural elements can be obtained in a number of ways. We have looked briefly at applications of the unit displacement theorem (section 4.5.1) and Castigliano's first theorem (section 4.5.2), and we obtained a general expression (eqn [4.49]) for element stiffness as

$$\boldsymbol{k} = \int \boldsymbol{B}^{\mathrm{T}}\boldsymbol{DB}\,\mathrm{d}vol$$

The stiffness matrix is then obtained by forming the matrix $\boldsymbol{B}$ through the relationship [4.46] and then integrating the triple product ($\boldsymbol{B}^{\mathrm{T}}\boldsymbol{DB}$). We shall see more of this in Chapter 8 when we study the stiffness method in more detail, but it will be useful at this stage to look ahead to the matrix formulation of the method when applied to the analysis of a complete structure.

Suppose the force–displacement relationship for a typical ith element in the structure is

$$\boldsymbol{S}_i = \boldsymbol{k}_i\boldsymbol{s}_i \qquad [4.52]$$

and suppose we have n elements in total. Then we can relate nodal forces and displacements separately but collectively as

$$\begin{bmatrix} \boldsymbol{S}_1 \\ \boldsymbol{S}_2 \\ \vdots \\ \boldsymbol{S}_n \end{bmatrix} = \begin{bmatrix} \boldsymbol{k}_1 & & & 0 \\ & \boldsymbol{k}_2 & & \\ & & \ddots & \\ 0 & & & \boldsymbol{k}_n \end{bmatrix} \begin{bmatrix} \boldsymbol{s}_1 \\ \boldsymbol{s}_2 \\ \vdots \\ \boldsymbol{s}_n \end{bmatrix}$$

which may be written more compactly as

$$\boldsymbol{S} = \boldsymbol{ks} \qquad [4.53]$$

Now we need to assemble the separate elements into the structure and express equilibrium of nodal forces as

$$\boldsymbol{Kr} = \boldsymbol{R} \qquad [4.54]$$

We now have two sets of forces and corresponding displacements: those for the unassembled structure described by the matrices in eqn [4.53], and those for the actual structure described by the matrices in eqn [4.54]. If we assume that the two sets of displacements $\boldsymbol{s}$ and $\boldsymbol{r}$ are related by

$$\boldsymbol{s} = \boldsymbol{Ar} \qquad [4.55]$$

then by the principle of contragredience (section 4.4.3) we have

$$\begin{aligned} \boldsymbol{R} &= \boldsymbol{A}^{\mathrm{T}}\boldsymbol{S} \\ &= \boldsymbol{A}^{\mathrm{T}}\boldsymbol{ks} \\ &= \boldsymbol{A}^{\mathrm{T}}\boldsymbol{kAr} \end{aligned}$$

using eqns [4.53] and [4.55]. From this and eqn [4.54] we see that

$$\boldsymbol{K} = \boldsymbol{A}^{\mathrm{T}}\boldsymbol{kA} \qquad [4.56]$$

and

$$\boldsymbol{S} = \boldsymbol{kAr} \qquad [4.57]$$

Thus the structure stiffness matrix $\boldsymbol{K}$ can be developed from the individual element stiffnesses using the transformation given in eqn [4.56].

The applied loads for matrix $\boldsymbol{R}$ must be nodal loads consistent with the choice and ordering of the nodal coordinates. Since applied loads will frequently be acting at points between nodes, a statical transfer of effects using fixed-end moments and shears is used to obtain nodal loads for the matrix $\boldsymbol{R}$. The fixed-end effects are then added to the member forces resulting from eqn [4.53] or eqn [4.57]. We shall illustrate the method in example 4.4, in which we reanalyse the five-span continuous beam of example 4.2 using the stiffness method.

Example 4.4

We refer to Fig. 4.5. The flexibility matrix of a typical prismatic beam has been found (eqn [4.21]), and we can use this to develop the corresponding stiffness matrix from the relationship $\boldsymbol{K} = \boldsymbol{F}^{-1}$. In doing this we need to be careful to match the two degrees of freedom for $\boldsymbol{K}$, in this case the end rotations of the beam, with the end forces defined for the flexibility matrix $\boldsymbol{F}$. Thus for each span, from eqn [4.21],

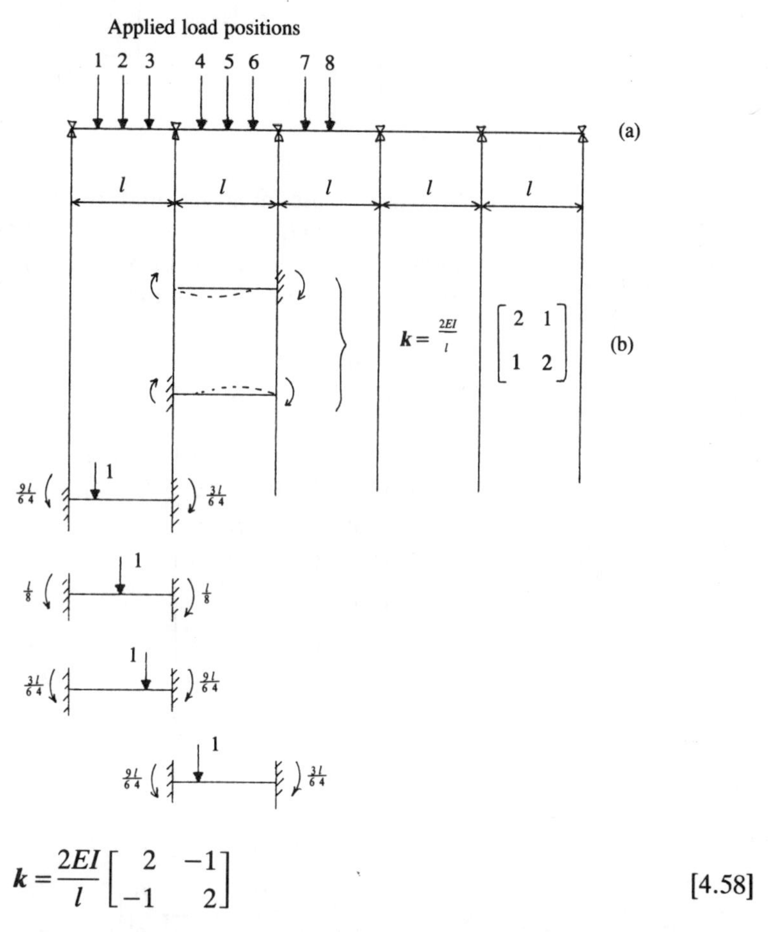

Figure 4.5 Example 4.4

$$\boldsymbol{k} = \frac{2EI}{l}\begin{bmatrix} 2 & -1 \\ -1 & 2 \end{bmatrix} \quad [4.58]$$

It will be convenient to adopt a change of sign convention from that used in obtaining eqn [4.21], where we used a curvature convention for the sign of the two bending moments. The effect of changing to a clockwise positive convention is to change the negative signs in $\boldsymbol{k}$ to positive. This will readily be seen in Fig. 4.5(b). Thus we shall use for each span

$$\boldsymbol{k} = \frac{2EI}{l}\begin{bmatrix} 2 & 1 \\ 1 & 2 \end{bmatrix} \quad [4.59]$$

Hence for all five spans,

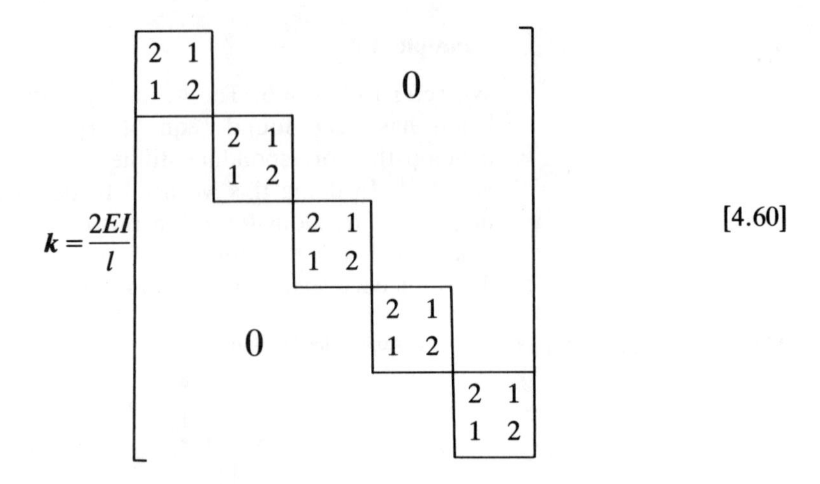

$$\boldsymbol{k} = \frac{2EI}{l}\begin{bmatrix} \begin{matrix}2 & 1\\1 & 2\end{matrix} & & & & 0 \\ & \begin{matrix}2 & 1\\1 & 2\end{matrix} & & & \\ & & \begin{matrix}2 & 1\\1 & 2\end{matrix} & & \\ & & & \begin{matrix}2 & 1\\1 & 2\end{matrix} & \\ 0 & & & & \begin{matrix}2 & 1\\1 & 2\end{matrix} \end{bmatrix} \qquad [4.60]$$

Now we relate member end rotations $\boldsymbol{s}$ to nodal (structure) rotations $\boldsymbol{r}$ by eqn [4.55], $\boldsymbol{s} = \boldsymbol{A}\boldsymbol{r}$. To obtain matrix $\boldsymbol{A}$ we set each nodal rotation equal to unity and write down, in columns, the corresponding member end rotations:

$$\boldsymbol{A} = \begin{bmatrix} 1 & 0 & 0 & 0 & 0 & 0 \\ 0 & 1 & 0 & 0 & 0 & 0 \\ \hdashline 0 & 1 & 0 & 0 & 0 & 0 \\ 0 & 0 & 1 & 0 & 0 & 0 \\ \hdashline 0 & 0 & 1 & 0 & 0 & 0 \\ 0 & 0 & 0 & 1 & 0 & 0 \\ \hdashline 0 & 0 & 0 & 1 & 0 & 0 \\ 0 & 0 & 0 & 0 & 1 & 0 \\ \hdashline 0 & 0 & 0 & 0 & 1 & 0 \\ 0 & 0 & 0 & 0 & 0 & 1 \end{bmatrix} \qquad [4.61]$$

Thus

$$\boldsymbol{kA} = \frac{2EI}{l}\begin{bmatrix} 2 & 1 & 0 & 0 & 0 & 0 \\ 1 & 2 & 0 & 0 & 0 & 0 \\ \hdashline 0 & 2 & 1 & 0 & 0 & 0 \\ 0 & 1 & 2 & 0 & 0 & 0 \\ \hdashline 0 & 0 & 2 & 1 & 0 & 0 \\ 0 & 0 & 1 & 2 & 0 & 0 \\ \hdashline 0 & 0 & 0 & 2 & 1 & 0 \\ 0 & 0 & 0 & 1 & 2 & 0 \\ \hdashline 0 & 0 & 0 & 0 & 2 & 1 \\ 0 & 0 & 0 & 0 & 1 & 2 \end{bmatrix} \qquad [4.62]$$

and

$$
\boldsymbol{K} = \boldsymbol{A}^{\mathrm{T}}\boldsymbol{k}\boldsymbol{A} = \frac{2EI}{l}\begin{bmatrix} 2 & 1 & & & & \\ 1 & 4 & 1 & & 0 & \\ & 1 & 4 & 1 & & \\ & & 1 & 4 & 1 & \\ & & & 1 & 4 & 1 \\ & 0 & & & 1 & 2 \end{bmatrix} \qquad [4.63]
$$

The applied nodal forces for matrix $\boldsymbol{R}$ can be obtained by calculating the fixed-end moments (Appendix 3). Thus

$$
\boldsymbol{R} = l \begin{bmatrix} 0.1406 & 0.125 & 0.0469 & 0 & 0 & 0 & 0 & 0 \\ -0.0469 & -0.125 & -0.1406 & 0.1406 & 0.125 & 0.0469 & 0 & 0 \\ 0 & 0 & 0 & -0.0469 & -0.125 & -0.1406 & 0.1406 & 0.125 \\ 0 & 0 & 0 & 0 & 0 & 0 & -0.0469 & -0.125 \\ 0 & 0 & 0 & 0 & 0 & 0 & 0 & 0 \\ 0 & 0 & 0 & 0 & 0 & 0 & 0 & 0 \end{bmatrix} \qquad [4.64]
$$

Solving the equations $\boldsymbol{Kr} = \boldsymbol{R}$ (eqn [4.54]), we produce

$$
\boldsymbol{r} = \frac{l^2}{2EI}\begin{bmatrix} 0.088\,430 & 0.091\,510 & 0.048\,830 & -0.023\,700 & -0.024\,520 & -0.013\,090 & 0.006\,354 & 0.006\,579 \\ -0.036\,260 & -0.058\,010 & -0.050\,760 & 0.047\,390 & 0.049\,040 & 0.026\,170 & -0.012\,710 & -0.013\,160 \\ 0.009\,720 & 0.015\,550 & 0.013\,610 & -0.025\,270 & -0.046\,650 & -0.044\,700 & 0.044\,480 & 0.046\,050 \\ -0.002\,617 & -0.004\,187 & -0.003\,663 & 0.006\,805 & 0.012\,560 & 0.012\,040 & -0.024\,600 & -0.046\,050 \\ 0.000\,7477 & 0.001\,196 & 0.001\,047 & -0.001\,944 & -0.00\,3589 & -0.003\,439 & 0.007\,029 & 0.013\,160 \\ -0.000\,3738 & -0.000\,598 & -0.000\,523 & 0.000\,972 & 0.001\,794 & 0.001\,719 & -0.003\,515 & -0.006\,579 \end{bmatrix} \qquad [4.65]
$$

and then by eqn [4.57],

$$
\boldsymbol{S} = \boldsymbol{kAr} = l\begin{bmatrix} 0.140\,600 & 0.125\,000 & 0.046\,900 & 0 & 0 & 0 & 0 & 0 \\ 0.015\,910 & -0.024\,500 & -0.052\,700 & 0.071\,080 & 0.073\,600 & 0.039\,300 & -0.019\,070 & -0.019\,740 \\ \hdashline -0.062\,800 & -0.100\,500 & -0.087\,900 & 0.069\,500 & 0.051\,400 & 0.007\,640 & 0.019\,060 & 0.019\,730 \\ -0.016\,820 & -0.026\,900 & -0.023\,540 & -0.003\,150 & -0.044\,260 & -0.063\,230 & 0.076\,250 & 0.078\,940 \\ \hdashline 0.016\,820 & 0.026\,900 & 0.023\,550 & -0.043\,740 & -0.080\,740 & -0.077\,400 & 0.064\,360 & 0.046\,050 \\ 0.004\,486 & 0.007\,176 & 0.006\,284 & -0.011\,660 & -0.021\,530 & -0.020\,620 & -0.004\,720 & -0.046\,050 \\ \hdashline -0.004\,486 & -0.007\,178 & -0.006\,280 & 0.011\,660 & 0.021\,530 & 0.020\,640 & -0.042\,170 & -0.078\,940 \\ -0.001\,122 & -0.001\,795 & -0.001\,570 & 0.002\,917 & 0.005\,380 & 0.005\,162 & -0.010\,540 & -0.019\,730 \\ \hdashline 0.001\,122 & 0.001\,795 & 0.001\,571 & -0.002\,916 & -0.005\,380 & -0.005\,159 & 0.010\,540 & 0.019\,740 \\ 0 & 0 & 0 & 0 & 0 & 0 & 0 & 0 \end{bmatrix} \qquad [4.66]
$$

The final bending moments are obtained by adding fixed-end moments from eqn [4.64] to those in eqn [4.66]. The results are in Table 4.1, and can be compared with the results of example 4.2 in eqn [4.44].

Table 4.1 Bending moments ($\times 10^2/l$) for example 4.4

	Position of unit load							
Section	1	2	3	4	5	6	7	8
1	0	0	0	0	0	0	0	0
2	6.28	10.05	8.79	7.11	7.36	3.93	−1.91	−1.97
3	−6.28	−10.05	−8.79	−7.11	−7.36	−3.93	1.91	1.97
4	−1.68	−2.69	−2.35	4.38	8.08	7.74	7.63	7.89
5	1.68	2.69	2.35	−4.38	−8.08	−7.74	−7.63	−7.89
6	0.45	0.718	0.628	−1.17	−2.15	−2.06	4.22	7.90
7	−0.45	−0.718	−0.628	1.17	2.15	2.06	−4.22	−7.90
8	−0.112	−0.18	−0.157	0.292	0.538	0.516	−1.054	−1.97
9	0.112	0.18	0.157	−0.292	−0.538	−0.516	1.054	1.97
10	0	0	0	0	0	0	0	0

We have actually carried out more work than is really necessary since we could take advantage of the fact that the bending moment is zero at each end of the beam under all applied loads. This means that we do not need to calculate the end rotations, and thus the size of the stiffness matrix can be reduced from 6×6 to 4×4. The reader is invited to show that this reduced stiffness matrix is

$$\boldsymbol{K} = \frac{EI}{l}\begin{bmatrix} 7 & 2 & & 0 \\ 2 & 8 & 2 & \\ & 2 & 8 & 2 \\ 0 & & 2 & 7 \end{bmatrix}$$

and to continue the analysis to produce the same results.

It will again be evident in following through this example that many of the matrices are sparsely populated. We met similar circumstances with the flexibility method. The sparseness of matrices is an important feature of the stiffness method and can be exploited in several ways to economize on space in the computer. Indeed, the handling of sparse matrices is a subject which has received considerable attention (Jennings 1966).

4.6 Sign convention for axes, forces and displacements

We conclude this chapter by looking at the subject of sign conventions, since we shall soon be immersed in theory and methods and will not want to compound any difficulties with additional problems of interpretation of signs. Sign conventions are not usually difficult until we encounter three-dimensional situations, where the difficulties tend to relate to 'visualization'. It has been traditional practice to introduce conventions in two dimensions and then extend these to three dimensions only when necessary. In spite of likely initial difficulties in visualization, it is now better practice to introduce the sign conventions in three dimensions *ab initio*.

If one analyses the sign conventions generally adopted by authors of textbooks and writers of computer packages for structural analysis, one finds that the most commonly used conventions are based on X, Y and Z coordinate axes directed as shown in Fig. 4.6(a). The X axis coincides with the longitudinal

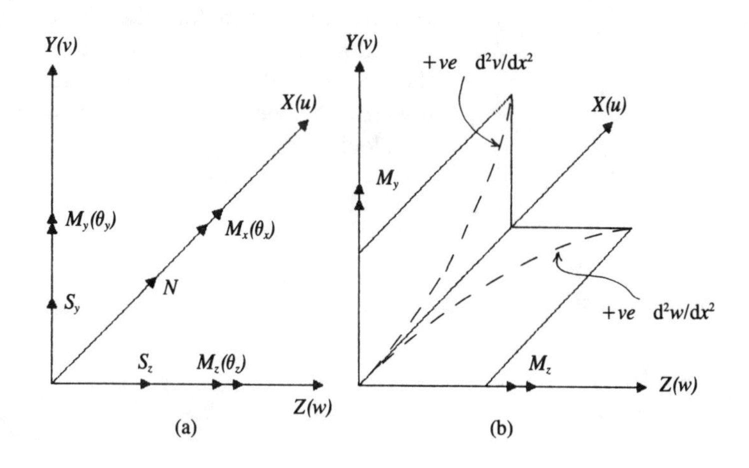

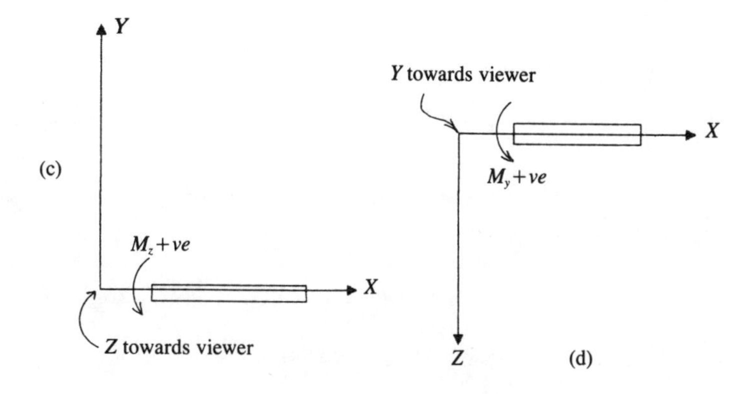

Figure 4.6 Sign conventions for rectangular axes, coordinates and stress resultants: (a) positive axes and stress resultants (b) positive curvatures (c) beam in X–Y plane (d) beam in X–Z plane

axis of the member, and the Z and Y axes are often specified to be principal axes of the cross-section. Linear displacements are u, v and w in the directions X, Y and Z respectively, and rotations are θ_x, θ_y, θ_z about axes X, Y and Z. Forces N, S_y and S_z are positive in the positive X, Y and Z directions respectively. N is the axial thrust. Moments are M_x, M_y and M_z, with M_x being the twisting moment about the longitudinal axis. We adopt a clockwise positive sign convention for the rotations and moments using a double-headed arrow convention as shown in (a). The right-hand screw rule is used to determine the signs of all moments; thus a clockwise (positive) moment will rotate a nut with a right-hand thread, causing movement in the positive direction of the axis.

Now if we consider the curvatures arising from bending in the X–Y and X–Z planes, these will be positive if as shown by the dotted lines in Fig. 4.6(b); thus positive moment M_z will produce negative curvature d^2v/dx^2, whereas positive M_y will produce positive curvature d^2w/dx^2. If we look at flexure of a beam in the X–Y plane, as in (c), a positive M_z moment will appear anticlockwise since the positive Z axis is directed towards the viewer, whereas in fact the positive moment is clockwise when viewed in the positive direction of the Z axis. Similarly if we view the beam in the X–Z plane, as in (d), the Y axis is positive towards the viewer so that again the positive moment appears anticlockwise.

CHAPTER

5

Stress resultant distributions

5.1 Introduction

Structural analysis provides information on the distribution of forces in structures and is often described by distribution diagrams showing the values of stress resultants throughout the structure. The construction of these diagrams and their interpretation is an extremely important facility to acquire and we shall devote this chapter to that purpose. We shall assume that sufficient analysis has been carried out on a structure so that all distributions of stress resultants can be determined using statical principles. This will generally mean that our analysis has provided us with the values of stress resultants at all the nodes of the structure.

5.2 The simply-supported beam

We shall consider the distributions of bending moment M, shear force S and thrust N in beams for which all the applied forces and reactions are known or can be determined by statical principles. We begin with the definitions of these stress resultants:

1. The bending moment at a transverse section in a beam is the algebraic sum of the moments of the forces to *either* side of the section taken about the centroid of the section.
2. The shear force at a transverse section is the algebraic sum of the components of the forces to *either* side of the section taken *parallel* to the section.
3. The thrust at a transverse section is the algebraic sum of the components of the forces to *either* side of the section taken *perpendicular* to the section.

These definitions will now be illustrated with reference to the beams in Fig. 5.1. In (a) the beam is horizontal and carries three concentrated loads at fixed positions as shown. The reactions are calculated by taking moments about one end, say the left-hand end to find R_2, and then expressing vertical equilibrium of the beam to find the other reaction R_1. Thus

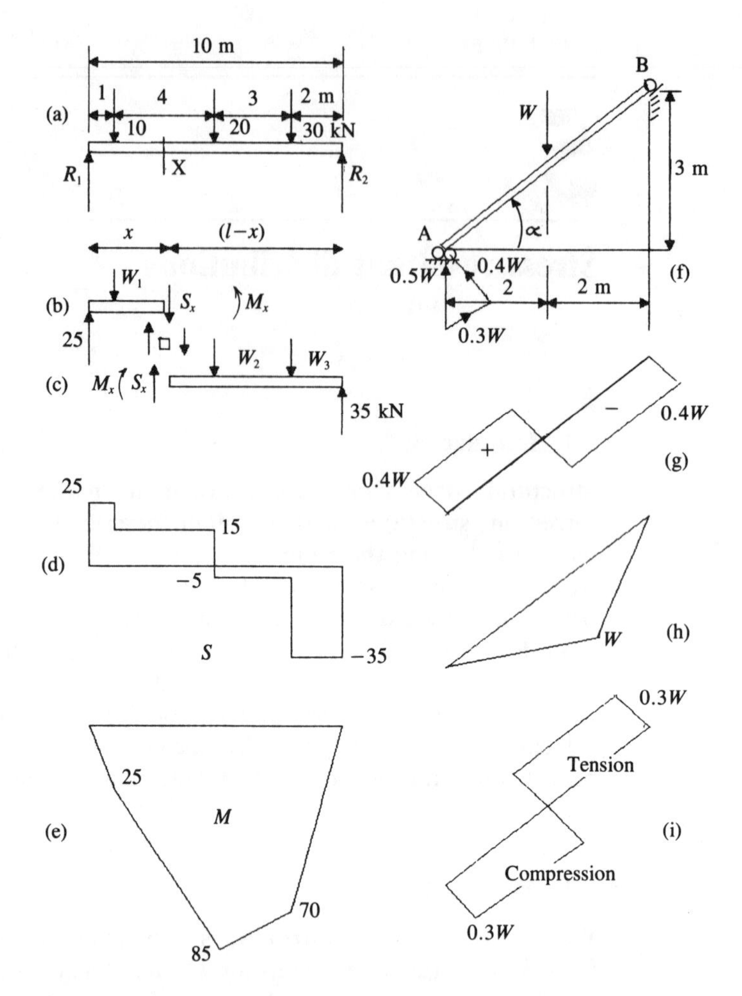

Figure 5.1 Stress resultant distribution diagrams for beams

$$R_2 \times 10 = 10 \times 1 + 20 \times 5 + 30 \times 8$$

So $R_2 = 35$ kN. Then

$$R_1 + R_2 = 10 + 20 + 30$$

Hence $R_1 = 25$ kN.

Consider section X at distance x from the left-hand end, as a typical section in the beam. Reminding ourselves of the definitions above, we can evaluate the shear stress resultant at x as

$$S_x = 25 - 10 = 15 \text{ kN}$$

It will help us in determining a sign for S_x to consider a thin slice of the beam as illustrated between diagrams (b) and (c). It is clear that the forces to the left of X produce a net upwards force S_x and the forces to the right of the slice produce a net downwards force S_x. We shall regard this situation as representing positive shear force at X. Now to construct the complete shear force diagram for the beam. Starting at the left-hand end, the shear force is equal to R_1 (25 kN) from $x = 0$ to $x = 1$ when we meet the 10 kN load. To the right of this load the shear force is $25 - 10 = 15$ kN, and this value holds from $x = 1$ to $x = 5$. In the same way we proceed along the beam, calculating the shear force between each pair of loads. Note that we finish with $S_x = -35$ kN at the right-hand end, which of course agrees with our value of the reaction R_2. The diagram representing the shear force distribution in the beam is given in Fig. 5.1(d), and we see that at each load position there is an instantaneous change in the value of the shear force equal to the magnitude of the load. Thus we need to identify two values of shear force at each load position, one considered immediately to the left and one immediately to the right. If these discontinuities in the shear force diagram are difficult to understand, the reader might like to imagine that each load is applied over a short length of beam, which will usually be the case anyway, and then the change in shear force from one side of the load to the other takes place gradually along the short length of the beam occupied by the load. It will then be apparent that the two values of shear force really do exist in practice.

We turn now to the calculation of bending moments in the beam. Let us start again with point X lying between the first pair of loads. In calculating the bending moment M_x at X we take moments of all forces parallel to the cross-section, about the centroid of the section. Now we can do this either to the left of X or to the right. The choice is arbitrary, so we usually take the easier route! Considering forces to the left,

$$M_x = 25x - 10(x - 1)$$

So we have a linear relationship for M_x between the first pair of loads and therefore, by deduction, between any pair of applied concentrated loads. In these circumstances we simply calculate bending moments under each load and join the ordinates with straight lines. Thus

$$M_{10} = 25 \times 1 = 25 \text{ kN m}$$
$$M_{20} = 25 \times 5 - 10 \times 4 = 85 \text{ kN m}$$
$$M_{30} = 25 \times 8 - 10 \times 7 - 20 \times 3 = 70 \text{ kN m}$$

or, more directly from the right-hand side,

$$M_{30} = 35 \times 2 = 70 \text{ kN m}$$

We must now consider a sign convention for bending moment. In the calculations we have just carried out we have expressed each bending moment as a positive quantity, and if we consider the sense of the curvature produced in the beam we can see that this is concave upwards throughout the length of the beam. If we consult section 4.6, where we introduced sign conventions, we will see that in the X–Y plane this curvature is positive. The sign convention proposed in section 4.6 for bending moments is unsuitable for use in the construction of bending moment diagrams, where we must use a convention based on the sense of the curvature produced in the beam. We have therefore implied that the bending moments we have calculated for this beam are all positive. Another way of describing this situation is to say that 'sagging' moments are positive; however, this breaks down when we come to vertical members. It is now common practice to avoid the need for positive and negative signs in bending moment diagrams by always drawing the diagrams on the 'tension' sides of the members. This avoids the difficulty with vertical members as well. The bending moment diagram in Fig. 5.1(e) has been drawn on the tension side of the beam and therefore does not need to be labelled with a sign.

Sign conventions are often difficult to apply correctly and so need to be thoroughly understood. Before we leave this diagram let us observe again that a clockwise/anticlockwise convention, which we use for signs for individual bending moments in the structural analysis, cannot be used for bending moment diagrams, which must be drawn on the basis of sense of curvature in the beam. If the moment M_x in Fig. 5.1(b) and (c) is studied, it can be seen that the moment is anticlockwise in (b) and clockwise in (c). (If the reader is still uncertain about this, then calculation, by taking moments about X to the left and then to the right, will clear the matter.) The two different signs of M_x in (b) and (c) are consistent with a sagging curvature and this determines how we draw the diagram.

To complete our study of the beam in Fig. 5.1(a), we observe that since all forces are purely vertical on the horizontal beam, no axial forces are present and so we do not need to draw a thrust diagram.

We turn now to the inclined beam of Fig. 5.1(f). The beam is hinged at B and supported on rollers on a horizontal plane at A; there can therefore be no horizontal component of reaction at A. Taking moments about B, we find the vertical reaction at A to be $W/2$; the vertical reaction at B is therefore also $W/2$. To find the M, V and N stress resultants we need to know forces parallel and perpendicular to the axis of the member; however, the external force at A is vertical so we shall need to resolve it. The component of $W/2$ in the AB (axial) direction is

$$(W/2)\sin\alpha = (W/2)(3/5) = (3/10)W$$

The component of $W/2$ perpendicular to the axis is

$$(W/2)\cos\alpha = (W/2)(4/5) = (2/5)W$$

These components are easily seen in the triangle of forces at A in diagram (f). It is now a straightforward matter to construct the diagrams as at (g), (h) and (i). For (g) we have used a sign convention that an upwards force to the left gives positive shear. For (h) the diagram is drawn on the tension side of the beam. In (i) to avoid any ambiguity, we have labelled the diagram 'compression' and 'tension'. This last diagram should be studied; the upper half of the beam is in tension and the lower half in compression. All the diagrams are drawn with the ordinates perpendicular to the axis of the member, and this is standard practice.

In this case we have worked entirely from the end A. If there remains any uncertainty about what we have done, the reader is advised to rework the example entirely from end B.

Example 5.1

We shall construct shear force and bending moment diagrams for the beam of Fig. 5.2(a). The overhang BE and the offset

Figure 5.2 Example 5.1

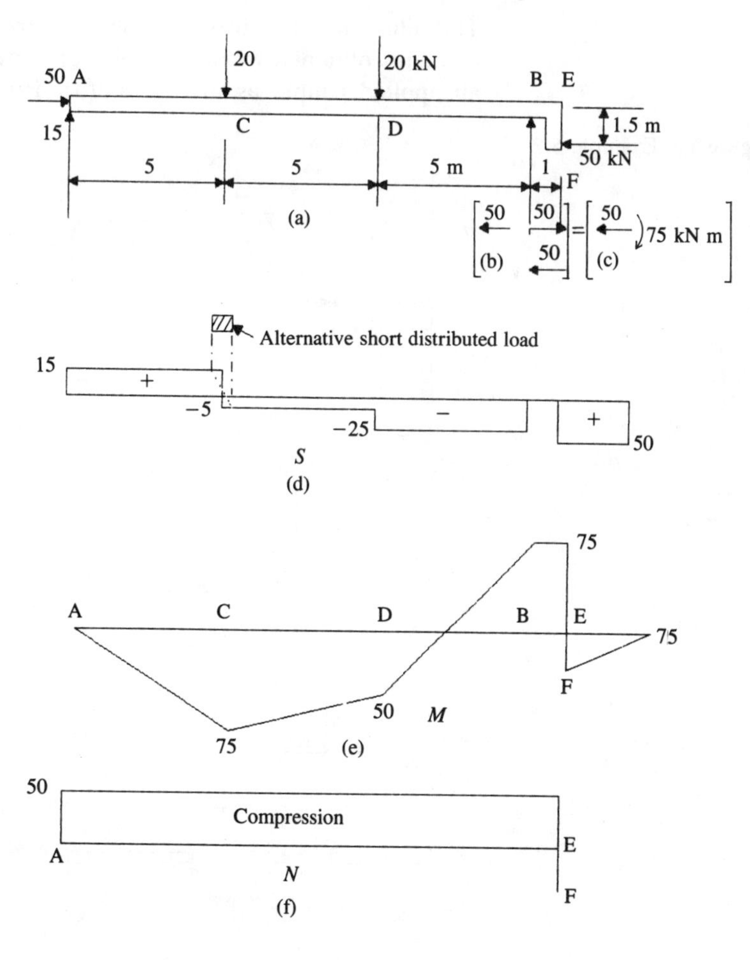

horizontal load at F should be noted. The effect of this latter force is unchanged if we add equal and opposite forces at E, as shown in (b). We then see in (c) that this is equivalent to a moment (couple) of 75 kN m and a horizontal force of 50 kN at E. Taking moments about B, the vertical reaction at A is

$$R_{VA} = (1/15)(20 \times 5 + 20 \times 10 - 50 \times 1.5) = 15\ \text{kN}$$

$$R_{VB} = 20 + 20 - 15 = 25\ \text{kN}$$

The shear force diagram is constructed at (d). There is no shear force on BE but there is an applied shear of 50 kN on EF. The bending moment diagram is shown at (e); note the constant value of bending moment of 75 kN m over BE and the triangular diagram for FE. With the horizontal restraint provided by the hinged support at A, it is easily seen that the whole beam ACDBE is in compression equal to 50 kN. There is no thrust in EF and the thrust diagram is therefore as at (f).

Example 5.2

This illustrates the forms of shear force and bending moment diagrams obtained when a simply-supported beam is subjected to an applied couple as in Fig. 5.3(a). By taking moments about

Figure 5.3 Example 5.2

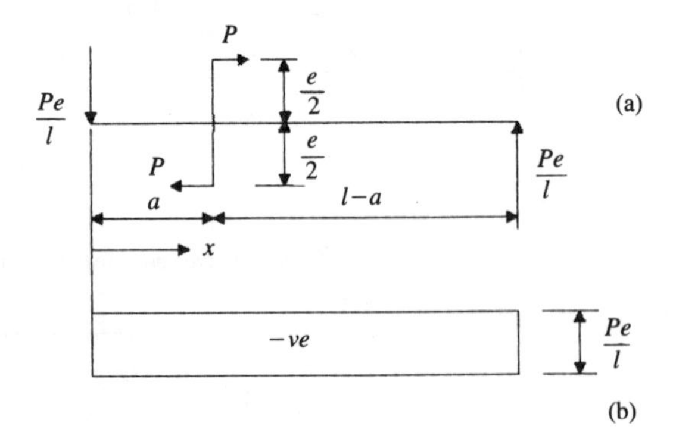

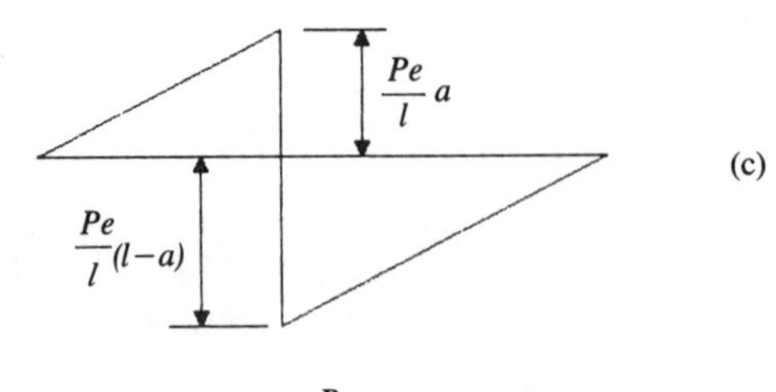

either end, we find that the reactions are both Pe/l, downwards at the left-hand end and upwards at the right. These are the only transverse, shear producing, forces on the beam, so the shear force is uniform and equal to Pe/l over the whole beam. This will be negative according to our convention, as in Fig. 5.3(b). The bending moment diagram is best approached from both ends, and we get

$$M_x = \begin{cases} \dfrac{Pe}{l}x & 0 \leqslant x \leqslant a \\ \dfrac{Pe}{l}(l-x) & a \leqslant x \leqslant l \end{cases}$$

The complete diagram in (c) is therefore composed of two parallel straight lines giving curvature which is concave down over the left-hand part and concave up over the right-hand part. There is a discontinuity in the diagram where the couple is applied. To fully understand this situation we need to see three moments at this point; $(Pe/l)a$ to the left, $(Pe/l)(l-a)$ to the right, and an externally applied moment Pe. The equilibrium of the beam at this point is easily verified if we consider all three moments and equate clockwise and anticlockwise moments, as in Fig. 5.3(d).

5.3 The relationship between load, shear and moment

We pause in our development of stress resultant diagrams to establish some fundamental relationships between them.

Consider an elemental length of beam δx (Fig. 5.4) at distance

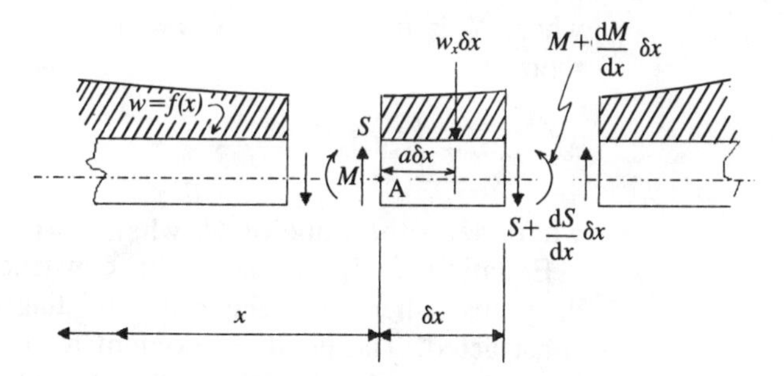

Figure 5.4 Relationships between load, shear and moment

x along the axis of the beam. The beam carries a load $w = f(x)$. The load on the element is $w_x\,\delta x$, where w_x is the average intensity of the load from x to $x + \delta x$ and the load is taken to act at a point distant $a\,\delta x$ from the section at x. For vertical equilibrium,

$$w_x\,\delta x + S + \frac{\mathrm{d}S}{\mathrm{d}x}\,\delta x = S$$

Hence

$$\frac{dS}{dx} = -w_x \qquad [5.1]$$

i.e. intensity of load is equal to minus rate of change of shear. Now, taking moments about A, on the centroid at x,

$$M + \left(S + \frac{dS}{dx}\delta x\right)\delta x + w_x \delta x a \delta x = M + \frac{dM}{dx}\delta x \qquad [5.2]$$

or

$$S + \frac{dS}{dx}\delta x + w_x a \delta x = \frac{dM}{dx}$$

As $\delta x \rightarrow 0$, this reduces to

$$S = \frac{dM}{dx} \qquad [5.3]$$

i.e. shear force is equal to rate of change of bending moment. Combining eqns [5.1] and [5.3],

$$w = -\frac{dS}{dx} = -\frac{d^2M}{dx^2}$$

From eqn [5.3], if $S = 0$, then $dM/dx = 0$ and M has a stationary value, either maximum or minimum.

Equations [5.1] and [5.3] can be expressed in an alternative form using integrals as follows. Since $w = -dS/dx$, then

$$S_x = S_0 - \int_0^x w \, dx \qquad [5.4]$$

where S_0 is the value of S_x when $x = 0$. Also, since $S = dM/dx$, then

$$M_x = M_0 + \int_0^x S \, dx \qquad [5.5]$$

where M_0 is the value of M_x when $x = 0$.

Equation [5.5] is useful in constructing bending moment diagrams when the shear force diagram has already been constructed. The bending moment at a section can be obtained by evaluating the area under the shear force diagram from $x = 0$ to $x = x$. For example, the bending moments at C and D in example 5.1 are

$$M_C = 0 + \int_0^5 15 \, dx = 75 \text{ kN m}$$

$$M_D = 75 + \int_5^{10} -5 \, dx = 75 - 25 = 50 \text{ kN m}$$

Equation [5.4] needs a little additional interpretation when applied to concentrated loads. As we have seen, these produce discontinuities in the shear force diagrams. If, however, the concentrated loads are understood to be distributed over short lengths of the beam, as illustrated in Fig. 5.2(d) for the load at C, the application of the principles is seen more clearly.

5.4 Distributed loads

We now look at the types of diagrams resulting from the application of distributed loading on beams. The intensity (force per unit length) of load is say w, and this may be constant (uniformly distributed load) or it may be a function of x along the beam. We take a very simple case first of all. In Fig. 5.5(a) we have a simply-supported beam of length l carrying a uniformly distributed load w. The total load is wl and, by symmetry, each reaction must be $wl/2$. At any point in the beam at x from the left-hand end,

$$S_x = (wl/2) - wx \qquad [5.6]$$

and $S_0 = wl/2$, $S_l = -wl/2$. The shear force diagram is shown in Fig. 5.5(b), and we note the negative slope of the line,

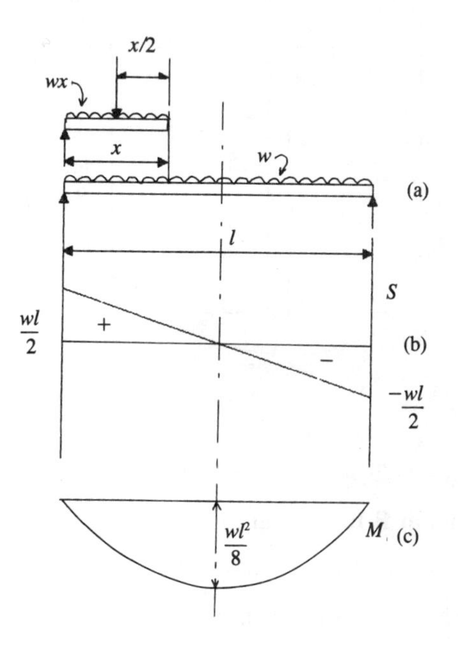

Figure 5.5 Uniformly distributed load on simply-supported beam

$dS/dx = -w$, as in eqn [5.1].

The bending moment at x is

$$M_x = (wl/2)x - wx(x/2)$$
$$= (w/2)(lx - x^2) \qquad [5.7]$$

Equation [5.7] represents a parabola with a maximum ordinate of $wl^2/8$ at $x = l/2$. The maximum bending moment exists where $S_x = 0$ in accordance with eqn [5.3]. The bending moment diagram is shown in Fig. 5.5(c) and is drawn on the tension side of the beam.

Example 5.3

The beam shown in Fig. 5.6 is simply-supported at A and B and overhangs to D. The reactions are easily shown to be $R_A = (4/9)wl$ and $R_B = (8/9)wl$. The shear force at end A is clearly (4/9) wl. The shear force at B to the left is

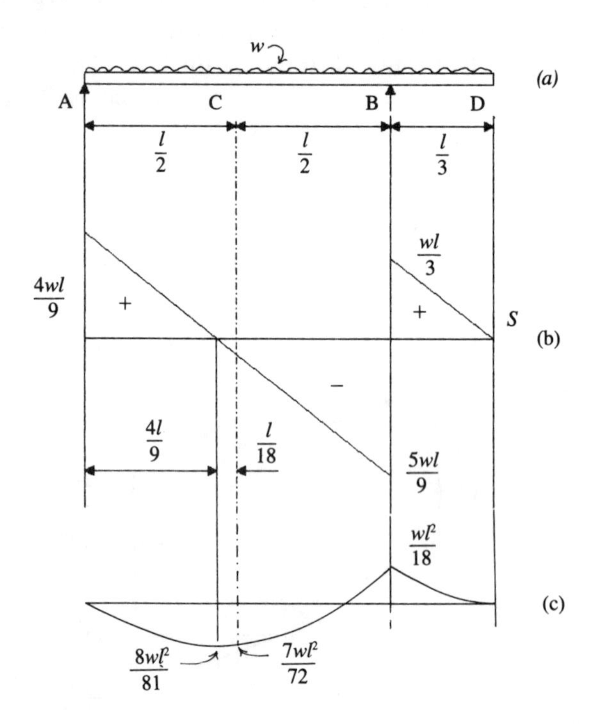

Figure 5.6 Example 5.3

$$S_B = (4/9)wl - wl = -(5/9)wl$$

and at B to the right is

$$S_B = -(5/9)wl + R_B = wl/3$$

The shear force at D is

$$S_D = wl/3 - wl/3 = 0$$

as expected,

The shear force diagram is shown in Fig. 5.6(b). The point of zero shear within AB can be found by expressing the shear force as a function of x:

$$S_x = (4/9)wl - wx$$
$$= 0 \qquad \text{for } x = (4/9)l$$

We shall use this value of x to find the maximum bending moment in AB. Before leaving the shear force diagram, observe that the two lines comprising the diagram in AB and BD are parallel and consider why this is so.

To construct the bending moment diagram we can locate specific values at points in the beam and draw the parabolic curves between them. Thus

$$M_0 = 0$$

$$M_{4l/9} = \frac{4wl}{9}\frac{4l}{9} - w\frac{4l}{9}\frac{1}{2}\frac{4l}{9} = \frac{8wl^2}{81}$$

Alternatively, using the area under the shear force diagram,

$$M_{4l/9} = \frac{4wl}{9}\frac{1}{2}\frac{4l}{9} = \frac{8wl^2}{81}$$

Next,

$$M_{l/2} = \frac{4wl}{9}\frac{l}{2} - \frac{wl}{2}\frac{l}{4} = \frac{7wl^2}{72}$$

and

$$M_B = \frac{wl}{3}\frac{l}{6} = \frac{wl^2}{18}$$

The last equation gives tension in the top at B.

The complete bending moment diagram is shown at Fig. 5.6(c).

Example 5.4

Here we have a uniformly distributed load which does not cover the whole span, shown in Fig. 5.7(a). The principle of superposition can be useful in cases like this. If we were to provide temporary supports to the beam at the ends of the load, the reactions at these supports would each be one-half of the total load, i.e. $5 \times 4 \times (1/2) = 10$ kN, as in (b). The bending moment diagram for the beam would then be a parabola of height $w(4^2/8) = 10$ kN m, as in (c). The temporary supports are now effectively removed by the application of downwards loads of 10 kN each at the ends of the distributed load, as in (d). The bending moment diagram corresponding to these loads is shown at (e), and to obtain the complete diagram we now add the parabola, shown by the dashed line. It helps in sketching in the parabolic curve to note that it has a common tangent with the bending moment diagrams in the outer parts of the beam.

Figure 5.7 Example 5.4

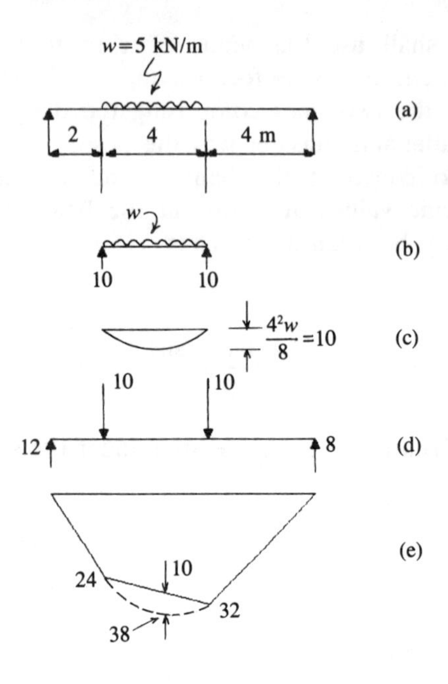

Example 5.5

A vertical beam is shown in Fig. 5.8. The load carried is a distributed load varying uniformly in intensity from zero at $x = 0$ to w_0 at $x = h$. We can find the reactions by taking moments about one end, say the lower end A.

Figure 5.8 Example 5.5

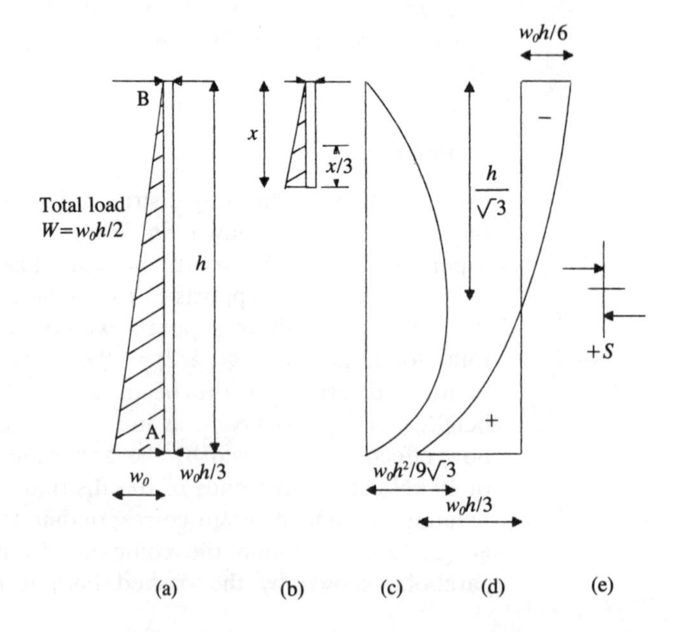

$$R_B = \frac{1}{h}\left(w_0 \frac{h}{2}\frac{h}{3}\right) = w_0 \frac{h}{6}$$

$$R_A = w_0\left(\frac{h}{2} - \frac{h}{6}\right) = w_0 \frac{h}{3}$$

The intensity of the load as a function of x is $w = w_0 x/h$. The shear force at x from B is

$$S_x = -w_0 \frac{h}{6} + w_0 \frac{x}{h}\frac{x}{2}$$

$$= -\frac{w_0}{6h}(h^2 - 3x^2)$$

The bending moment at x is

$$M_x = w_0 \frac{h}{6} x - w_0 \frac{x}{h}\frac{x}{2}\frac{x}{3} = \frac{w_0 x}{6h}(h^2 - x^2)$$

Fig. 5.8(b) will help with this. We see that the bending moment is a cubic function of x along the beam. The bending moment and shearing force diagrams are shown at (c) and (d) respectively. The bending moment diagram is drawn on the tension side of the member and the sign convention used for the shear force diagram is shown at (e). The maximum bending moment occurs where $S_x = 0$, i.e. when $3x^2 = h^2$ or $x = h/\sqrt{3}$. Hence

$$M_{max} = \frac{w_0 h^2}{9\sqrt{3}}$$

5.5 Hinged beams

The presence of a hinge (moment release) in a structure means that no bending moment can exist at that point. This information is used when evaluating reactions and when drawing bending moment diagrams. We examine first of all a very simple case of a beam ABCD in Fig. 5.9, with internal hinges at B and C and fully restrained at A and D. Studying the part beam BC in diagram (b), we can see that vertical shear forces of $W/2$ are needed at B and C to maintain vertical equilibrium. Consequently, equal and opposite forces are applied by BC to the outer parts of the beam at B and C, as in diagram (c). We can now draw the shear force diagram as at (d) and the bending moment diagram as at (e). The following observations should be made:

1. There is no change of shear across the hinges at B and C since no load is applied at these positions.
2. The bending moment is zero at the hinges B and C.

Figure 5.9 Hinged beam

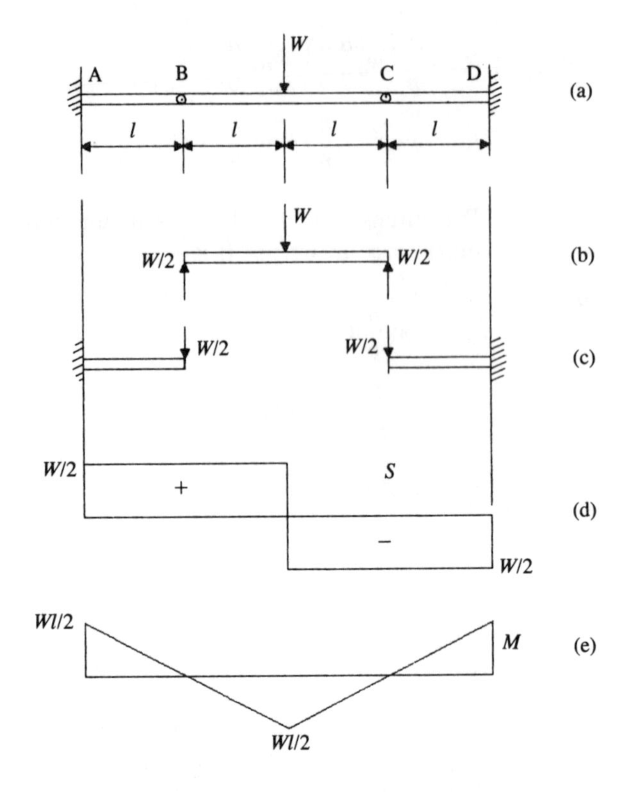

It is of interest in this particular case that the shear and bending are unaffected by the presence of the hinges at B and C, since these would have been points of inflexion had the beam been continuous from A to D. The reader is invited to show this, and also to study the effect on the maximum moment in AB and CD if the hinge positions are moved along the beam, keeping them symmetrically disposed about the centre of AD.

Example 5.6

The beam ABCDEF shown in Fig. 5.10 is continuous over the support at E, and we wish to construct shear force and bending moment diagrams. We need to determine the reactions first, so we take moments about the hinge to the left and get $R_A = W/4$ (diagram (b) should also be studied). Then we take moments about F for the part CDEF, remembering the downwards shear force at C (diagram (c)), and get $R_E = 3W/2$. R_F is therefore $2W - W/4 - 3W/2 = W/4$. The shear force diagram is at (d) and the bending moment diagram at (e). The maximum bending moment is at E and is $WL/4$ (hogging); there is a maximum sagging bending moment at B of $WL/16$; and in EF there is a turning point at $L/4$ from F.

Figure 5.10 Example 5.6

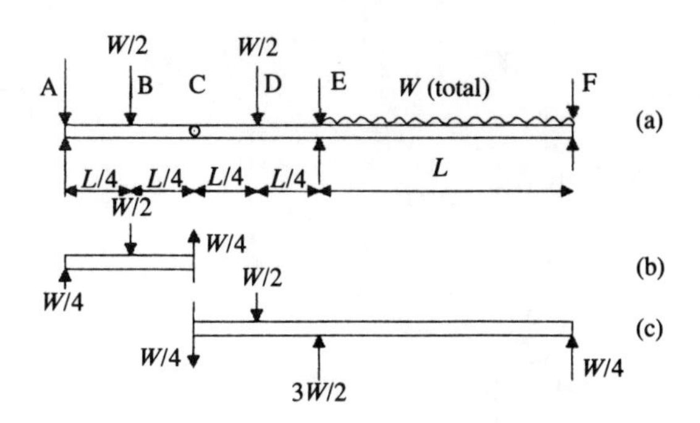

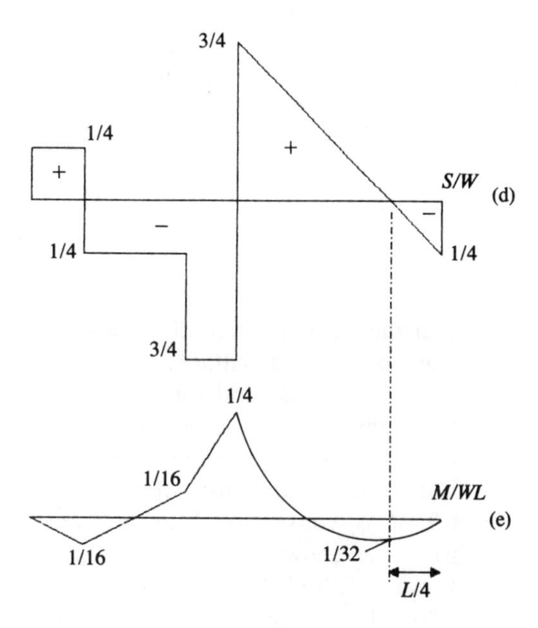

5.6 Twisting moment diagrams and the three-dimensional force system

In order to introduce the twisting moment diagram we need to consider a general three-dimensional force system as in Fig. 5.11. The rectangular axes through the centroid C of the cross-section are X, Y and Z, and it is conventional to take the X axis as coincident with the member axis. In order to adhere to this convention we shall need to reorientate our axes as we move from member to member in a structure – from horizontal members to vertical members, for example. Such a 'movable' system of axes is called a 'local' system, as distinct from the 'global' system which applies to the complete structure and which will not be changed once we have started the analysis.

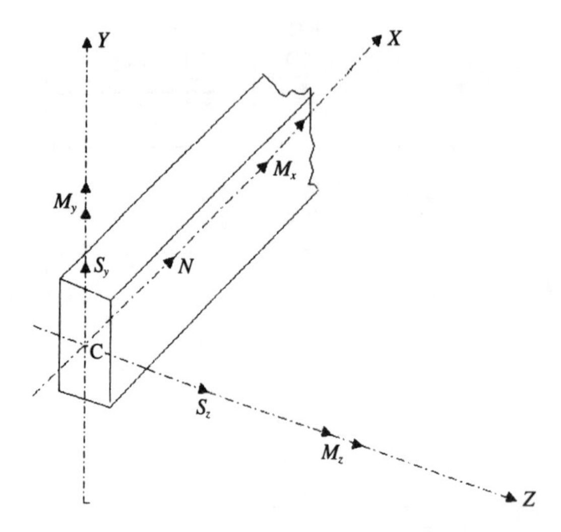

Figure 5.11 Three-dimensional force system

In the general three-dimensional situation we need to recognize six stress resultants: moments about each of the three rectangular axes, and forces in each of the three perpendicular directions. These are all shown in Fig. 5.11. The moments M_x, M_y and M_z are positive if acting in a clockwise sense when viewed in the positive directions of the axes. As we mentioned in section 4.6, it is conventional to indicate these with double-headed arrows as shown. The moment M_x is the 'twisting' moment or 'torque'. S_y and S_z are the shear forces, and N is the thrust.

Now let us see how this works in practice. In Fig. 5.12(a) we have a structure with two members at right angles. The vertical member is rigidly constrained at its lower end. The local coordinate systems and resulting diagrams are shown at (b) for the horizontal member and at (c) for the vertical member. In both cases the twisting moment is M_x. The signs need careful study. The bending moment diagrams are drawn on the tension sides of the members.

5.7 Complete structures

We now consider the construction of stress resultant distribution diagrams for complete structures. We continue to make the assumption that sufficient information is available to us, from analysis or otherwise, to enable the construction of the diagrams from statical principles.

Figure 5.12 Stress resultants in three dimensions

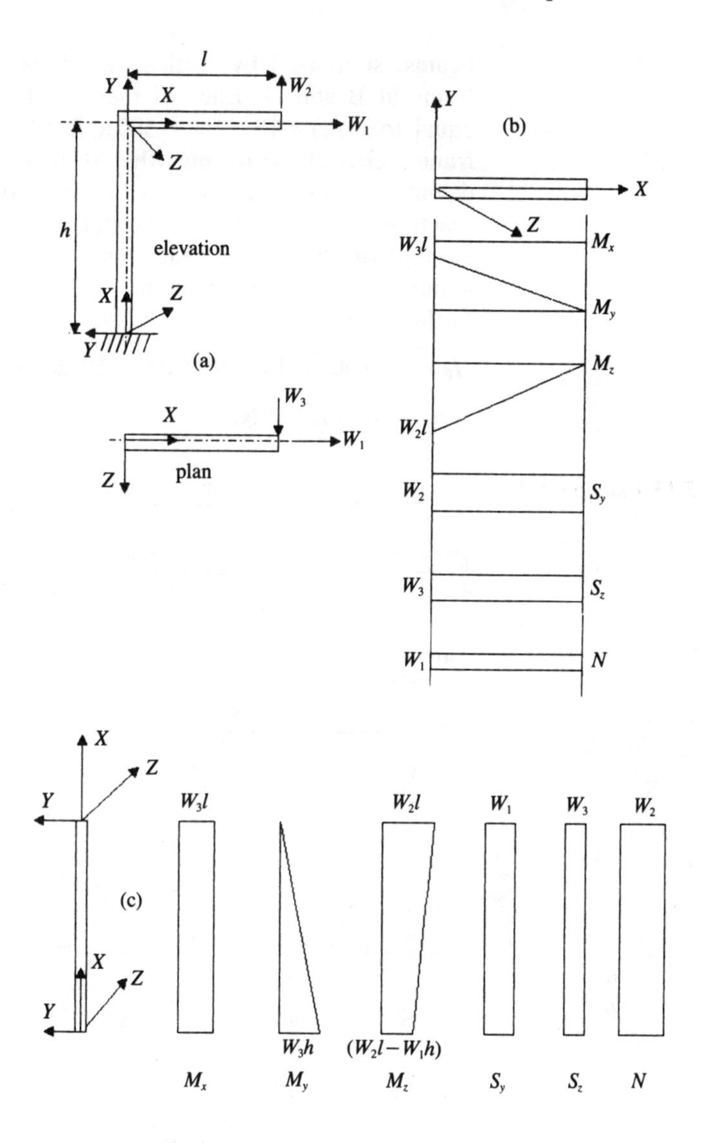

Example 5.7

Consider the structure shown in Fig. 5.13(a). We wish to draw the bending moment diagram corresponding to the uniformly distributed load w. Before commencing problems like this it is advisable to give some thought to the order in which the work will be approached. A judicious choice of order will often lead to a quicker solution than where one 'jumps in at the deep end'!

The side spans AB and DE are really just simply-supported

beams, supported by vertical reactions at A and E and by the frame at B and D. Each of these vertical reactions is therefore equal to $(1/2) \times 10 \times 100 = 500$ kN. We should now analyse the frame GBCDF carrying the distributed load on BD and downwards loads of 500 kN at B and D. Thus the vertical reactions at G and F are each $(1/2)(20 \times 100) + 500 = 1500$ kN. To find the horizontal reactions at G and F, we take moments about the hinge at C either to the left or the right. For example, to the right of C,

$$H_F \times 10 + 500 \times 10 + 10 \times 100 \times 10/2 = 1500 \times 15$$

Hence $H_F = 1250$ kN.

Figure 5.13 Example 5.7

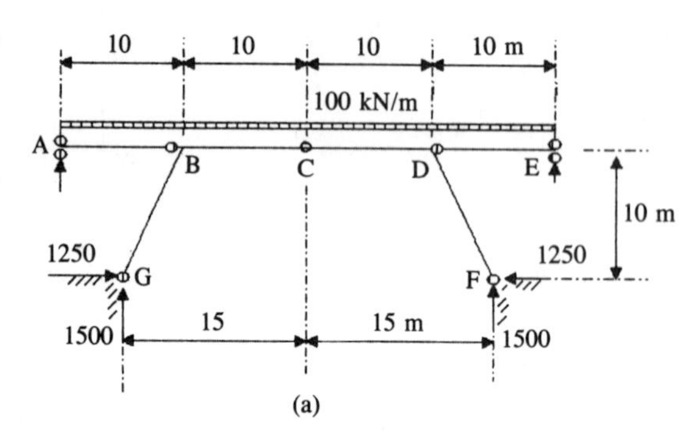

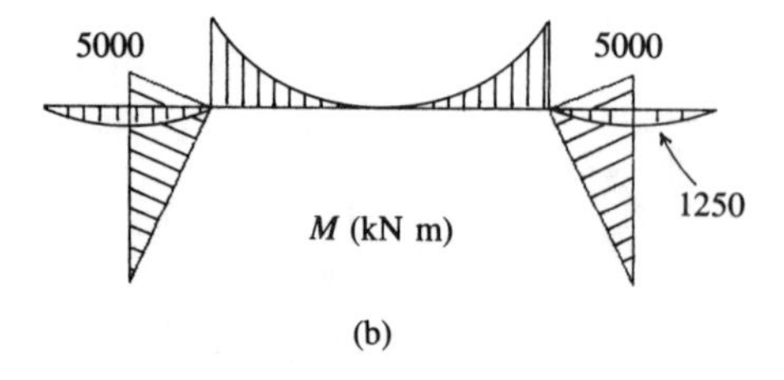

The bending moment diagram can now be drawn as in Fig. 5.13(b). For each member the diagram is drawn on the tension side and the ordinates are taken at right angles to the member axis.

Example 5.8

The drawing of bending moment diagrams for continuous beams, such as that in Fig. 5.14(a), is best carried out span by span once the bending moments at the ends of each span are known. This is information which will normally be available from the structural analysis. Suppose we are given the end moments in each span as in diagram (b), the sign convention being that clockwise moments are positive. The first thing we must do is to examine the signs of these moments to determine the sense of the curvatures produced.

Figure 5.14 Example 5.8

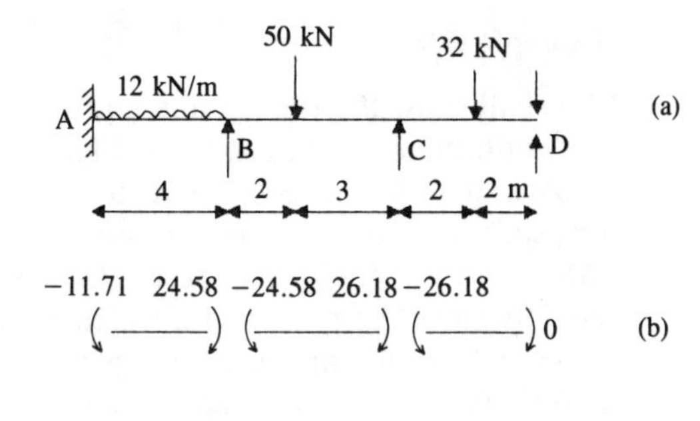

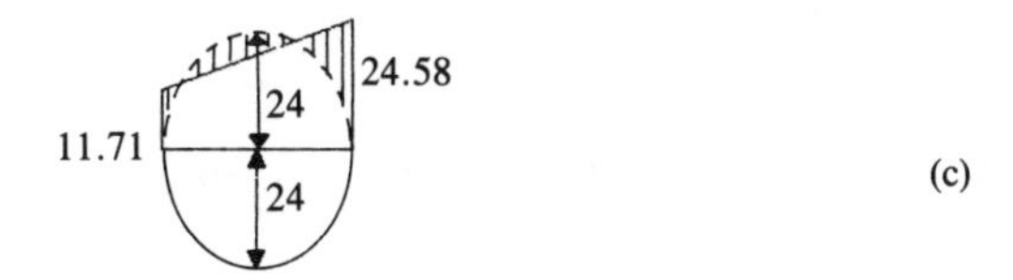

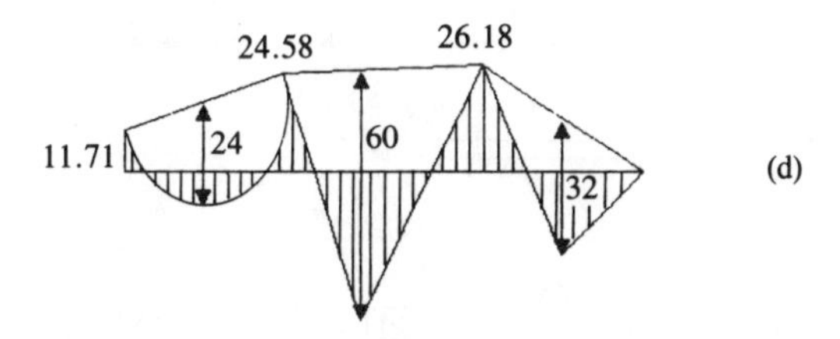

Take span AB as an example. The moment at A is negative, therefore producing curvature which is concave downwards (hogging); the moment at B to the left is positive, and this also produces curvature which is concave downwards. Since both moments produce the same kind of curvature, we plot both on the same side of the member as in diagram (c). We now draw a construction line between these plotted points to assist us in completing the diagram for this span. The bending moment diagram for the span if it were simply-supported is known as the 'free' bending moment diagram (FBMD), and for this span it will be a parabola with a mid-span ordinate of $wl^2/8 = (12 \times 4^2)/8 =$ 24 kN m. Since this produces curvature which is concave upwards (sagging), we draw this below the line.

Now we need to subtract these diagrams geometrically, and this is easily done by inverting one of them. Suppose we invert the parabola and get the dashed line in diagram (c); then the resulting bending moment diagram is shown shaded. We can obtain a more convenient geometrical subtraction if we construct the parabola using the line joining the end moment ordinates as the base. This allows us to retain the axis of the beam as a datum line. On this basis we get the diagram as at (d). One advantage is that we retain our convention of drawing the diagram on the tension side of the member axis.

Example 5.9

We shall draw the shear force and bending moment diagrams for the continuous beam shown in Fig. 5.15. The bending moments at A and B have been found to be 142.4 kN m and 75.3 kN m respectively, both producing curvature which is concave down. Whereas we could construct the bending moment diagram directly from the given data, we cannot construct the shear force diagram without first calculating the vertical reactions at A, B and C. We do that first, and again it helps if we proceed span by span.

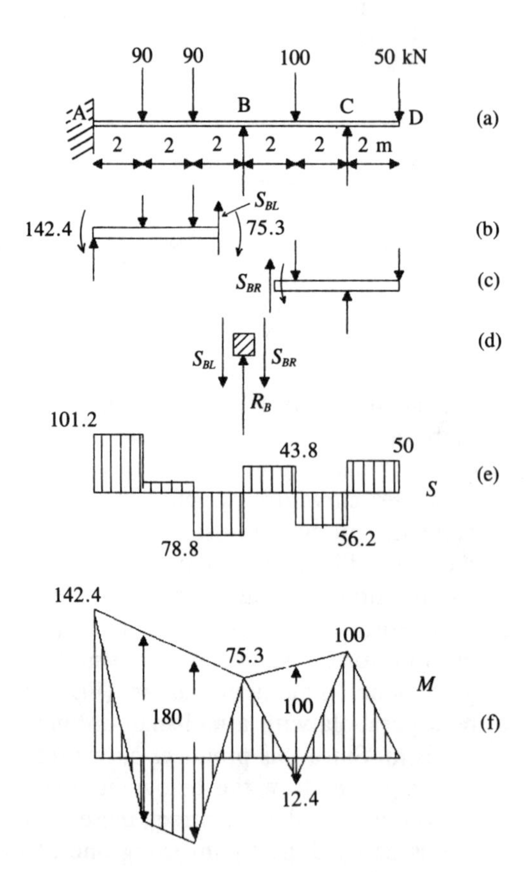

Figure 5.15 Example 5.9

Consider separately span AB in diagram (b), and span BCD in diagram (c). The internal moment of 75.3 kN m (hogging) at B acts clockwise on BA and anticlockwise on BCD. Now it will help our understanding of what is happening at B if we imagine a cut to be made to the left of B as in diagram (b) and to the right of B in diagram (c). In this way the shear forces at B are more readily seen; we label them S_{BL} and S_{SR}. We turn to diagram (d) and see a section of the beam exactly at B. Look at the directions of the shear forces S_{BL} and S_{BR}, the external vertical reaction at B, and the internal moment of 75.3. Do not proceed until this situation is thoroughly understood.

If we now take moments about B for the part BCD in diagram (c), and note that S_{BR} has no moment about B, we get

$$R_C = \frac{1}{4}(100 \times 2 + 50 \times 6 - 75.3) = 106.2 \text{ kN}$$

Now take moments about B for BA in diagram (b):

$$R_A = \frac{1}{6}(90 \times 2 + 90 \times 4 + 142.4 - 75.3) = 101.2 \text{ kN}$$

We can now find R_B by expressing vertical equilibrium of the whole structure:

$$R_B = 90 + 90 + 100 + 50 - 106.2 - 101.2 = 122.6 \text{ kN}$$

However, if we have made a mistake in calculating R_C or R_A this will not be detected. It will be better to evaluate R_B without using the vertical equilibrium condition, which can then be reserved for a final check. Take moments about A for the whole beam:

$$R_B = \frac{1}{6}(90 \times 2 + 90 \times 4 + 100 \times 8 + 50 \times 12$$

$$- 106.2 \times 10 - 142.4) = 122.6 \text{ kN}$$

which agrees with our vertical equilibrium equation.

We have not yet evaluated S_{BL} or S_{BR}, but these will come out quite naturally when we draw the shear force diagram as at (e):

$$S_{BL} = 78.8 \text{ kN}$$

$$S_{BR} = 43.8 \text{ kN}$$

Adding these as in diagram (d) gives

$$R_B = S_{BL} + S_{BR} = 78.8 + 43.8 = 122.6 \text{ kN}$$

The bending moment diagram is quite easily drawn as at (f).

Example 5.10

For the three-storey rigidly jointed frame shown in Fig. 5.16 we wish to draw the bending moment and thrust diagrams given the following results of an analysis which neglects axial shortening of the members:

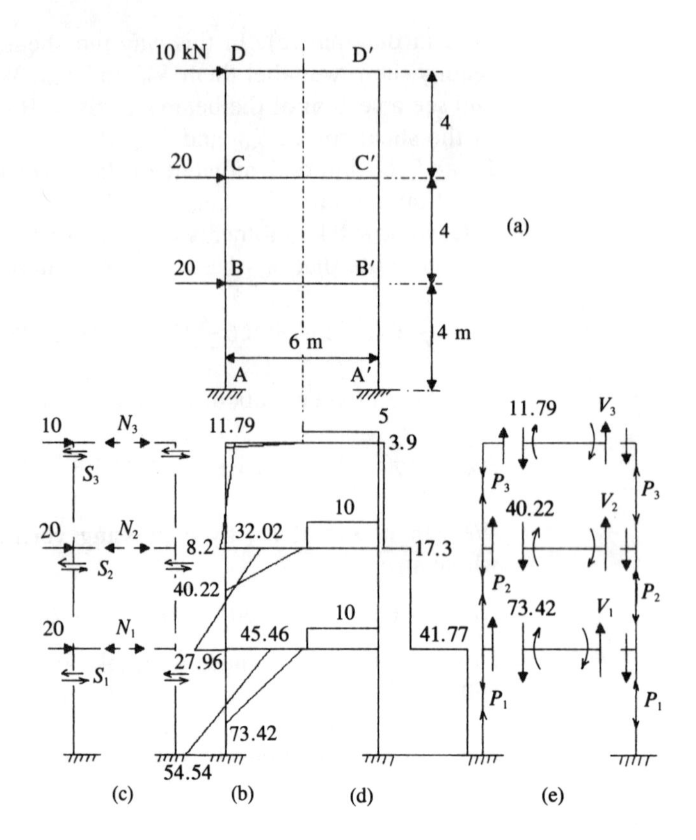

Figure 5.16 Example 5.10

M_{AB}	M_{BA}	M_{BC}	M_{CB}
−54.54	−45.46	−27.96	−32.02

M_{CD}	M_{DC}	
−8.2	−11.79	kN m

At first sight we seem to be short of information for this structure, since the given numerical values apply only to the left-hand side of the structure. However, we can deduce the remaining results from the skew-symmetrical behaviour of this structure because points of inflexion will exist at the mid-span of each beam. This allows us to construct the bending moment diagram as in Fig. 5.16(b). This diagram applies over the left-hand half of the frame. The signs of the bending moments (rotational convention) are the same on the right-hand side according to skew symmetry.

Given the bending moment diagram we can now calculate any

internal forces we wish. Let us first find the axial forces in the beams. We do this by calculating the shear forces in the columns, as in (c). In both columns of the upper storey,

$$S_3 = (11.79 + 8.2)/4 = 5 \text{ kN}$$

Now we could have deduced this result from the type of symmetry exhibited by this structure under this particular form of loading. The shear in the two columns of any storey must be equal, and in total equal to the applied external shear at the storey level immediately above. Thus, in the second storey,

$$S_2 = (27.96 + 30.02)/4 = 14.5 \text{ kN}$$

This should be 15 kN, so we suspect some rounding errors in the given data for this storey. We shall proceed with $S_2 = 15$ kN. Similarly $S_1 = 25$ kN, and this is obviously correct since it equals one-half of the total applied load. The axial forces in the beams are therefore

$$N_3 = S_3 = 5 \text{ kN}$$
$$N_2 = S_2 - S_3 = 10 \text{ kN}$$
$$N_1 = S_1 - S_2 = 10 \text{ kN}$$

Just as we obtained beam axial forces from column shears, we calculate column axial forces from beam shears. From diagram (e),

$$V_3 = 11.79 \times (2/6) = 3.9 \text{ kN}$$
$$V_2 = 40.22 \times (2/6) = 13.4 \text{ kN}$$
$$V_1 = 73.42 \times (2/6) = 24.47 \text{ kN}$$

We sum these shears down the columns to arrive at the axial forces:

$$P_3 = 3.9 \text{ kN}$$
$$P_2 = 3.9 + 13.4 = 17.3 \text{ kN}$$
$$P_1 = 17.3 + 24.47 = 41.77 \text{ kN}$$

The thrust diagram is shown at (d) for the right-hand side. The diagram for the left-hand side is similar except that the columns are in tension, of course.

This type of structure is amenable to an approximate form of analysis which we shall explore in Chapter 12.

Exercises

5.1 A fixed-end beam with symmetrically disposed internal hinges is shown in Fig. 5.17. Find the value of x such that the end moments are each $wl^2/12$.

Figure 5.17 Exercise 5.1

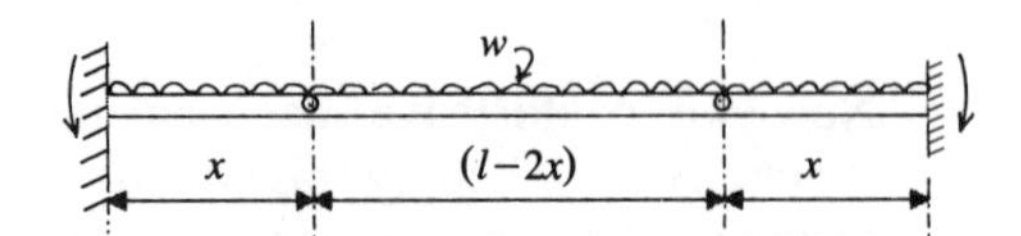

5.2 Draw bending moment, shear and thrust diagrams for the structure shown in Fig. 5.18.

Figure 5.18 Exercise 5.2

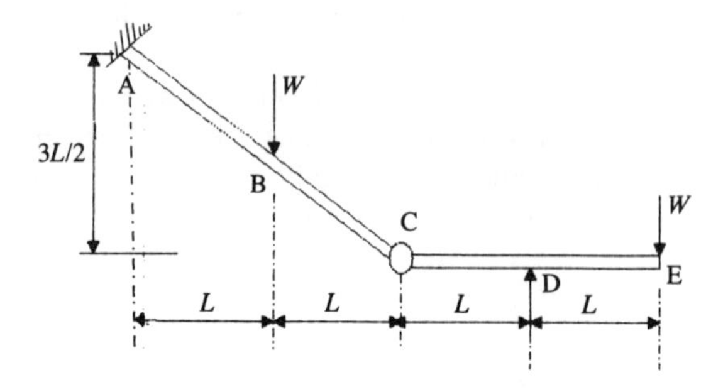

5.3 Figure 5.19 shows a 'cantilever and suspended span' arrangement ABCDEF. If the part BCD carries a uniformly distributed load w, calculate M_B.

Figure 5.19 Exercise 5.3

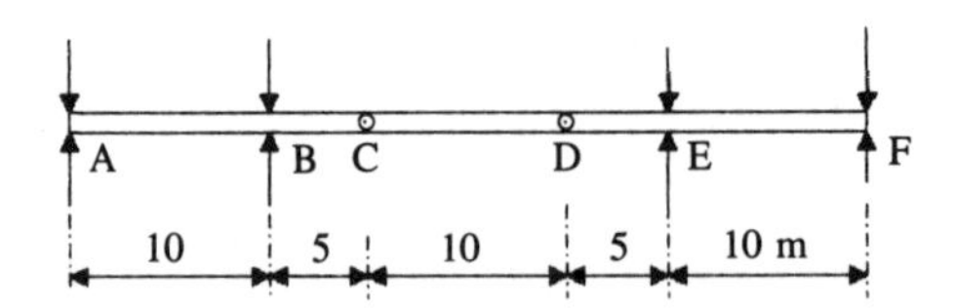

5.4 (a) Draw the bending moment and shear force diagrams for the structure shown in Fig. 5.20.
(b) If member BD is inserted and a hinge is introduced at C, determine the maximum bending moment in BC and compare it with that from (a).

Figure 5.20 Exercise 5.4

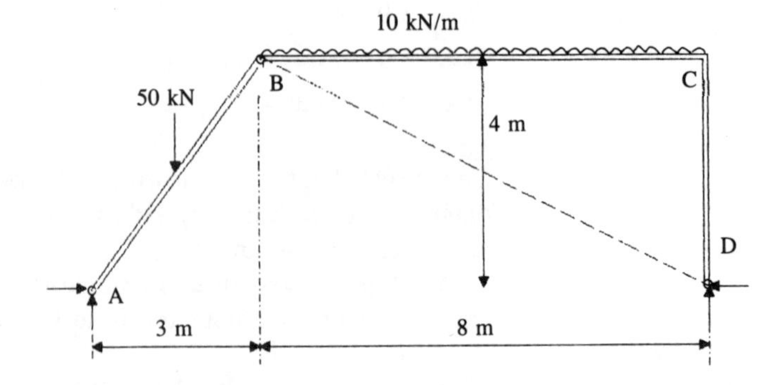

5.5 In Fig. 5.21 the structure ABDCD′B′A′ lies in a horizontal plane.

Figure 5.21 Exercise 5.5

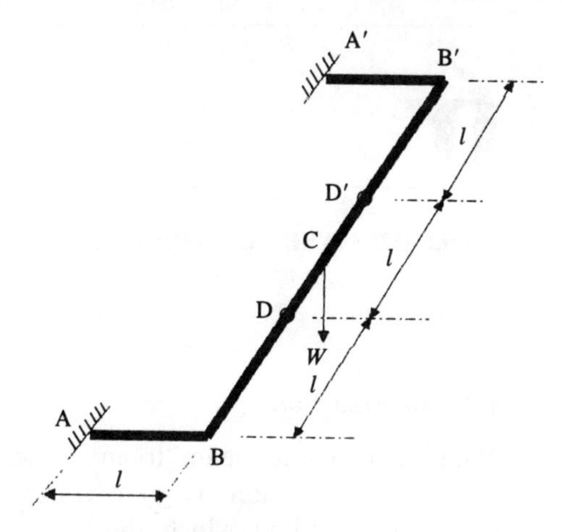

A and A′ are *encastré*. Draw bending moment and twisting moment diagrams corresponding to a single vertical load W at C, the centre of DD′.

CHAPTER

6

The analysis of beams

6.1 Introduction

We give the name 'beam' to any structural element whose length, in an axial direction, is great compared with its cross-sectional dimensions and in which the predominant structural action is bending; however, this will generally be accompanied by transverse shear, axial force and sometimes twisting actions. The beam is one of the most important structural elements owing to its widespread use in many types of structure, and its importance is such as to warrant a detailed treatment. The 'column' differs from the beam in that the structural action is predominantly in response to axial compressive forces although the other actions of bending, shear and twisting may also be present. The theory of beams is usually developed on the understanding that the flexural behaviour is unaffected by the presence of any axial force even if the stresses due to axial force are included in the analysis. Conversely, in the treatment of columns it is often assumed that the response to axial load is unaffected by the presence of (small) bending moments. If both flexure and axial force are significant then we have what is known as the 'beam-column', which requires special treatment in which the interaction of axial force and bending is properly represented.

If the beam has a constant cross-section throughout its length, it is said to be 'prismatic'. Non-prismatic beams are used and require some additional numerical analysis which we shall examine later in this chapter. The shape and dimensions of the cross-section of a beam are important as we shall see. The cross-section of the beam may be solid or hollow and, it may be 'thin-walled' or 'thick-walled'. The cross-section may have one or more axes of symmetry or none at all. If the cross-section is hollow it will be important to know whether it is 'open' or 'closed', especially if it is required to resist a twisting moment.

Since cross-sectional geometry and mathematical properties are of central importance to the development of beam theory we shall examine these first and will then go on to develop the general theory of pure bending of beams. We shall then consider special problems of thin-walled beams and non-prismatic beams. The column and the beam-column will be examined in Chapter 7.

6.2 Cross-sectional properties

Here we list the mathematical definitions of the properties of cross-sections. We do not introduce proofs of the various relationships but refer the reader to standard texts (Budynas 1977; Elias 1986). We shall direct positive Z and Y axes as shown in Fig. 6.1(a); the X axis will be reserved for use as the longitudinal axis of the beam. We will generally employ the Y and Z axes as vertical and horizontal axes respectively in the beam cross-section. The origin of axes will be taken at C, the geometrical centroid of the cross-section; thus X will be the centroidal axis of the member.

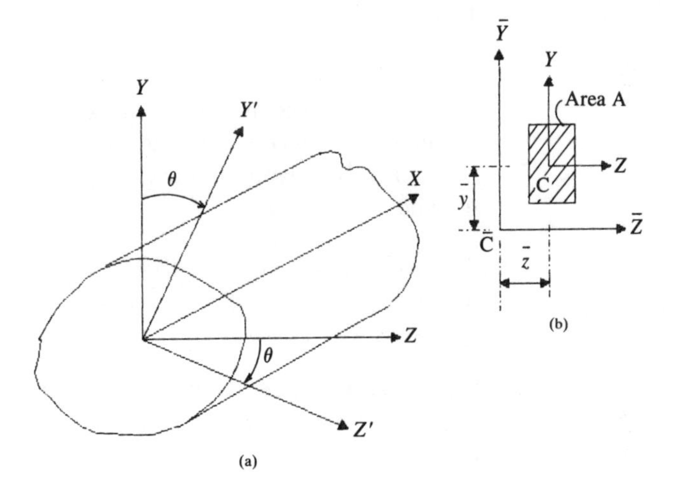

Figure 6.1 Geometrical properties of cross-sections

The cross-sectional area is given by

$$A = \int_A \mathrm{d}A \qquad [6.1]$$

The first moments of area are

$$Q_y = \int_A z \, \mathrm{d}A \qquad [6.2]$$

$$Q_z = \int_A y \, \mathrm{d}A \qquad [6.3]$$

At the centroid C,

$$Q_{y\mathrm{C}} = Q_{z\mathrm{C}} = 0 \qquad [6.4]$$

and thus C will lie on any axis of symmetry. Equation [6.4] is useful for determining the position of the centroid.

The second moments of area are

$$I_y = \int_A z^2 \, dA \qquad [6.5]$$

$$I_z = \int_A y^2 \, dA \qquad [6.6]$$

and the product moment of area is

$$I_{zy} = \int_A zy \, dA \qquad [6.7]$$

The 'polar' second moment is

$$I_P = \int_A r^2 \, dA \qquad [6.8]$$

where r is measured from the X axis. Now since $r^2 = z^2 + y^2$, then

$$I_P = I_y + I_z \qquad [6.9]$$

In calculating second moments of area we make use of the 'parallel axis' theorem. In Fig. 6.1(b), if $\bar{Z}$ and $\bar{Y}$ are parallel to Z and Y respectively and if C is the centroid of the shaded area A, then

$$I_{\bar{y}} = I_y + A\bar{z}^2 \qquad [6.10]$$

$$I_{\bar{z}} = I_z + A\bar{y}^2 \qquad [6.11]$$

$$I_{\bar{z}\bar{y}} = I_{zy} + A\bar{z}\bar{y} \qquad [6.12]$$

If the Y and Z axes are rotated through angle θ, as in Fig. 6.1(a), then the following relationships apply (*Civil Engineer's Reference Book*):

$$I_{y'} = \frac{1}{2}(I_y + I_z) + \frac{1}{2}(I_y - I_z)\cos 2\theta - I_{zy}\sin 2\theta \qquad [6.13a]$$

$$I_{z'} = \frac{1}{2}(I_y + I_z) - \frac{1}{2}(I_y - I_z)\cos 2\theta + I_{zy}\sin 2\theta \qquad [6.13b]$$

$$I_{z'y'} = \frac{1}{2}(I_y - I_z)\sin 2\theta + I_{zy}\cos 2\theta \qquad [6.13c]$$

Now when

$$\tan 2\theta = \frac{-2I_{zy}}{I_y - I_z} \qquad [6.14]$$

then $I_{z'y'} = 0$ and

$$I_{max,min} = \frac{I_z + I_y}{2} \pm \sqrt{\left[\left(\frac{I_y - I_z}{2}\right)^2 + I_{zy}^2\right]} \qquad [6.15]$$

For every section there is a particular value of the angle θ for which the product of inertia $I_{z'y'}$ is zero. The corresponding orthogonal axes are called the 'principal' axes of the cross-section. Geometrical properties for some of the standard cross-sections are given in Appendix 1.

6.3 Pure bending of an arbitrary section beam: axes of symmetry

The basic assumption made in the development of the engineer's theory of bending (ETB) is that a plane section through the beam normal to the longitudinal axis will remain plane and normal to the axis under an applied bending moment. Longitudinal stresses will be developed and any arbitrary plane section will therefore displace in such a way that tensile stresses will develop on one side and compressive stresses on the other side of a particular axis called the 'neutral' axis. There is therefore no longitudinal strain at points on this axis. It is further assumed that the material of the beam is homogeneous, isotropic and linearly elastic.

Consider the arbitrary cross-section beam shown in Fig. 6.2(a).

Figure 6.2 Pure bending of arbitrary section beam

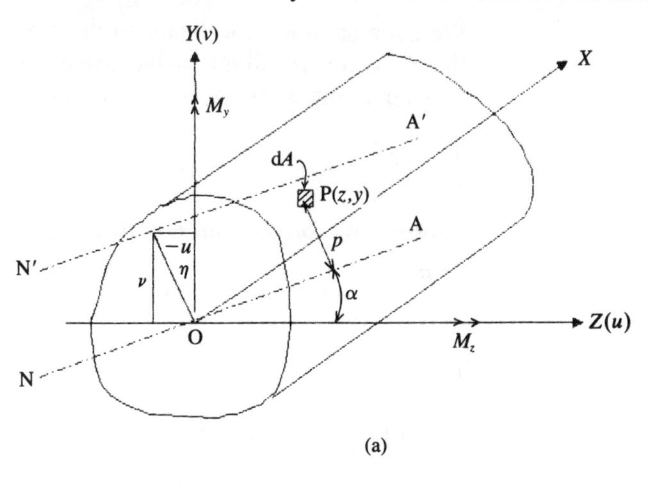

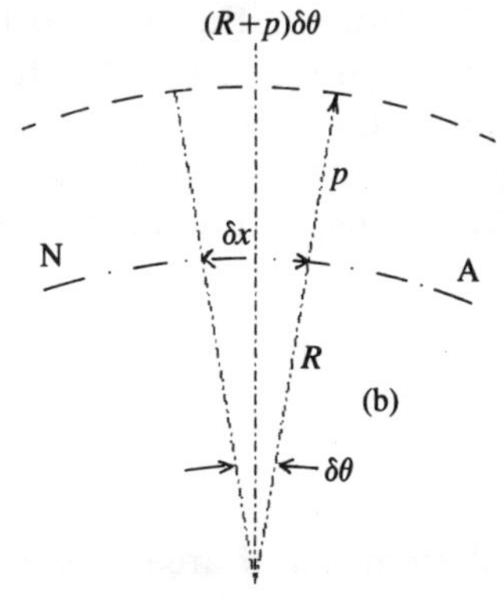

The longitudinal axis is X and any arbitrary pair of orthogonal axes Y and Z lie in the plane of the cross-section. Suppose that the origin of plane Y–Z lies on the neutral axis at point O. We shall show that O is the centroid of the cross-section. Element δA at point P in the cross-section has coordinates (z, y) and is distant p from the neutral axis. The normal stress on the element δA (with tension assumed positive) is

$$\sigma_{x,\mathrm{P}} = E\varepsilon_{x,\mathrm{P}} \qquad [6.16]$$

where $\varepsilon_{x,\mathrm{P}}$ is the strain in the X direction at P.

Now, if the unstrained length of the element is δx and the beam is bent to a radius of curvature R from the neutral axis, the strained length of the element (see Fig. 6.2(b)) will be $(R+p)\delta\theta$. Hence the strain at p from the neutral axis is

$$\varepsilon_{x,\mathrm{P}} = \frac{(R+p)\,\delta\theta - R\,\delta\theta}{R\,\delta\theta} = p/R$$

The stress is therefore, from eqn [6.16],

$$\sigma_{x,\mathrm{P}} = \frac{E}{R}p = \frac{E}{R}(y\cos\alpha - z\sin\alpha) \qquad [6.17]$$

We have assumed the beam to be subjected to pure bending, so there is no resultant axial force on the cross-section. The corresponding stress resultant is therefore zero; hence

$$\int_A \sigma_x\,\mathrm{d}A = 0 \qquad [6.18]$$

Substituting for σ_x from eqn [6.17],

$$\int_A \frac{E}{R}p\,\mathrm{d}A = 0$$

i.e.

$$\int_A p\,\mathrm{d}A = 0. \qquad [6.19]$$

since E/R is constant. Equation [6.19] tells us that the neutral axis passes through the centroid of the cross-section.

Now, the bending moments M_z and M_y can be obtained by integrating the moments of the force on the element about the relevant axes. Thus, with tension assumed positive,

$$M_z = \int_A \sigma_x y\,\mathrm{d}A \qquad [6.20]$$

$$M_y = -\int_A \sigma_x z\,\mathrm{d}A \qquad [6.21]$$

The second moments of area I_z and I_y and product of inertia I_{zy}

are given by eqns [6.5], [6.6] and [6.7]. So, substituting eqn [6.17] in eqns [6.20] and [6.21] we obtain

$$M_z = \frac{E}{R}\int_A (y \cos\alpha - z \sin\alpha) y \, dA \qquad [6.22]$$

$$M_y = -\frac{E}{R}\int_A (y \cos\alpha - z \sin\alpha) z \, dA \qquad [6.23]$$

Noting eqns [6.5], [6.6] and [6.7] we obtain

$$M_z = \frac{E}{R}(I_z \cos\alpha - I_{zy} \sin\alpha) \qquad [6.24]$$

$$M_y = -\frac{E}{R}(I_{zy} \cos\alpha - I_y \sin\alpha) \qquad [6.25]$$

We can now use eqns [6.24] and [6.25] to eliminate $(E/R) \sin\alpha$ and $(E/R) \cos\alpha$ and obtain an expression for the normal stress σ_x:

$$\sigma_x = \frac{(M_z I_y + M_y I_{zy}) y}{I_y I_z - I_{zy}^2} - \frac{(M_y I_z + M_z I_{zy}) z}{I_y I_z - I_{zy}^2} \qquad [6.26]$$

If axes Y and Z are principal axes, then $I_{zy} = 0$ and eqn [6.26] takes the form

$$\sigma_x = \frac{M_z y}{I_z} - \frac{M_y z}{I_y} \qquad [6.27]$$

It is known that an axis of symmetry always coincides with a principal axis; hence if either Z or Y is an axis of symmetry then Z and Y are principal axes and eqn [6.27] applies. For asymmetrical cross-sections, if the principal axes are known then eqn [6.27] can be used. If this is not so then eqn [6.26] is advised, where Z and Y are convenient non-principal axes.

The angle α, which orientates the neutral axis, can be obtained by putting $\sigma_x = 0$ in eqn [6.26]:

$$y/z = \tan\alpha \qquad [6.28]$$

$$\tan\alpha = \frac{M_y I_z + M_z I_{zy}}{M_z I_y + M_y I_{zy}} \qquad [6.29]$$

6.4 Stress and deformation in pure bending

We now develop relationships between the applied bending moments, cross-section and material properties, and the displacements occurring in the beam; in fact, we shall derive the general differential equation of bending. The neutral axis (NA) passes through the centroid as already shown (eqn [6.4]). If there are no shear stresses and there is no twisting of the beam, then the transverse displacement η is perpendicular to the neutral axis. Suppose the neutral axis moves into position N′A′, as shown in Fig. 6.2(a). The curvature of the beam measured to the neutral axis is $1/R$, and

$$\frac{1}{R}=\frac{d^2\eta/dx^2}{[1+(d\eta/dx)^2]^{3/2}}$$

which is usually approximated to

$$\frac{1}{R}=\frac{d^2\eta}{dx^2} \qquad [6.30]$$

Now, let the displacements of the centroid be $-u$ in the Z direction and v in the Y direction, as shown in Fig. 6.2(a). Hence

$$-u=\eta\sin\alpha; \qquad \eta=-u/\sin\alpha \qquad [6.31]$$

$$v=\eta\cos\alpha; \qquad \eta=v/\cos\alpha \qquad [6.32]$$

Now, from eqns [6.24] and [6.25], and noting that positive applied moments will induce negative curvatures in the η–X plane,

$$M_z=\frac{E}{R}(I_z\cos\alpha-I_{zy}\sin\alpha)=-E\frac{d^2\eta}{dx^2}(I_z\cos\alpha-I_{zy}\sin\alpha)$$

$$M_y=-\frac{E}{R}(I_{zy}\cos\alpha-I_y\sin\alpha)=E\frac{d^2\eta}{dx^2}(I_{zy}\cos\alpha-I_y\sin\alpha)$$

Combining eqns [6.30], [6.31] and [6.32],

$$\frac{\sin\alpha}{R}=-\frac{d^2u}{dx^2} \qquad [6.33]$$

$$\frac{\cos\alpha}{R}=\frac{d^2v}{dx^2} \qquad [6.34]$$

Hence

$$M_z=-E\left(I_z\frac{d^2v}{dx^2}+I_{zy}\frac{d^2u}{dx^2}\right) \qquad [6.35]$$

$$M_y=E\left(I_{zy}\frac{d^2v}{dx^2}+I_y\frac{d^2u}{dx^2}\right) \qquad [6.36]$$

Equations [6.35] and [6.36] are now solved for the curvatures.

The horizontal curvature is

$$\frac{d^2u}{dx^2} = \frac{M_y I_z + M_z I_{zy}}{E(I_y I_z - I_{zy}^2)} = \frac{1}{R_y} \qquad [6.37]$$

where R_y is the radius of curvature about axis Y, and the vertical curvature is

$$\frac{d^2v}{dx^2} = \frac{-(M_z I_y + M_y I_{zy})}{E(I_y I_z - I_{zy}^2)} = \frac{1}{R_z} \qquad [6.38]$$

where R_z is the radius of curvature about axis Z. Again, if Y and z are principal axes, then $I_{zy} = 0$ and eqns [6.37] and [6.38] simplify to

$$\frac{d^2u}{dx^2} = \frac{M_y}{EI_y} \qquad [6.39]$$

$$\frac{d^2v}{dx^2} = \frac{-M_z}{EI_z} \qquad [6.40]$$

Summary

This section may be summarized as follows, with equation numbers repeated for convenience.

1. In general, assuming tensile stress positive, the normal stress is given by

$$\sigma_x = \frac{(M_z I_y + M_y I_{zy})y - (M_y I_z + M_z I_{zy})z}{I_y I_z - I_{zy}^2} \qquad [6.26]$$

The slope of the neutral axis is given by

$$\tan \alpha = \frac{M_y I_z + M_z I_{zy}}{M_z I_y + M_y I_{zy}} \qquad [6.29]$$

2. For Y and Z principal axes,

$$\sigma_x = \frac{M_z y}{I_z} - \frac{M_y z}{I_y} \qquad [6.27]$$

The curvatures in the Z–X and Y–X planes respectively are

$$\frac{1}{R_y} = \frac{d^2u}{dx^2} = \frac{M_y}{EI_y} \qquad [6.39]$$

$$\frac{1}{R_z} = \frac{d^2v}{dx^2} = -\frac{M_z}{EI_z} \qquad [6.40]$$

3. The relationships between the signs of the applied bending moments and the curvatures are determined by the conventions in Fig. 6.3, with σ_x positive if tensile.

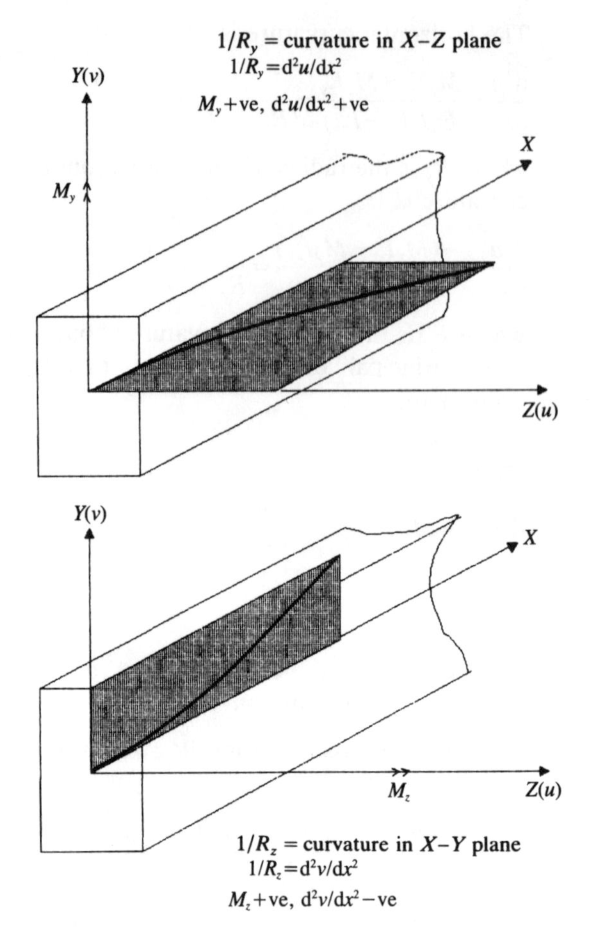

Figure 6.3 Sign conventions for curvatures

Example 6.1

A beam has a cross-section as shown in Fig. 6.4. A bending moment $M_z = -78.125$ kN m is applied; determine the maximum tensile and compressive stresses induced in the cross-section.

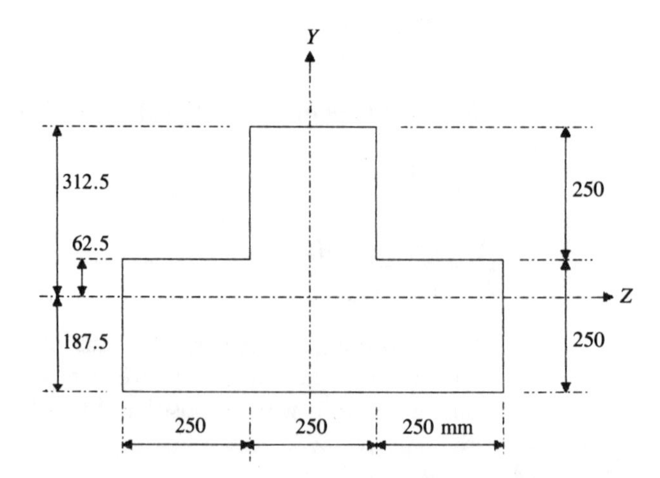

Figure 6.4 Example 6.1

The Y axis is an axis of symmetry and is therefore a principal axis. The Z axis is therefore the other principal axis and can be located by observing that the first moment of area about a principal axis is zero. Thus since $Q_z = \int y \, dA = 0$ (eqn [6.3]) we can take first moments of area about any convenient axis parallel to Z and equate this to $A\bar{y}$, where A is the total cross-sectional area and $\bar{y}$ is the distance from the chosen axis to the Z axis. Taking moments about the base of the section,

$$A\bar{y} = 250 \times 750 \times 125 + 250 \times 250 \times 375$$

As $A = 250 \times 1000$, then $\bar{y} = 187.5$ mm.

Now to find I_z we break the section into rectangles and use the standard result from Appendix 1 for second moment of area about one edge ($bd^3/3$). Hence

$$I_z = 750 \times \frac{187.5^3}{3} + 500 \times \frac{62.5^3}{3} + 250 \times \frac{312.5^3}{3}$$

$$= 4.232 \times 10^9 \text{ mm}^4$$

Now eqn [6.27] is relevant, with $M_y = 0$; thus

$$\sigma_x = \frac{M_z y}{I_z} = \frac{-78.125(\text{kN m})y}{4.232 \times 10^9(\text{mm}^4)}$$

$$= -0.01\,846y \text{ N/mm}^2$$

The maximum compressive stress will be along the top edge ($y = 312.5$) and is -5.76 N/mm^2. The maximum tensile stress will be along the bottom edge ($y = -187.5$) and is 3.46 N/mm^2.

Example 6.2

The L-shaped beam shown in Fig. 6.5 carries a bending moment

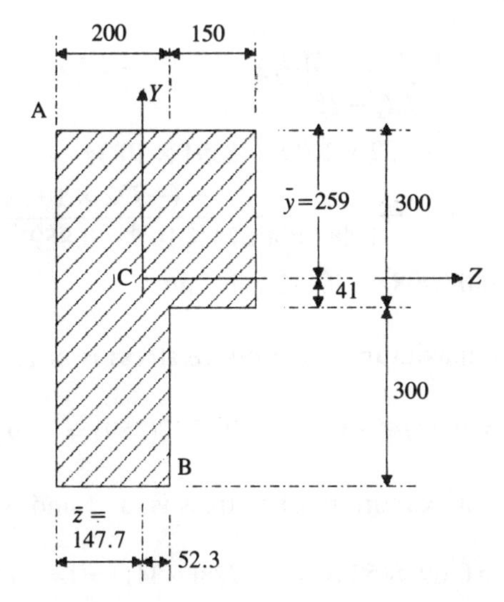

Figure 6.5 Example 6.2

$M_z = -722$ kN m. Determine the maximum compressive and tensile stresses and their locations in the cross-section.

We need first to determine the position of the centroid C:

$$\bar{y} = \frac{350 \times 300 \times 150 + 200 \times 300 \times 450}{350 \times 300 + 300 \times 200} = 259$$

$$\bar{z} = \frac{600 \times 200 \times 100 + 300 \times 150 \times 275}{165\,000} = 147.7$$

Then

$$I_z = \int y^2 \, dA = 350 \times \frac{259^3}{3} + 200 \times \frac{341^3}{3} + 150 \times \frac{41^3}{3}$$

$$= 4.674 \times 10^9 \text{ mm}^4$$

$$I_y = \int z^2 \, dA = 600 \times \frac{147.7^3}{3} + 300 \times \frac{52.3^3}{3} + 300 \times \frac{202.3^3}{3}$$

$$= 1.487 \times 10^9 \text{ mm}^4$$

$$I_{zy} = \int zy \, dA = 150 \times 300 \times 127.3 \times 109 + 600$$

$$\times 200(-47.7)(-41) = 0.8591 \times 10^9 \text{ mm}^4$$

Now $M_z = -722$ kN m and $M_y = 0$. Hence

$$\tan \alpha = \frac{M_z I_{zy}}{M_z I_y} = \frac{I_{zy}}{I_y} = \frac{0.8591 \times 10^9}{1.487 \times 10^9} = 0.5777$$

$$\alpha = 30.016°$$

and

$$\sigma_x = \frac{-M_z I_{zy} z + M_z I_y y}{(I_y I_z - I_{zy}^2)}$$

$$= \frac{-(-722 \times 10^6 \times 0.8591 \times 10^9)z + (-722 \times 10^6 \times 1.487 \times 10^9)y}{1.487 \times 4.674 \times 10^{18} - 0.8591^2 \times 10^{18}}$$

$$= 0.0998z - 0.1728y \text{ N/mm}^2$$

The maximum compressive stress is at A and is equal to

$$\sigma_x = 0.0998(-147.7) - 0.1728(259) = -59.5 \text{ N/mm}^2$$

The maximum tensile stress is at B and is equal to

$$\sigma_x = 0.0998(52.3) - 0.1728(-341) = 64.14 \text{ N/mm}^2$$

6.5 Displacements due to bending

The calculation of displacements in beams proceeds from the solution of the differential eqns [6.37] and [6.38]:

$$\frac{d^2u}{dx^2} = \frac{M_y I_z + M_z I_{zy}}{E(I_y I_z - I_{zy}^2)}$$

$$\frac{d^2v}{dx^2} = \frac{-(M_z I_y + M_y I_{zy})}{E(I_y I_z - I_{zy}^2)}$$

where $d^2u/dx^2 = 1/R_y$ is the curvature in the Z–X plane, and $d^2v/dx^2 = 1/R_z$ is the curvature in the Y–X plane.

In the majority of practical cases the beam cross-section will be symmetrical about the Y axis and the loading will be vertical. In these circumstances we need use only eqn [6.38], and the differential equation simplifies to

$$\frac{d^2v}{dx^2} = -\frac{M_z}{EI_z} \qquad [6.41]$$

We shall now spend some time exploring the use of this equation. Before going further the reader should revise the sign conventions attached to v and M_z in eqn [6.41]; Fig. 6.3 should help in this. M_z is positive if acting clockwise when viewed in the positive direction of the z axis. The displacement v is in the Y direction and is positive if upwards. The negative sign in eqn [6.41] is needed because positive moment M_z tends to produce negative curvature d^2v/dx^2. We can usually solve eqn [6.41] by direct integration, determining the constants of integration from the displacement boundary conditions of the particular problem.

6.5.1 Standard cases of beam displacements

Here we consider four cases of a uniform section beam with a symmetrical cross-section, so eqn [6.41] applies for vertical loading:

$$\frac{d^2v}{dx^2} = -\frac{M_z}{EI_z}$$

The four cases (a) to (d) are illustrated in Fig. 6.6(a) to (d) respectively.

Figure 6.6 Standard cases of beam displacements

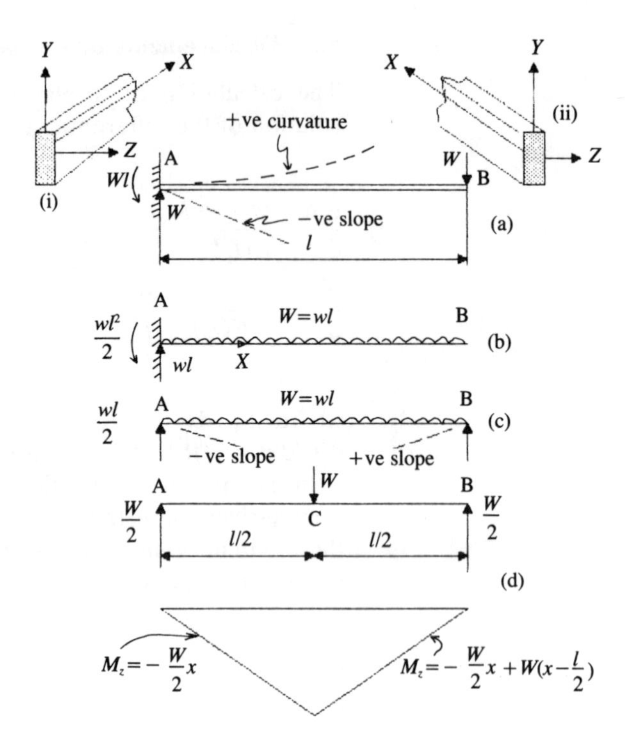

Case (a)

With the origin at A, as in Fig. 6.6(a)(i), we have

$$M_z = f(x) = Wl - Wx = W(l - x)$$

Thus

$$EI_z \frac{d^2v}{dx^2} = -W(l - x)$$

Integrating,

$$EI_z \frac{dv}{dx} = \frac{W}{2}(l - x)^2 + A$$

This is zero for $x = 0$; thus $A = -Wl^2/2$. Hence

$$EI_z \frac{dv}{dx} = \frac{W}{2}(l - x)^2 - \frac{Wl^2}{2}$$

The maximum slope occurs at $x = l$:

$$\left(\frac{dv}{dx}\right)_{max} = -\frac{Wl^2}{2EI_z}$$

Integrating again,

$$EI_z v = -\frac{W}{6}(l - x)^3 - \frac{Wl^2}{2}x + B$$

This is zero for $x = 0$; thus $B = Wl^3/6$. Hence

$$EI_z v = -\frac{W}{6}(l-x)^3 - \frac{Wl^2}{2}x + \frac{Wl^3}{6} \qquad [6.46]$$

The maximum deflection occurs at $x = l$:

$$v_{max} = -\frac{Wl^3}{3EI_z} \qquad [6.47]$$

The negative sign indicates downwards deflection.

Great care is needed in setting up eqn [6.42] in order to ensure correct signs. The reader should view the situation from the origin for both X and Z axes. Let us see if we can handle the signs correctly if we change origin to B and direct the X axis towards A, as in Fig. 6.6(a)(ii). We again have eqn [6.41]:

$$\frac{d^2v}{dx^2} = -\frac{M_z}{EI_z}$$

Now $M_z = +Wx$, i.e. clockwise about the Z axis. Therefore

$$EI_z \frac{d^2v}{dx^2} = -Wx$$

$$EI_z \frac{dv}{dx} = -\frac{W}{2}x^2 + A$$

This is zero for $x = l$; thus $A = Wl^2/2$. Hence

$$EI_z v = -\frac{W}{6}x^3 + \frac{Wl^2}{2}x + B$$

This is zero for $x = l$; thus $B = -Wl^3/3$. Hence

$$EI_z v = -\frac{W}{6}x^3 + \frac{Wl^2}{2}x - \frac{Wl^3}{3} \qquad [6.48]$$

The maximum deflection is clearly at $x = 0$, and is as before

$$v_{max} = -\frac{Wl^3}{3EI_z}$$

When integrating the bracketed terms we retain the brackets at each stage. The reader should verify that if the bracketed terms are expanded, the integration constants are changed but the final result is the same.

Case (b)

The moment at A is $wl^2/2$ and the vertical reaction is wl. Hence, with the origin at A,

$$M_z = \frac{wl^2}{2} - wlx + \frac{w}{2}x^2 = \frac{w}{2}(l-x)^2$$

Hence, in eqn [6.41],

$$EI_z\frac{d^2v}{dx^2} = -\frac{w}{2}(l-x)^2 \qquad [6.49]$$

$$EI_z\frac{dv}{dx} = \frac{w}{6}(l-x)^3 + A$$

This is zero for $x = 0$; thus $A = -wl^3/6$. Hence

$$EI_z\frac{dv}{dx} = \frac{w}{6}(l-x)^3 - \frac{wl^3}{6} \qquad [6.50]$$

$$EI_z v = -\frac{w}{24}(l-x)^4 - \frac{wl^3}{6}x + B$$

This is zero for $x = 0$; thus $B = wl^4/24$. Hence

$$EI_z v = -\frac{w}{24}(l-x)^4 - \frac{wl^3}{6}x + \frac{wl^4}{24} \qquad [6.51]$$

The maximum slope is at $x = l$:

$$\left(\frac{dv}{dx}\right)_{max} = -\frac{wl^3}{6EI_z} \qquad [6.52]$$

The maximum deflection is at $x = l$:

$$v_{max} = -\frac{wl^4}{8EI_z} \qquad [6.53]$$

Case (c)

Now we have a beam, simply-supported at each end, carrying a uniformly distributed load w. Again we take the origin at A:

$$M_z = -\frac{wl}{2}x + \frac{w}{2}x^2 \qquad [6.54]$$

$$EI_z\frac{d^2v}{dx^2} = \frac{wl}{2}x - \frac{w}{2}x^2$$

$$EI_z\frac{dv}{dx} = \frac{wl}{4}x^2 - \frac{w}{6}x^3 + A$$

This is zero for $x = l/2$ (from symmetry); thus $A = -wl^3/24$. Hence

$$EI_z\frac{dv}{dx} = \frac{wl}{4}x^2 - \frac{w}{6}x^3 - \frac{wl^3}{24} \qquad [6.55]$$

$$EI_z v = \frac{wl}{12}x^3 - \frac{w}{24}x^4 - \frac{wl^3}{24}x + B$$

This is zero for $x = 0$; thus $B = 0$. Hence

$$EI_z v = \frac{wl}{12}x^3 - \frac{w}{24}x^4 - \frac{wl^3}{24}x \qquad [6.56]$$

The maximum slope is at $x = 0$ and is, from eqn [6.55],

$$\left(\frac{dv}{dx}\right)_{max,x=0} = -\frac{wl^3}{24EI_z} \qquad [6.57]$$

The slope at $x = l$ is also maximum but of opposite sign to that at $x = 0$:

$$\left(\frac{dv}{dx}\right)_{max,x=l} = \frac{wl^3}{24EI_z}$$

The maximum deflection is clearly, from symmetry, at $x = l/2$:

$$v_{max} = -\frac{5wl^4}{384EI_z} \qquad [6.58]$$

Case (d)

This beam is simply-supported at its ends and carries a central concentrated load. With the origin at A,

$$M_z = -\frac{W}{2}x = -EI_z\frac{d^2v}{dx^2} \qquad [6.59]$$

However, this expression is valid only for the left-hand half of the beam $(0 \leqslant x \leqslant l/2)$. For the right-hand half of the beam $(l/2 \leqslant x \leqslant l)$ we need a new expression:

$$M_z = -\frac{W}{2}x + W\left(x - \frac{l}{2}\right) = -EI_z\frac{d^2v}{dx^2} \qquad [6.60]$$

Owing to the discontinuity in the bending moment diagram, as seen in Fig. 6.6(d), we cannot represent the bending moment M_z by a single algebraic expression. We can however proceed by integrating eqns [6.59] and [6.60] separately. In AC,

$$EI_z\frac{d^2v}{dx^2} = \frac{W}{2}x$$
$$EI_z\frac{dv}{dx} = \frac{W}{4}x^2 + A_1 \qquad [6.61]$$

In CB,

$$EI_z\frac{d^2v}{dx^2} = \frac{W}{2}x - W\left(x - \frac{l}{2}\right)$$
$$EI_z\frac{dv}{dx} = \frac{W}{4}x^2 - \frac{W}{2}\left(x - \frac{l}{2}\right)^2 + A_2 \qquad [6.62]$$

We can relate the two constants A_1 and A_2 by noting that the slope of the beam at C is the same in AC and CB. Hence, with

$x = l/2$ in eqns [6.61] and [6.62],

$$\frac{Wl^2}{16} + A_1 = \frac{Wl^2}{16} + A_2$$

Hence $A_1 = A_2$, and since the slope at C is zero owing to symmetry,

$$A_1 = A_2 = -\frac{Wl^2}{16}$$

The equality of the constants is a convenient result and is due to the fact that we integrated eqn [6.62] without expanding the bracketed term. If the bracket is expanded on integration, the constants A_1 and A_2 are not equal. Integrating again, in AC,

$$EI_z v = \frac{W}{12}x^3 - \frac{Wl^2}{16}x + B_1$$

and in CB,

$$EI_z v = \frac{W}{12}x^3 - \frac{W}{6}\left(x - \frac{l}{2}\right)^3 - \frac{Wl^2}{16}x + B_2$$

The deflection at C is the same whichever expression is used. So, observing that $v = 0$ at $x = 0$ and $x = l$, we obtain $B_1 = B_2 = 0$, and we note that again the constants of integration are the same providing we do not expand the bracketed term. Hence

$$EI_z v = \begin{cases} \frac{W}{48}(4x^3 - 3l^2x) & 0 \leqslant x \leqslant \frac{l}{2} \\ \frac{W}{48}(4x^3 - 3l^2x) - \frac{W}{6}\left(x - \frac{l}{2}\right)^3 & \frac{l}{2} \leqslant x \leqslant l \end{cases} \quad [6.63]$$

The maximum deflection is clearly at the centre. So, putting $x = l/2$ in either of eqns [6.63] we obtain the same result:

$$v_{\max} = -\frac{Wl^3}{48EI_z} \quad [6.64]$$

The maximum slope, at the ends, is

$$\left.\begin{aligned} \left(\frac{dv}{dx}\right)_{\max, x=0} &= -\frac{Wl^2}{16EI_z} \\ \left(\frac{dv}{dx}\right)_{\max, x=l} &= \frac{Wl^2}{16EI_z} \end{aligned}\right\} \quad [6.65]$$

The above procedure is the basis of the 'Macaulay' method of integrating the differential equation. It allows us to use a single expression for the bending moment distribution in the beam where a concentrated load causes a discontinuity in the expression. Bracketed terms are not expanded during integration and are only included in an evaluation when the term within the brackets is positive. In order to draw attention to this convention, square brackets are used for such terms in the remainder of section 6.5. Applying the Macaulay method to the current problem (d) we combine eqns [6.59] and [6.60] into a single equation for the whole beam:

$$M_z = -\frac{W}{2}x + W\left[x - \frac{l}{2}\right] \qquad [6.66]$$

The bracketed term in eqn [6.66] is included only when $x - l/2$ is positive, i.e. in the right-hand side of the beam. Thus

$$EI_z\frac{d^2v}{dx^2} = \frac{W}{2}x - W\left[x - \frac{l}{2}\right]$$

$$EI_z\frac{dv}{dx} = \frac{W}{4}x^2 - \frac{W}{2}\left[x - \frac{l}{2}\right]^2 + A \qquad [6.67]$$

This is zero for $x = l/2$; thus $A = -Wl^2/16$. Hence

$$EI_z v = \frac{W}{12}x^3 - \frac{W}{6}\left[x - \frac{l}{2}\right]^3 - \frac{Wl^2}{16}x + B$$

This is zero for $x = 0$; thus $B = 0$. (Note that the bracketed term is excluded when putting $x = 0$ since $x - l/2$ is then negative.)

The results so far obtained are useful standard cases of beam displacements and are included in Appendix 2. We now study some further examples of the calculation of beam displacements.

Example 6.3

The beam with a single, variable position, concentrated load is a useful general case to have available (Fig. 6.7). With the origin at A,

Figure 6.7 Example 6.3

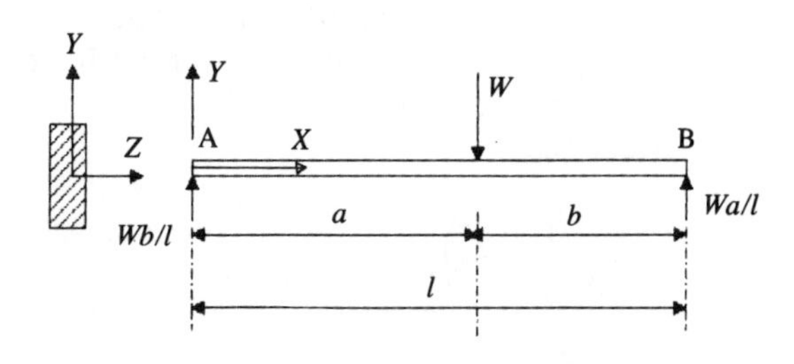

$$M_z = -\frac{Wb}{l}x + W[x - a] \qquad [6.68]$$

$$EI_z \frac{d^2v}{dx^2} = \frac{Wb}{l}x - W[x-a]$$

$$EI_z \frac{dv}{dx} = \frac{Wb}{2l}x^2 - \frac{W}{2}[x-a]^2 + A \qquad [6.69]$$

No information is available about slope in the beam, so we cannot at this stage determine the constant A. We have

$$EI_z v = \frac{Wb}{6l}x^3 - \frac{W}{6}[x-a]^3 + Ax + B$$

This is zero for $x = 0$; thus $B = 0$. It is also zero for $x = l$; thus $A = (Wb/6l)(b^2 - l^2)$. Hence

$$EI_z v = \frac{Wb}{6l}x^3 - \frac{W}{6}[x-a]^3 + \frac{Wb}{6l}(b^2 - l^2)x$$

$$= \frac{Wb}{6l}x(x^2 + b^2 - l^2) - \frac{W}{6}[x-a]^3 \qquad [6.70]$$

The deflection under the load is

$$v_W = -\frac{Wa^2b^2}{3EI_z l} \qquad [6.71]$$

This is not the maximum deflection in the beam. To obtain the maximum deflection we put $dv/dx = 0$. Now we have to decide which version of eqn [6.69] to use, the one with the bracketed term ($x \geqslant a$) or the one without the bracketed term ($x \leqslant a$). If $a > l/2$ then it would be reasonable to assume that the maximum deflection will occur to the left of the load. In this case,

$$0 = \frac{Wb}{2l}x^2 + \frac{Wb}{6l}(b^2 - l^2)$$

$$x = \sqrt{\left(\frac{l^2 - b^2}{3}\right)} \qquad [6.72]$$

If $a < l/2$ then the maximum deflection will be to the right of the load point. The simplest way to obtain the result for this case will be to change the origin to the right-hand end. By induction, the maximum deflection will be at
From B:

$$x = \sqrt{\left(\frac{l^2 - a^2}{3}\right)}$$

From A:

$$x = l - \sqrt{\left(\frac{l^2 - a^2}{3}\right)} \qquad [6.73]$$

Example 6.4

We wish to find the maximum deflection in the beam shown in Fig. 6.8, for which $EI_z = 25\,800$ kN m². We have

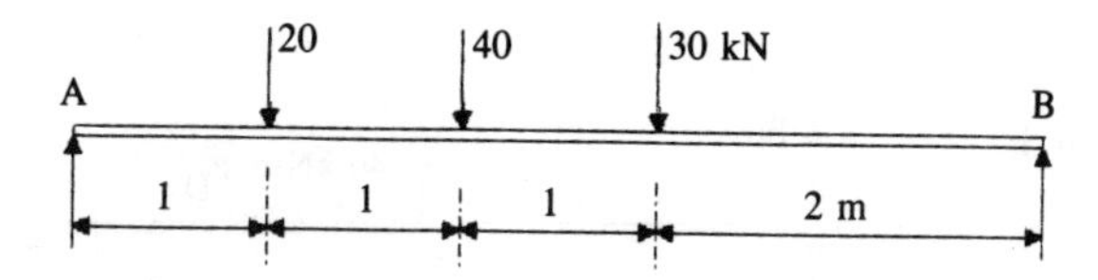

Figure 6.8 Example 6.4

$$R_A = \frac{4}{5} \times 20 + \frac{3}{5} \times 40 + \frac{2}{5} \times 30 = 52 \text{ KN}$$

$$M_z = -52x + 20[x-1] + 40[x-2] + 30[x-3]$$

$$= -EI_z \frac{d^2v}{dx^2} \quad [6.74]$$

$$EI_z \frac{dv}{dx} = 26x^2 - 10[x-1]^2 - 20[x-2]^2 - 15[x-3]^2 + A \quad [6.75]$$

$$EI_z v = \frac{26}{3}x^3 - \frac{10}{3}[x-1]^3 - \frac{20}{3}[x-2]^3$$
$$- 5[x-3]^3 + Ax + B$$

This is zero for $x = 0$; thus $B = 0$. It is also zero for $x = 5$; thus $A = -130$. Hence

$$EI_z v = \frac{26}{3}x^3 - \frac{10}{3}[x-1]^3 - \frac{20}{3}[x-2]^3 - 5[x-3]^3 - 130x \quad [6.76]$$

To find the position of maximum deflection we put $dv/dx = 0$ in eqn [6.75]. However, we do not know which bracketed terms to include until we know the position. It is likely that the maximum deflection will occur between $x = 2$ and $x = 3$, near the centre of the beam; so we shall use

$$EI_z \frac{dv}{dx} = 0 = 26x^2 - 10[x-1]^2 - 20[x-2]^2 - 130$$

Solving this equation we obtain $x = 2.44$ m, which is in the assumed range. Hence $v_{max} = 7.8$ mm by substitution in eqn [6.76].

Example 6.5

The simply-supported beam ACDB shown in Fig. 6.9(a) carries a uniformly distributed load over a 1 m length from C to D. We cannot, at first sight, use a continuous expression for bending moment since the load terminates at an intermediate position. We can, however, continue the load to the end of the beam and then cancel the unwanted load by applying an upward load between D and B, as shown in Fig. 6.9(b). We can then write a Macaulay-type expression for the whole beam:

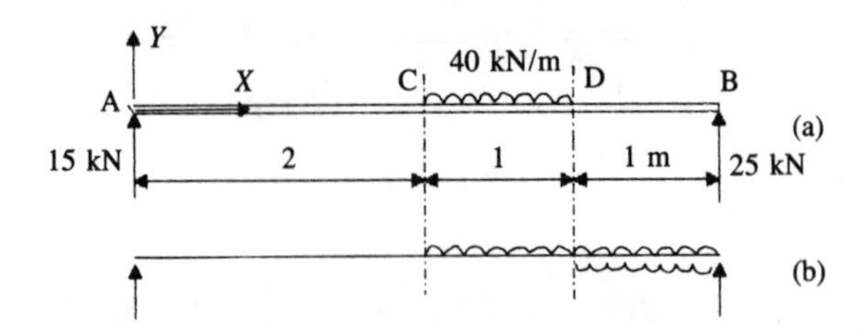

Figure 6.9 Example 6.5

$$M_z = -15x + \frac{40}{2}[x-2]^2 - \frac{40}{2}[x-3]^2 = -EI_z\frac{d^2v}{dx^2} \qquad [6.77]$$

$$EI_z\frac{dv}{dx} = \frac{15}{2}x^2 - \frac{40}{6}[x-2]^3 + \frac{40}{6}[x-3]^3 + A \qquad [6.78]$$

$$EI_z v = \frac{5}{2}x^3 - \frac{10}{6}[x-2]^4 + \frac{10}{6}[x-3]^4 + Ax + B$$

This is zero for $x = 0$; thus $B = 0$. It is also zero for $x = 4$; thus $A = -33.75$. Hence

$$EI_z v = \frac{5}{2}x^3 - \frac{10}{6}[x-2]^4 + \frac{10}{6}[x-3]^4 - 33.75x \qquad [6.79]$$

To determine v_{max} we put $dv/dx = 0$ in eqn [6.78]. However, we must know in which part of the beam the maximum deflection occurs. It is virtually certain that this is in CD, so we use

$$EI_z\frac{dv}{dx} = 0 = \frac{15}{2}x^2 - \frac{40}{6}[x-2]^3 - 33.75$$

or

$$x^3 - 7.125x^2 + 12x - 2.94 = 0$$

From this the appropriate solution is $x = 2.13$, which is in the assumed range. Hence, substituting in eqn [6.79], $v_{max} = -47.7/EI_z$.

The beams we have studied so far have been statically determinate. We can, however, use the same approach via the differential equation with statically indeterminate beams. We shall find that for each additional statically indeterminate force there will exist an additional displacement-related boundary condition. The following examples will illustrate the principles.

Example 6.6

The beam shown in Fig. 6.10 is what is called a 'propped' cantilever. The statical indeterminacy is one since the introduction of a vertical release at B, or a moment release at A, will leave the structure statically determinate.

Figure 6.10 Example 6.6

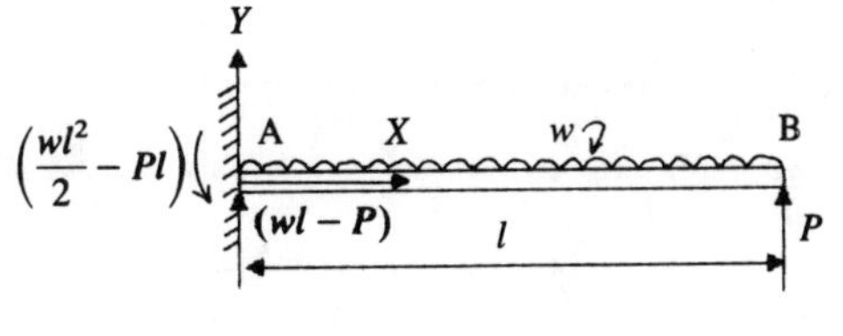

Taking the origin at A,

$$M_z = -EI_z \frac{d^2v}{dx^2} = \left(\frac{wl^2}{2} - Pl\right) + \frac{w}{2}x^2 - (wl - P)x \qquad [6.80]$$

$$EI_z \frac{dv}{dx} = -\left(\frac{wl^2}{2} - Pl\right)x - \frac{w}{6}x^3 + \frac{1}{2}(wl - P)x^2 + A \qquad [6.81]$$

$$= 0 \quad \text{for} \quad x = 0; \qquad A = 0 \qquad [6.82]$$

Hence

$$EI_z v = -\frac{1}{2}\left(\frac{wl^2}{2} - Pl\right)x^2 - \frac{w}{24}x^4 + \frac{1}{6}(wl - P)x^3 + B \qquad [6.83]$$

$$= 0 \quad \text{for} \quad x = 0; \qquad B = 0 \qquad [6.84]$$

$$= 0 \quad \text{for} \quad x = l \qquad [6.85]$$

Hence

$$0 = -\frac{1}{2}\left(\frac{wl^2}{2} - Pl\right)l^2 - \frac{w}{24}l^4 + \frac{1}{6}(wl - P)l^3$$

from which

$$P = 3wl/8 \qquad [6.86]$$

Notice how the statically indeterminate part of the problem is solved by using the third boundary condition [6.85], the first two having been used to evaluate the constants of integration [6.82] and [6.84]. The bending moment at A is $(wl^2/2) - Pl = wl^2/8$. Slope and deflection can now be found using eqns [6.81] and [6.83] respectively.

Example 6.7

The beam shown in Fig. 6.11 is 'fixed-ended'. The end moments, the 'fixing' moments, are of interest to us. (Appendix 3 contains values of fixed-end moments for several standard cases.) We first use statical principles to express the vertical reactions at A and B in terms of M_A and M_B and the applied load W. With the origin at A,

Figure 6.11 Example 6.7

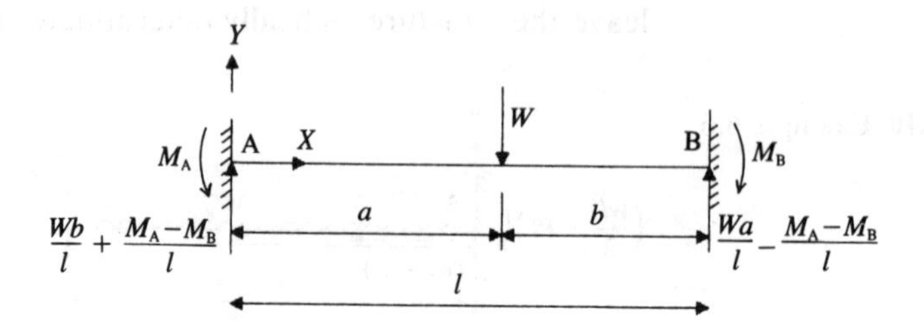

$$M_z = -EI_z\frac{d^2v}{dx^2} = M_A - \left(\frac{Wb}{l} + \frac{M_A - M_B}{l}\right)x + W[x-a] \quad [6.87]$$

$$EI_z\frac{dv}{dx} = -M_A x + \left(\frac{Wb}{l} + \frac{M_A - M_B}{l}\right)\frac{x^2}{2} - \frac{W}{2}[x-a]^2 + A \quad [6.88]$$

$$= 0 \quad \text{for} \quad x = 0; \qquad A = 0 \quad [6.89]$$

$$= 0 \quad \text{for} \quad x = l \quad [6.90]$$

Hence

$$M_A + M_B = Wab/l \quad [6.91]$$

Then

$$EI_z v = -\frac{M_A}{2}x^2 + \frac{1}{6}\left(\frac{Wb}{l} + \frac{M_A - M_B}{l}\right)x^3 - \frac{W}{6}[x-a]^3 + B \quad [6.92]$$

$$= 0 \quad \text{for} \quad x = 0; \qquad B = 0 \quad [6.93]$$

$$= 0 \quad \text{for} \quad x = l \quad [6.94]$$

Hence

$$2M_A + M_B = Wb(1 - b^2/l^2) \quad [6.95]$$

Solving eqns [6.91] and [6.95], we obtain

$$M_A = Wab^2/l^2 \qquad M_B = Wa^2b/l^2 \quad [6.96]$$

The reactions R_A and R_B can be found from

$$R_A = \frac{Wb}{l} + \frac{M_A - M_B}{l} = \frac{Wb^2}{l^3}(l + 2a) \quad [6.97]$$

$$R_B = W - R_A = \frac{Wa^2}{l^3}(l + 2b) \quad [6.98]$$

The four boundary conditions in this problem allow us to eliminate two constants of integration and determine two (statically indeterminate) reactive forces.

Example 6.8

A cantilever beam (Fig. 6.12) has a cross-section as shown in Fig. 6.5 and carries a point load at the free end as shown. Determine the components of transverse displacement of the load point if $E = 10^4\ \text{N/mm}^2$.

Figure 6.12 Example 6.8

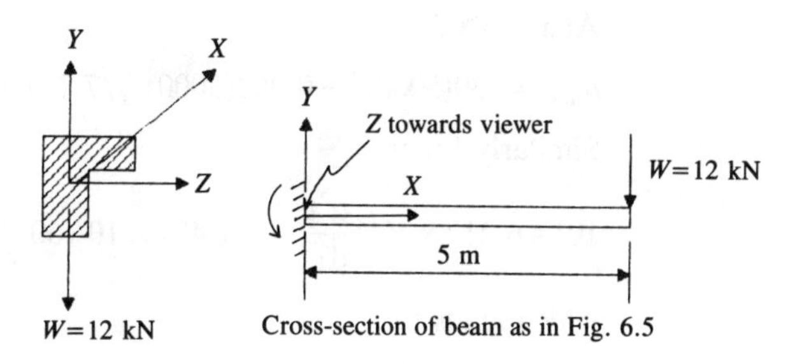

The relevant equations for the curvatures are eqns [6.37] and [6.38], i.e.

$$\frac{d^2u}{dx^2} = \frac{M_y I_z + M_z I_{zy}}{E(I_y I_z - I_{zy}^2)}$$

$$\frac{d^2v}{dx^2} = \frac{-(M_z I_y + M_y I_{zy})}{E(I_y I_z - I_{zy}^2)}$$

Now we already know the following properties from example 6.2:

$$I_z = 4.674 \times 10^9\ \text{mm}^4$$

$$I_y = 1.487 \times 10^9\ \text{mm}^4$$

$$I_{zy} = 0.8591 \times 10^9\ \text{mm}^4$$

The applied loading is on the Y axis only. Hence, with the origin at the fixed end,

$$M_z = 60\ \text{kN m at } x = 0;$$

$$M_z = (60 \times 10^6 - 12 \times 10^3 x)$$

$$M_y = 0$$

$$I_y I_z - I_{zy}^2 = 6.212 \times 10^{18}$$

$$M_z I_{zy} = 0.8591 \times 10^9 (60 \times 10^6 - 12 \times 10^3 x)$$

$$M_z I_y = 1.487 \times 10^9 (60 \times 10^6 - 12 \times 10^3 x)$$

Hence

$$10^4 \times 6.212 \times 10^{18} \frac{d^2u}{dx^2} = 0.8591 \times 10^9 (60 \times 10^6 - 12 \times 10^3 x)$$

$$7.231 \times 10^7 \frac{d^2u}{dx^2} = 60 - 0.012x$$

$$7.231 \times 10^7 \frac{du}{dx} = 60x - 0.006x^2 + A$$

$$= 0 \quad \text{for} \quad x = 0; \qquad A = 0$$

$$7.231 \times 10^7 u = 30x^2 - 0.002x^3 + B$$

$$= 0 \quad \text{for} \quad x = 0; \qquad B = 0$$

At $x = 5000$,

$u_{max} = \{30(5000)^2 - 0.002(5000)^3\}/7.231 \times 10^7 = 6.9$ mm

Similarly for v,

$$10^4 \times 6.212 \times 10^{18} \frac{d^2v}{dx^2} = -1.487 \times 10^9 (60 \times 10^6 - 12 \times 10^3 x)$$

which leads to

$v_{max} = -11.97$ mm

6.6 Shear in beams

The engineer's theory of bending (ETB) which we have examined in sections 6.3–6.5 is based on the existence of pure bending in the beam, and under these circumstances the assumption that plane sections remain plane is valid. If shear forces are present then the plane sections assumption is strictly invalid. The behaviour of a beam under the combined actions of bending and shear is adequately described by the St Venant approximation, which assumes that the longitudinal stress distribution and the curvatures due to bending are unaffected by the presence of shear stresses. The effect of shear on beam displacements is usually small and can generally be ignored; however, the shear stresses developed are of importance and always need to be assessed.

Imagine an elemental cube of material $\delta x \times \delta y \times \delta z$, as shown in Fig. 6.13. In general shear stresses will be acting on all six faces of the cube and these stresses are designated according to the convention described in section 3.3. Direct stresses will also exist (we have already been much involved with σ_x) and these could be represented as acting along the centre lines of the elemental cube. However, they are not going to affect what we are going to do, and so have been omitted from Fig. 6.13.

We take moments about an axis through the centre of the cube parallel to the X axis:

$$2(\tau_{zy}\, \delta y\, \delta x)(\delta z/2) = 2(\tau_{yz}\, \delta z\, \delta x)(\delta y/2)$$

Hence

$$\tau_{zy} = \tau_{yz} \qquad [6.99]$$

Figure 6.13 Complementary shear stresses

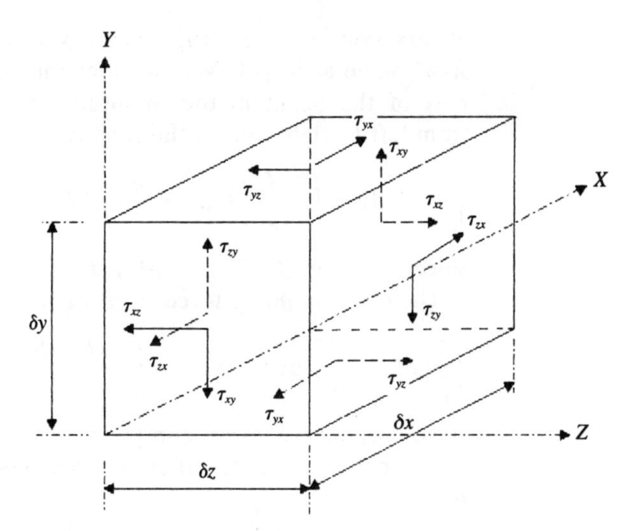

In a similar way, if we take moments about axes parallel to the Y and Z axes, we obtain

$$\tau_{zx} = \tau_{xz} \qquad [6.100]$$

$$\tau_{xy} = \tau_{yx} \qquad [6.101]$$

The pairs of shear stresses in eqns [6.99], [6.100] and [6.101] are called 'complementary' shear stresses.

To avoid undue complication we shall develop our theory for shear stress distribution on the assumption that the beam has an axis of symmetry coinciding with the Y axis and that the cross-section has a width b which is a function of y. It is assumed that the beam carries transverse loads in the Y direction only. This situation will be found to cover the majority of practical circumstances involving solid section beams. Such a beam is shown in Fig. 6.14(a). The Z axis passes through the centroid O

Figure 6.14 Shear stresses in beam

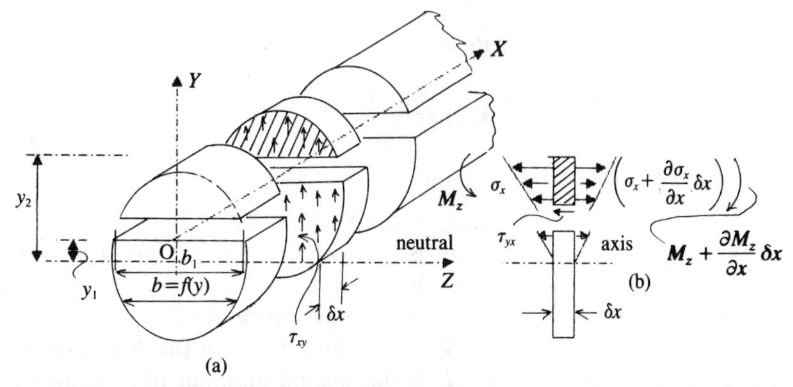

and is the neutral axis. Suppose we make a cut through the beam in a plane parallel to the X–Z plane and then cut out a slice of the beam of thickness δx as shown in Fig. 6.14(b). The shear stresses acting over the face of the slice will be τ_{xy} as shown, but we first consider the complementary shear stress τ_{yx} acting on the X–Z plane as in (b). We consider the equilibrium of the shaded part of the beam in the longitudinal (X) direction. The force from left to right due to the longitudinal stress σ_x is

$$\int_{y_1}^{y_2} \sigma_x b \, \mathrm{d}y = \int_{y_1}^{y_2} \frac{M_z y}{I_z} b \, \mathrm{d}y = \frac{M_z}{I_z} \int_{y_1}^{y_2} by \, \mathrm{d}y \qquad [6.102]$$

since from eqn [6.27] $\sigma_x = M_z y / I_z$.

The corresponding force from right to left is

$$\int_{y_1}^{y_2} \left(M_z + \frac{\mathrm{d}M_z}{\mathrm{d}x} \delta x \right) \frac{y}{I_z} b \, \mathrm{d}y = \int_{y_1}^{y_2} \frac{M_z + S_y \, \delta x}{I_z} by \, \mathrm{d}y \qquad [6.103]$$

since $\mathrm{d}M_z / \mathrm{d}x = S_y$ from eqn [5.3].

The net force is the difference between these two and is equal to

$$\frac{S_y}{I_z} \delta x \int_{y_1}^{y_2} by \, \mathrm{d}y$$

This must be equal to the shear force acting on the plane parallel to X–Z, i.e.

$$\tau_{yx} b_1 \, \delta x = \frac{S_y}{I_z} \delta x \int_{y_1}^{y_2} by \, \mathrm{d}y$$

Hence

$$\tau_{yx} = \frac{S_y}{b_1 I_z} \int_{y_1}^{y_2} by \, \mathrm{d}y \qquad [6.104]$$

The integral in eqn [6.104] is the first moment of the shaded area about the neutral axis. Putting

$$\int_{y_1}^{y_2} by \, \mathrm{d}y = \bar{A}\bar{y}$$

then

$$\tau_{yx(i)} = \frac{S_y \bar{A}\bar{y}}{b_i I_z} \qquad [6.105]$$

where

$\tau_{yx(i)}$ is the shear stress at level i
b_i is the width of the section at level i
$\bar{A}\bar{y}$ is the first moment of area of cross-section above level i about the neutral axis
S_y is the shear force in the Y direction
I_z is the second moment of area about axis Z

The complementary shear stress on vertical planes, τ_{xy}, is also given by eqn [6.105].

If we apply eqn [6.105] to analyse the shear stress distribution in a rectangular cross-section beam (Fig. 6.15) then

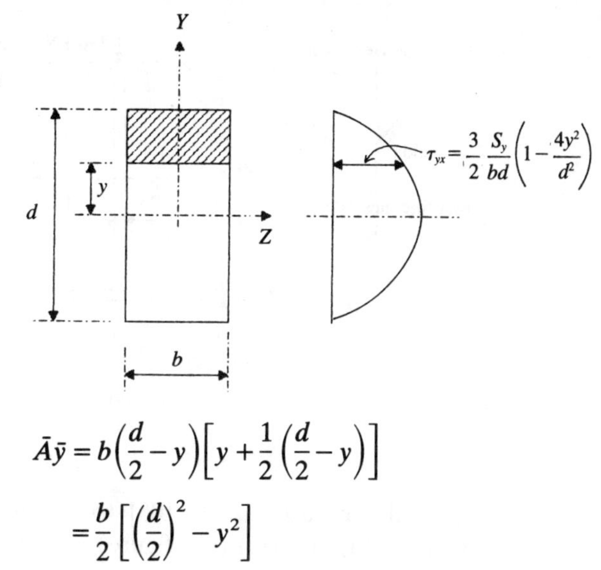

Figure 6.15 Distribution of shear stress in rectangular section beam

$$\bar{A}\bar{y} = b\left(\frac{d}{2} - y\right)\left[y + \frac{1}{2}\left(\frac{d}{2} - y\right)\right]$$

$$= \frac{b}{2}\left[\left(\frac{d}{2}\right)^2 - y^2\right]$$

Hence

$$\tau_{yx} = \frac{S_y b[(d/2)^2 - y^2]}{2bI_z}$$

$$= \frac{S_y}{2I_z}\left[\left(\frac{d}{2}\right)^2 - y^2\right]$$

and, since $I_z = bd^3/12$,

$$\tau_{yx} = \frac{3}{2}\frac{S_y}{bd}\left(1 - \frac{4y^2}{d^2}\right) \qquad [6.106]$$

At the top and bottom of the section, $y = \pm d/2$, and

$$\tau_{yx} = 0$$

At the neutral axis, $y = 0$, and

$$\tau_{yx} = \frac{3}{2}\left(\frac{S_y}{bd}\right)$$

Thus we see that the maximum shear stress is 1.5 times the average shear stress and occurs at the neutral axis. The shear stress distribution according to eqn [6.106] is parabolic, as shown in Fig. 6.15.

Consider now the shear stress distribution in an I-section beam as in Fig. 6.16(a). We can illustrate this situation best with a numerical example. We first calculate

$$I_z = 180 \times \frac{460^3}{12} - 166 \times \frac{400^3}{12} = 5.747 \times 10^8\ \text{mm}^4$$

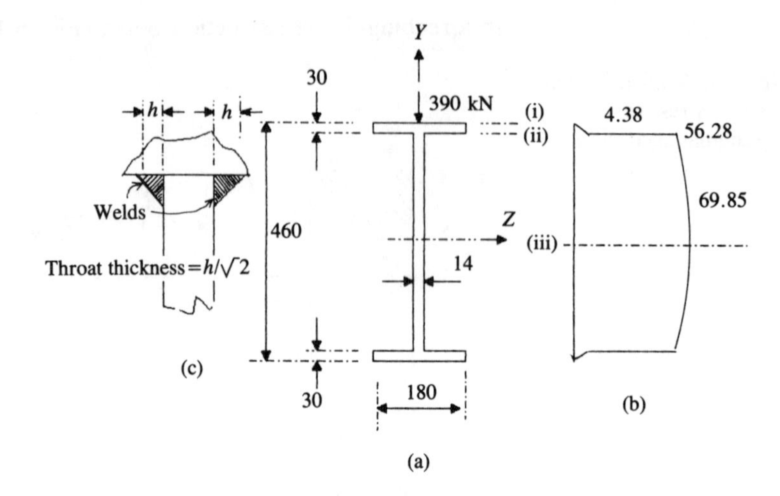

Figure 6.16 Distribution of shear stress in I-section beam

We shall now use eqn [6.105] to calculate the shear stress at levels (i), (ii) and (iii) (see figure). At level (i), $\tau_{yx} = 0$. At level (ii) in the flange,

$$\tau_{yx} = \frac{390 \times 10^3}{5.747 \times 10^8 \times 180}(180 \times 30 \times 215) = 4.38\ \mathrm{N/mm^2}$$

At level (ii) in the web,

$$\tau_{yx} = \frac{390 \times 10^3}{5.747 \times 10^8 \times 14}(180 \times 30 \times 215) = 56.28\ \mathrm{N/mm^2}$$

Thus we have two values of shear stress apparently coexisting at the same level in the cross-section. The smaller value, theoretically that in the flange, is unreliable owing to neglect of the sudden change in the width b from 180 to 14 mm. The theoretical value of 56.28 $\mathrm{N/mm^2}$ for the shear stress in the web at level (ii) is a reliable assessment of the actual shear stress in the web at the junction of the web and flange, and would be used if required to design a weld connecting web and flange. At level (iii), the neutral axis,

$$\tau_{yx} = \frac{390 \times 10^3}{5.747 \times 10^8 \times 14}(180 \times 30 \times 215 + 14 \times 200 \times 100)$$
$$= 69.85\ \mathrm{N/mm^2}$$

The distribution of shear stress in the web is parabolic as shown in Fig. 6.16(b), the maximum value being at the centroidal axis. The shape of the distribution in (b) suggests that a good approximation may be obtained by ignoring the flanges and assuming that all the shear is carried by the web and that there is a uniform, rectangular distribution. On this basis the average shear stress in the web is

$$\tau_{yx}(\text{average}) = \frac{390 \times 10^3}{400 \times 14} = 69.64\ \text{N/mm}^2$$

and this is obviously a good approximation to the more accurate value of $69.85\ \text{N/mm}^2$.

It is important to use only the web area in calculating this approximation. If the complete cross-sectional area is used, we obtain

$$\frac{390 \times 10^3}{400 \times 14 + 2 \times 180 \times 30} = 23.78\ \text{N/mm}^2$$

which result is seriously misleading.

The shear stress distribution in the flange is not without interest since it can be shown that a horizontal shear stress τ_{xz} exists which is of significance in thin-walled beams subject to bending and torsion.

We take the opportunity to design a weld to connect the web and flange of the beam in Fig. 6.16. The longitudinal shear stress τ_{yx} at the junction of the web and flange is

$$\tau_{yx} = \frac{S_y \bar{A} \bar{y}}{b I_z}$$

Hence the shear force Q per unit length of beam is

$$Q = \tau_{yx} b = S_y \bar{A} \bar{y} / I_z = 788\ \text{N/mm}$$

This is the shear force to be transmitted by a pair of fillet welds, as shown in Fig. 6.16(c). The 'nominal' size of the weld is h and the effective thickness is the 'throat' thickness $h/\sqrt{2}$. Thus the shear transmitting capacity of the pair of welds is $2p_w h/\sqrt{2}$, where p_w is the 'design' strength of the weld (N/mm^2). Taking $p_w = 215\ \text{N/mm}^2$, then $2 \times 215h/\sqrt{2} = 788\ \text{N/mm}$, from which $h = 2.6\ \text{mm}$. However, in practice the weld specified would probably be a minimum of 5 mm.

6.7 Torsion in beams

Again we need to remind ourselves that the engineer's theory of bending developed in sections 6.3 and 6.4 is based on the assumption that plane sections remain plane. If we were to look at any point in the cross-section of a beam, then this point will lie in the same transverse plane after bending as it does before bending. When a structural member is subjected to a twisting moment or 'torque', M_x in our notation, plane sections will not remain plane except under certain conditions. Plane transverse sections will generally suffer distortion in the axial direction; such distortion is called 'warping'. An exception to this is the circular section which does not warp when twisted – a result which could actually be deduced from the geometry of the circle!

The important stresses produced by torsion are shear stresses. Under certain circumstances, torque produces longitudinal stresses in addition to shear stresses. This occurs when the natural warping of the beam cross-section is restrained in some way. These are difficult situations analytically and we shall not pursue them here. Our principal concern is the shear stresses, and the corresponding deformations, produced by torsion.

6.7.1 Torsion of circular sections

In Fig. 6.17 a uniform bar of circular cross-section is twisted by a

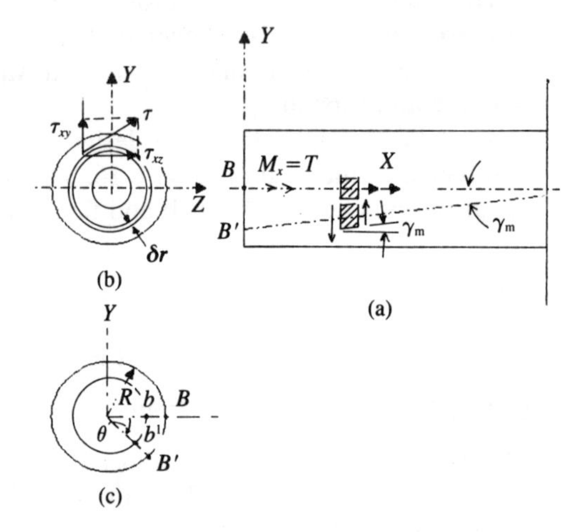

Figure 6.17 Circular torsion

pure torque $T\ (= M_x)$. It is assumed that the stresses are within the elastic limit, that any point on a radial line before twisting remains on the same radial line after twisting, and that the shear stress produced is proportional to the radius r to the point considered. Thus the maximum shear stress is at the surface of the bar; let us call this τ_m. Then at radius r,

$$\tau = \frac{r}{R}\tau_m \qquad [6.107]$$

and the tangential shear force over an elemental ring of thickness δr and area dA is

$$\frac{r}{R}\tau_m\,\mathrm{d}A$$

The moment of this force is

$$\delta T = \frac{r}{R}\tau_m\,\mathrm{d}A(r)$$

Hence the total torque is

$$T = \frac{\tau_m}{R}\int_0^R r^2\,dA \qquad [6.108]$$

Now $\int_0^R r^2\,dA$ is the polar second moment of area I_p (eqn [6.8]), and

$$I_p = \int_0^R r^2(2\pi r)\,dr = \frac{\pi R^4}{2} = \frac{\pi D^4}{32} = J \qquad [6.109]$$

where J is the section constant for torsion, which in this case is equal to the polar second moment of area. By combining eqns [6.107], [6.108] and [6.109],

$$\frac{T}{J} = \frac{\tau_m}{R} = \frac{\tau}{r} \qquad [6.110]$$

If the torque T is constant along the member then the rotation of any transverse plane about the X axis will be proportional to its distance along the X axis. The angle of twist is defined as θ and the angle γ is the shear strain at radius r; see Fig. 6.17(a) and (c). Now $bb' = r\theta = l\gamma$ and $BB' = R\theta = l\gamma_m$, where γ_m is the shear strain at the surface of the bar. Hence

$$\gamma_m = R\theta/l \qquad [6.111]$$

$$\gamma = r\theta/l \qquad [6.112]$$

Now the shear strain $\gamma = \tau/G$, where G is the modulus of rigidity (shear modulus). Hence

$$\frac{r\theta}{l} = \frac{\tau}{G} \qquad [6.113]$$

Combining eqns [6.110] and [6.113],

$$\frac{T}{J} = \frac{\tau}{r} = \frac{G\theta}{l} \qquad [6.114]$$

Equation [6.114] is the governing relationship for torsion.

For a tubular circular section of outside diameter D_o and internal diameter D_i, eqn [6.108] is modified to

$$T = \frac{\tau_m}{R_o}\int_{R_i}^{R_o} r^2\,dA \qquad [6.115]$$

where R_o and R_i are the outer and inner radii respectively. Thus

$$T = 2\pi\frac{\tau_m}{R_o}\int_{R_i}^{R_o} r^3\,dr$$

$$= \frac{\pi}{2}\frac{\tau_m}{R_o}(R_o^4 - R_i^4)$$

$$= \frac{\pi}{16} \frac{\tau_m}{D_o} (D_o^4 - D_i^4)$$

and

$$J = \frac{\pi}{2} (R_o^4 - R_i^4)$$

$$= \frac{\pi}{32} (D_o^4 - D_i^4) \qquad [6.116]$$

If the circular tube has a thin wall of thickness t, then we can approximate the torsion constant, from eqn [6.116], as follows:

$$J = \frac{\pi}{32} (D_o^2 - D_i^2)(D_o^2 + D_i^2)$$

$$= \frac{\pi}{32} (D_o - D_i)(D_o + D_i)(D_o^2 + D_i^2)$$

$$= \frac{\pi}{32} (2t)(2D)(2D^2)$$

$$= \frac{\pi}{4} tD^3 \qquad [6.117]$$

where D is the mean diameter of the tube.

Example 6.9

A circular tube, 3 m long, has an outside diameter of 250 mm and a wall thickness of 15 mm. A pure torque of 50 kN m is applied. $G = 80$ kN/mm^2. Calculate the maximum shear stress and the total angle of twist (a) using the exact expression for J, eqn [6.116], and (b) using the approximate expression for J, eqn [6.117].

(a)

$$J = \frac{\pi}{32} (D_o^4 - D_i^4) = \frac{\pi}{32} (250^4 - 220^4) = 1.5351 \times 10^8 \text{ mm}^4$$

$$\frac{\tau}{r} = \frac{T}{J}; \qquad \tau = \frac{50 \times 10^6 \times 125}{1.5351 \times 10^8} = 40.71 \text{ N/mm}^2$$

$$\frac{G\theta}{l} = \frac{T}{J}; \qquad \theta = \frac{50 \times 10^6 \times 3000}{80 \times 10^3 \times 1.5351 \times 10^8} = 0.01\,221 \text{ rad}$$

(b)

$$J = \frac{\pi}{4} D^3 t = \frac{\pi}{4} \times 235^3 \times 15 = 1.5289 \times 10^8 \text{ mm}^4$$

$$\tau = 40.71 \times \frac{1.5351}{1.5289} = 40.88 \text{ N/mm}^2$$

$$\theta = 0.01\,221 \times \frac{1.5351}{1.5289} = 0.01\,226 \text{ rad}$$

6.7.2 Torsion of non-circular sections

With torsion, as we have already seen, plane sections do not in general remain plane. The circular cross-section is an exception, and there are others. However, for the majority of the structural cross-sections we shall encounter there will be deformation of the cross-section in the axial direction. These 'warping' deformations are small in solid cross-sections and usually negligible in thin-walled 'closed' sections, but may be comparatively large in thin-walled 'open' sections. We shall look at this important distinction between open and closed sections shortly.

The theory of non-circular torsion is due to St Venant, the underlying assumption being that the rate of twist $d\theta/dx$ is proportional to the applied torque T. The constant of proportionality is the torsional rigidity GJ. Thus

$$T = GJ\frac{d\theta}{dx} \qquad [6.118]$$

This theory applies to 'uniform' torsion, where T is uniform along the member and where the warping displacements (if any) are unconstrained. Non-uniform torsion is a complex topic and will not be discussed here.

The torsional constant J is a function only of the geometry of the cross-section and has the units (length)^4. It is difficult to express in general terms except for certain simple shapes as quoted in Appendix 1. A convenient numerical method for calculating J for any cross-section is by finite differences. The four cross-sections shown in Fig. 6.18 all have the same cross-sectional area. However, their torsional constants calculated using the standard results from Appendix 1 are

$$J_{(a)} = 8.76 \times 10^6 \text{ mm}^4$$

$$J_{(b)} = 80 \times 10^6 \text{ mm}^4$$

$$J_{(c)} = 0.217 \times 10^6 \text{ mm}^4$$

$$J_{(d)} = 130 \times 10^6 \text{ mm}^4$$

It is evident that cross-sections in which the material is disposed as far away as possible from the centroid have larger torsional constants and hence greater torsional rigidity and smaller angles of twist for the same torque. Further, the 'closed' section (b) has a very significantly higher torsional constant than the 'open' section (c). The circular tube is clearly an ideal torsional cross-section.

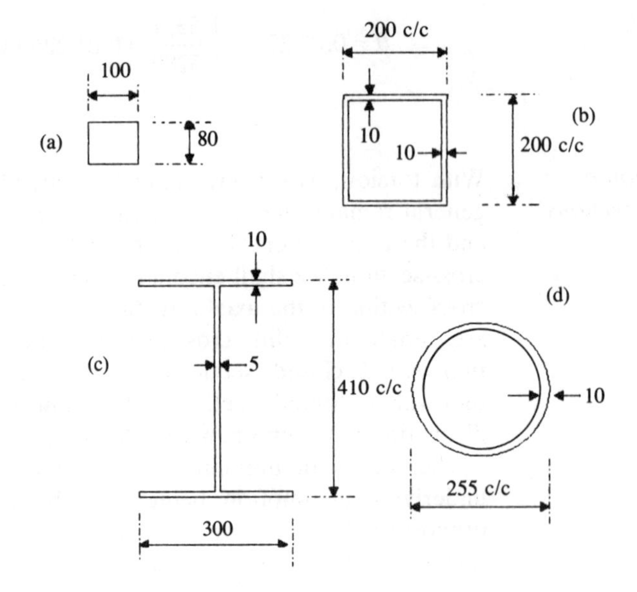

Figure 6.18 Example cross-sections for comparison of torsional constants

The maximum shear stress τ_m can be related to the applied torque T by a relationship

$$\tau_m = T/k \qquad [6.119]$$

where k is a function of the geometry of the cross-section. Values of k for standard geometrical shapes are quoted in reference books (e.g. *Civil Engineer's Reference Book*).

6.8 Thin-walled beams

6.8.1 Sectorial coordinate and shear centre

Many structures are made from 'thin' steel plate or even comparatively thin concrete sections. The steel box girder and the concrete core of a tall building are examples of thin-walled structures (Zbirohowski–Koscia 1967; Megson 1974). Even if a structure is not strictly thin-walled, the application of thin-walled theory may give a good initial approximation to its behaviour. A strict definition of 'thin-walled' is difficult to formulate; however, if t is a representative thickness of the material and if a is a representative cross-sectional dimension, then the structure can be deemed to be thin-walled if

$$a/t > 10 \qquad [6.120]$$

We can then analyse the structure assuming uniform stress conditions across the thickness of the material.

The calculation of section properties of thin-walled members can be simplified by omitting terms in t^2 and higher. For example if we evaluate I_z, I_y and I_{zy} for the inclined thin plate of Fig. 6.19, we get

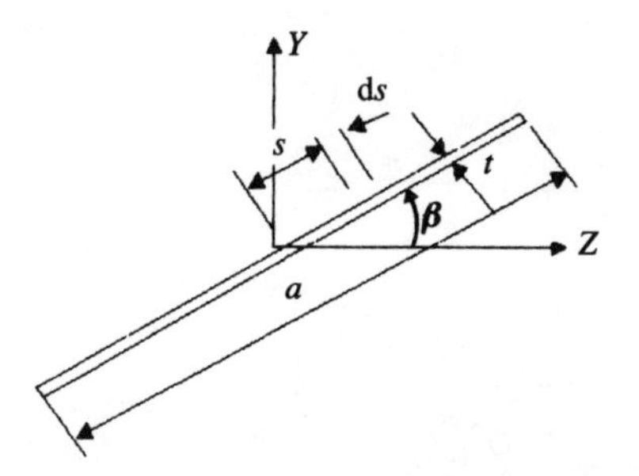

Figure 6.19 Section properties of thin plate

$$I_z = 2\int_0^{a/2} y^2 t \, ds = 2t\int_0^{a/2} (s \sin \beta)^2 \, ds$$

$$= \frac{a^3 t}{12} \sin^2 \beta \qquad [6.121]$$

Similarly,

$$I_y = \frac{a^3 t}{12} \cos^2 \beta \qquad [6.122]$$

Finally,

$$I_{zy} = 2\int_0^{a/2} zyt \, ds = 2t\int_0^{a/2} (s \cos \beta)(s \sin \beta) \, ds$$

$$= \frac{a^3 t}{24} \sin 2\beta \qquad [6.123]$$

We note that

1. If $\beta = 90°$: $I_z = a^3 t/12$, $I_y = 0$, $I_{zy} = 0$.
2. If $\beta = 0°$: $I_z = 0$, $I_y = a^3 t/12$, $I_{zy} = 0$.

When analysing thin-walled cross-sections it is convenient to introduce a new type of coordinate called the 'sectorial' coordinate. This coordinate is helpful in describing deformations and stresses when plane cross-sections no longer remain plane, which happens frequently in thin-walled structures when shear and torsion are present. In Fig. 6.20(a) we see part of the cross-section of a thin-walled member with a longitudinal axis in the X direction. The curvilinear coordinate s is measured along the middle surface of the section from any suitable generator. The sectorial coordinate of any typical point P on the mid-surface is referred to two reference points, the origin of s (O_s) and an arbitrary pole O_A, located at some suitable position in the plane of the cross-section (a Y–Z plane). The sectorial coordinate (ω) of point P is twice the area swept out by a radius vector O_AO_s in moving to O_AP:

$$\omega = \int_0^s h \, ds \qquad [6.124]$$

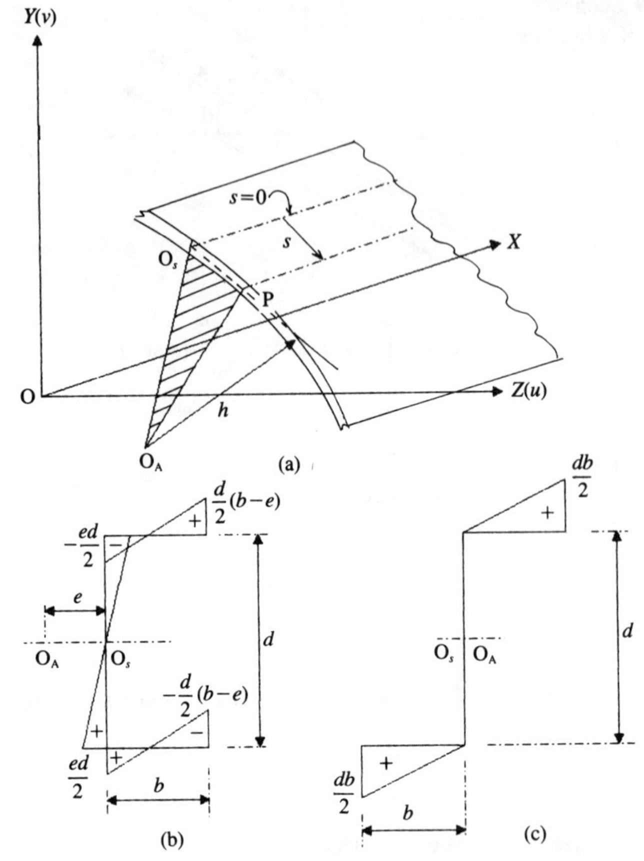

Figure 6.20 Sectorial coordinates for thin-walled sections

If the radius vector O_AP rotates in a clockwise sense in moving in the positive direction of s, then ω is positive. Two examples of ω distribution diagrams are shown in Fig. 6.20: (b) a channel section and (c) a Z-section.

Corresponding to the section properties defined in section 6.2 in cartesian coordinates, we now introduce four additional 'sectorial' properties:

Sectorial static moment:

$$Q_\omega = \int \omega \, \mathrm{d}A \qquad [6.125]$$

Sectorial product second moment of area with respect to the Z axis:

$$I_{\omega z} = \int \omega y \, \mathrm{d}A \qquad [6.126]$$

Sectorial product second moment of area with respect to the Y axis:

$$I_{\omega y} = \int \omega z \, \mathrm{d}A \qquad [6.127]$$

Sectorial second moment of area:

$$I_\omega = \int \omega^2 \, \mathrm{d}A \qquad [6.128]$$

It is useful to compare these cross-sectional properties with those defined in eqns [6.1] to [6.7] and to collect all the results into a compact form where the relationships are more evident. We note that, with the introduction of a factor of unity where required, all the section properties are integrals of the form

$$\iint f_1 f_2 \, \mathrm{d}z \, \mathrm{d}y = \int f_1 f_2 \, \mathrm{d}A$$

and thus we can generalize them in Table 6.1. If we choose

Table 6.1 Section properties in general coordinates $\int f_1 f_2 \, \mathrm{d}A$

f_2	f_1			
	1	z	y	ω
1	A	Q_y	Q_z	Q_ω
z	Q_y	I_y	I_{zy}	$I_{\omega y}$
y	Q_z	I_{zy}	I_z	$I_{\omega z}$
ω	Q_ω	$I_{\omega y}$	$I_{\omega z}$	I_ω

principal coordinates, then Table 6.1 transforms to Table 6.2.

Table 6.2 Section properties in principal coordinates $\int f_1 f_2 \, \mathrm{d}A$

f_2	f_1			
	1	z	y	ω
1	A	0	0	0
z	0	I_y	0	0
y	0	0	I_z	0
ω	0	0	0	I_ω

A further advantage of this form of presentation of section properties is that, if the individual distributions can be set out in graphical form, as in Fig. 6.20(b) and (c), then the area integrals can be computed using the table of standard product integrals in Appendix 4.

In cartesian coordinates, the location of the centroid is determined using Tables 6.1 and 6.2 or eqn [6.4]:

$$Q_z = \int 1y \, \mathrm{d}A = 0$$

$$Q_y = \int 1z \, dA = 0$$

In sectorial coordinates, the corresponding integral

$$Q_\omega = \int 1\omega \, dA = 0$$

locates a point in the plane of the cross-section called the 'principal' pole. This has importance for our future work in torsion when we shall recognize this pole as being the 'shear centre' of the cross-section.

We shall now introduce a method of locating the principal pole in an arbitrary thin-walled open section. This method was put forward by Zbirohowski–Koscia (1967) and further developed by Prof H Tottenham. Let the principal pole be O_P (z_{OP}, y_{OP}) and let O_A (z_{OA}, y_{OA}) be an arbitrary convenient pole. O_s is the sectorial origin ($s = 0$). Now if ω_{OP} is the sectorial coordinate related to the principal pole and ω_{OA} the sectorial coordinate related to the arbitrary pole, then

$$\omega_{OP} = \omega_{OA} - \alpha_y z + \alpha_z y - \alpha_O \qquad [6.129]$$

where

$$\alpha_y = (y_{OP} - y_{OA}) \qquad [6.130]$$

$$\alpha_z = (z_{OP} - z_{OA}) \qquad [6.131]$$

$$\alpha_0 = \alpha_z y_{OA} - \alpha_y z_{OA} \qquad [6.132]$$

To find α_0, α_z and α_y we solve the equations

$$\begin{bmatrix} A & Q_y & -Q_z \\ Q_y & I_y & -I_{zy} \\ Q_z & I_{zy} & -I_z \end{bmatrix} \begin{bmatrix} \alpha_0 \\ \alpha_y \\ \alpha_z \end{bmatrix} = \begin{bmatrix} Q_{\omega A} \\ I_{\omega A y} \\ I_{\omega A z} \end{bmatrix} \qquad [6.133]$$

If Z and Y are the principal axes, then eqns [6.133] take the simpler form

$$\begin{bmatrix} A & 0 & 0 \\ 0 & I_y & 0 \\ 0 & 0 & -I_z \end{bmatrix} \begin{bmatrix} \alpha_O \\ \alpha_y \\ \alpha_z \end{bmatrix} = \begin{bmatrix} Q_{\omega A} \\ I_{\omega A y} \\ I_{\omega A z} \end{bmatrix} \qquad [6.134]$$

from which α_O, α_y and α_z are obtained directly. The above equations relate to thin-walled open cross-sections. Similar equations can be derived for closed sections.

Example 6.10

We wish to locate the shear centre of the channel section shown in Fig. 6.21(a). We start by choosing a position for the arbitrary pole O_A as shown. Now since the Z axis is an axis of symmetry, the shear centre lies on this axis. Hence from eqn [6.130] $\alpha_y = 0$, and from eqn [6.132] $\alpha_O = 0$. Now

Figure 6.21 Example 6.10

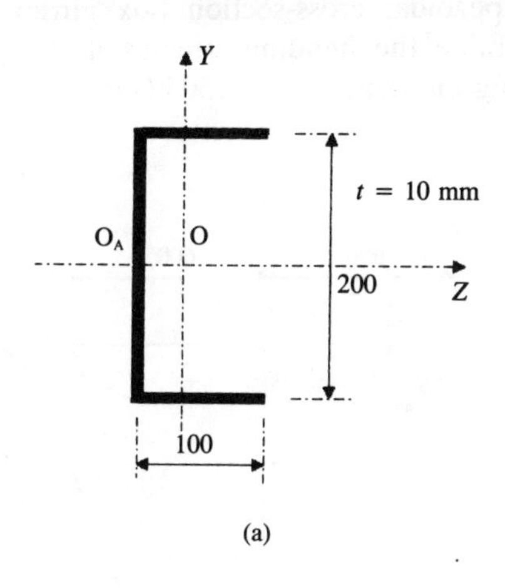

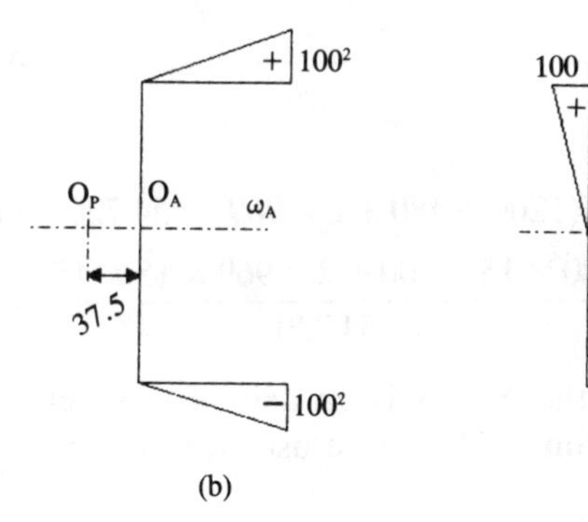

$$\alpha_z = -I_{\omega_A z}/I_z$$

$$I_z = 2 \times 10 \times 100 \times 100^2 + 10 \times 200^3/12 = 26.7 \times 10^6$$

$$I_{\omega_A z} = \int \omega_A y \, dA$$

The ω_A diagram is shown in Fig. 6.21(b) and the y diagram in (c). Hence, using the table of product integrals (Appendix 4),

$$I_{\omega_A z} = 100/2 \times 100^2 \times 100 \times 2 \times 10 = 10^9$$

Hence

$$\alpha_z = -10^9/26.7 \times 10^6 = -37.5 \text{ mm}$$

The position of the principal pole O_P is shown in (b).

6.8.2 Bending of thin-walled beams

For the calculation of bending stresses, eqn [6.26] is valid for Z and Y being non-principal axes, and eqn [6.27] for Z and Y being principal axes. The calculation of second moments of area can be simplified according to eqns [6.121]–[6.123].

Example 6.11

A trapezoidal cross-section box girder is shown in Fig. 6.22. Determine the bending stresses if the girder is subjected to a bending moment $M_z = -950$ kN m.

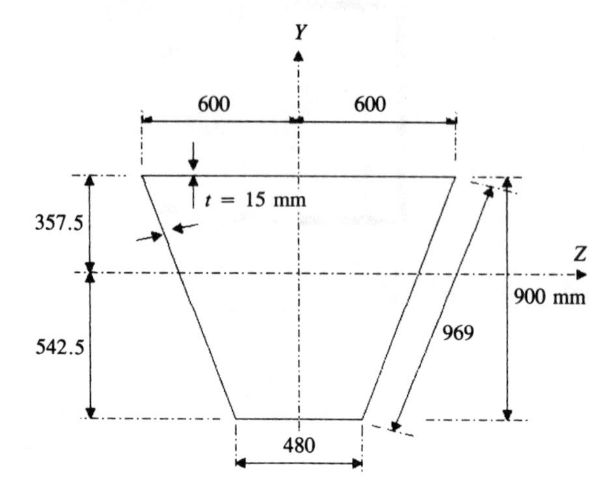

Figure 6.22 Example 6.11

$$A = 15(1200 + 480 + 2 \times 969) = 54\,720 \text{ mm}^2$$

$$\bar{y} = \frac{480 \times 15 \times 900 + 2 \times 969 \times 15 \times 450}{54\,720} = 357.5 \text{ mm}$$

Since the Y axis is an axis of symmetry, Z and Y are principal axes. Since $M_y = 0$ we use eqn [6.27]:

$$\sigma_x = \frac{M_z y}{I_z}$$

Now

$$I_z = 1200 \times 15 \times 357.5^2 + 480 \times 15 \times 542.5^2$$
$$+ 2 \times 15 \times \frac{969^3}{12} \times \left(\frac{900}{969}\right)^2 + 969 \times 15 \times 92.5^2 \times 2$$
$$= 6.630 \times 10^9 \text{ mm}^4$$

Hence the maximum tension at $y = -542.5$ is

$$-\frac{950 \times 10^6 \times (-542.5)}{6.630 \times 10^9} = 77.7 \text{ N/mm}^2$$

and the maximum compression at $y = 357.5$ is

$$-\frac{950 \times 10^6 \times 357.5}{6.630 \times 10^9} = -51.2 \text{ N/mm}^2$$

The analysis of a box girder will of course include considerations of several other aspects of behaviour such as shear stresses and stability of thin plates.

6.8.3 Torsion of thin-walled open sections

If a thin rectangular strip as in Fig. 6.23(a) is subjected to a uniform torque T, the maximum shear stresses are on the surfaces of the long sides of the rectangle. The distribution of shear stresses is as shown in Fig. 6.23(a), the shear stress being zero on the mid-surface. The torsion constant J may be obtained from the appropriate expression in Appendix 1, with t/a small:

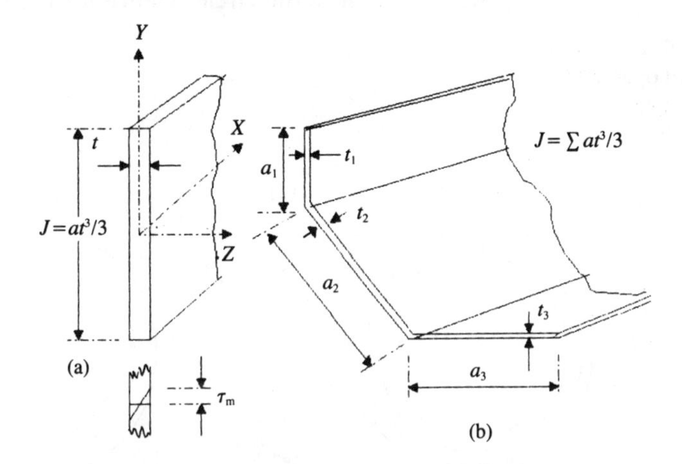

Figure 6.23 Torsion of thin-walled open section

$$J = at^3/3 \tag{6.135}$$

If strips are connected together to form an 'open' section as in Fig. 6.23(b), then

$$J = \sum at^3/3 \tag{6.136}$$

The applied torque, the torsional rigidity and the rate of twist are still related through eqn [6.118], i.e.

$$T = GJ\frac{\mathrm{d}\theta}{\mathrm{d}x}$$

The maximum value of the shear stress, on both surfaces, is

$$\tau_{\mathrm{m}} = Gt\frac{\mathrm{d}\theta}{\mathrm{d}x} \tag{6.137}$$

Combining the last two equations we have

$$T/J = \tau_{\mathrm{m}}/t \tag{6.138}$$

6.8.4 Torsion of thin-walled closed sections

We must first of all establish the fundamental difference between 'open' and 'closed' sections. In Fig. 6.24(a) we have a circular tube with a slit. If a pure torque is applied to any member, the shear stress system produced must form a closed path otherwise an out-of-balance resultant force is produced. In a closed section this loop develops around the section in a continuous path as in

Fig. 6.24(b), and the resulting shear stress is (nearly) uniform across the thickness. In the open section in (a) the closed loop can only form within the thickness of the material, and the resulting shear stress distribution is as shown with a change of sign from one side to the other. Now the torsion constant of a thin strip will not be altered if we bend it into a circle providing we do not close the circle. Hence for the slit tube,

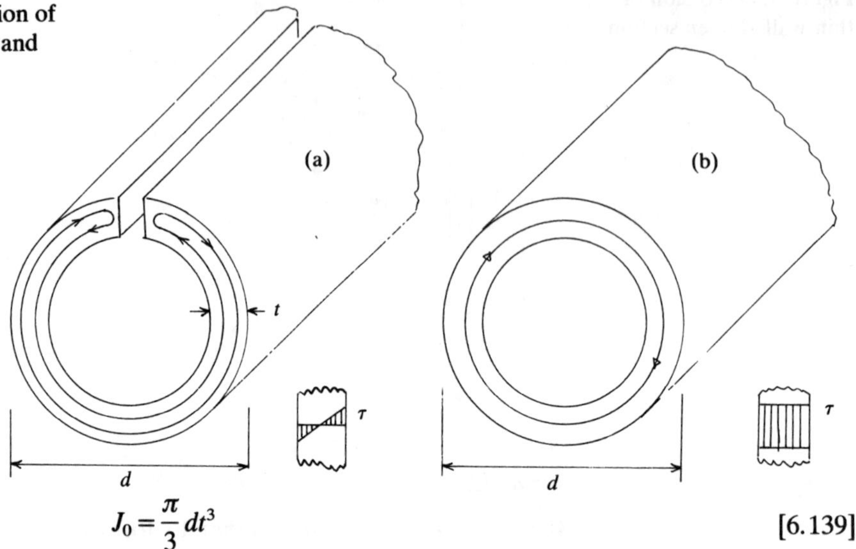

Figure 6.24 Torsion of thin-walled open and closed sections

$$J_0 = \frac{\pi}{3} dt^3 \qquad [6.139]$$

Comparing this with the value for the closed section in (b), we have from eqn [6.117]

$$J_c = \frac{\pi}{4} d^3 t$$

Thus

$$J_c / J_0 = \frac{3}{4} \frac{d^2}{t^2} \qquad [6.140]$$

This ratio will be large for practical values of d/t; for example, if $d/t = 20$ then $J_c/J_0 = 300$.

A useful concept in describing shear in thin-walled closed sections is the 'shear flow' q, which is defined as

$$q = \tau t \qquad [6.141]$$

The fundamental theory of the torsional behaviour of closed sections was established in the Bredt-Batho theory, which states that the shear flow in a closed cell is constant, i.e.

$$q = \tau t = \text{constant} \qquad [6.142]$$

The relationship between torque and shear flow is

$$T = 2Aq \qquad [6.143]$$

where A is the area enclosed by the middle surface of the tube. Thus for the circular tube, using eqns [6.142] and [6.143],

$$T = 2\pi \frac{d^2}{4} \tau t = \frac{\pi}{2} d^2 \tau t$$

This is consistent with eqns [6.110] and [6.117].

The general expression for the torsion constant for a thin-walled closed section is

$$J = \frac{4A^2}{\oint \mathrm{d}s/t} \qquad [6.144]$$

the integral being evaluated around the mid-surface.

Example 6.12

In the closed rectangular thin-walled tube shown in Fig. 6.25, the maximum shear stress is limited to $50\,\mathrm{N/mm^2}$. Determine the maximum allowable uniform twisting moment and the resulting twist per unit length. $G = 80\,\mathrm{kN/mm^2}$.

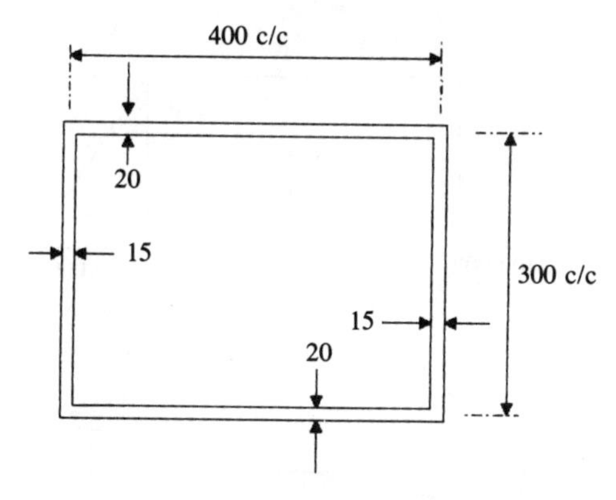

Figure 6.25 Example 6.12

$$q = \tau t = \text{constant} = T/2A$$

The maximum shear stress will occur in the thinner plate; thus

$$50 \times 15 = T/(2 \times 400 \times 300)$$

$$T = 180\ \mathrm{kN\,m}$$

From eqn [6.144] or Appendix 1,

$$J = \frac{4 \times 400^2 \times 300^2}{2 \times (400/20) + 2 \times (300/15)} = 7.2 \times 10^8\ \mathrm{mm}^4$$

and from eqn [6.118],

$$\frac{\mathrm{d}\theta}{\mathrm{d}x} = \frac{T}{GJ} = \frac{180 \times 10^6}{80 \times 10^3 \times 7.2 \times 10^8} = 3.125 \times 10^{-6}$$

Hence the twist per unit length is 3.125×10^{-3} rad/m.

Example 6.13

The cross-section of a thin-walled beam is shown in Fig. 6.26(a). Each plate is 15 mm thick. First, find the position of the shear centre. Then calculate the shear stress produced by a pure twisting moment of 5 kN m about a longitudinal axis through the shear centre.

Figure 6.26 Example 6.13

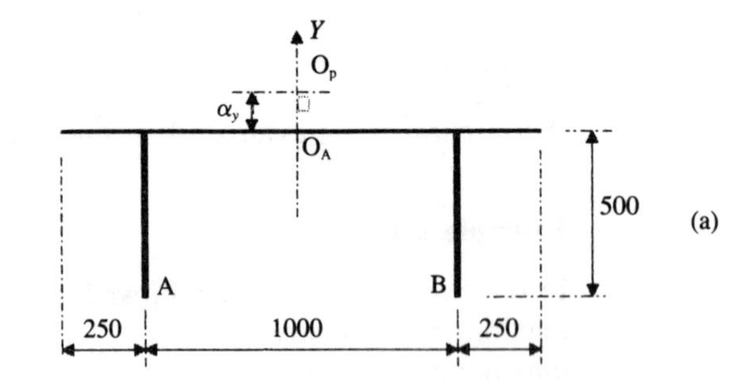

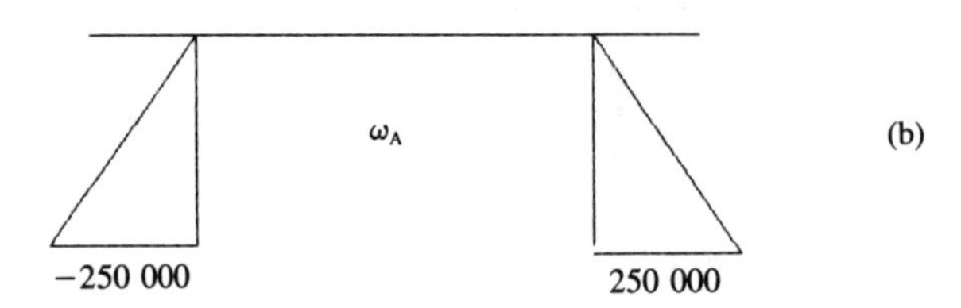

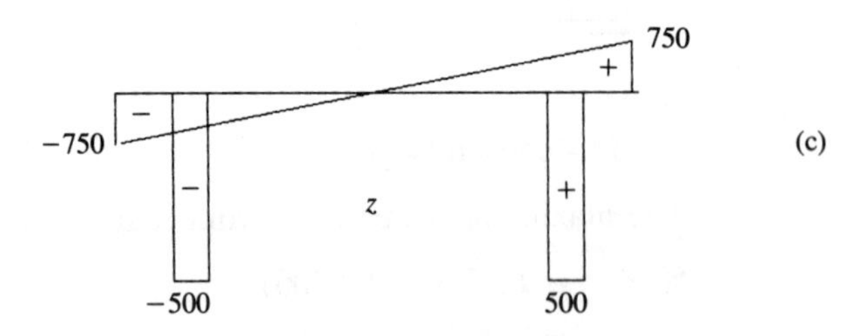

With an arbitrary pole located at O_A as in Fig. 6.26(a), we obtain an ω diagram as shown in (b). The Y axis is an axis of symmetry, so $\alpha_z = 0$; and from eqn [6.132] $\alpha_0 = 0$. Now

$$\alpha_y = I_{\omega_A y}/I_y$$

Using the tables of product integrals in Appendix 4,

$$I_y = 2 \times 500 \times 15 \times 500^2 + 15 \times 1500^3/12 = 7.97 \times 10^9$$

$$I_{\omega_A y} = \int \omega_A z \, dA = \frac{500}{2} \times 250\,000 \times 500 \times 15 \times 2$$

$$= 9.375 \times 10^{11}$$

Hence

$$\alpha_y = \frac{9.375 \times 10^{11}}{7.97 \times 10^9} = 118 \text{ mm}$$

Now from eqn [6.136],

$$J = \frac{1}{3} \sum at^3 = \frac{1}{3}(1500 + 500 + 500)15^3$$

$$= 2\,812\,500 \text{ mm}^4$$

and from eqn [6.138],

$$\tau_m = \frac{Tt}{J} = \frac{5 \times 10^6 \times 15}{2\,812\,500} = 26.7 \text{ N/mm}^2$$

6.9 Non-prismatic beams

In section 6.5 we considered the evaluation of displacements in beams due to bending. This involved the integration of a differential equation which we expressed in different forms depending on the circumstances of the cross-section of the beam. The relevant equations are [6.37] to [6.40] in section 6.4. In applying these equations to actual problems, we assumed constant conditions in the beam in respect of E and all values of I and were thus able to carry out the integrations without much difficulty. In a non-prismatic beam, the cross-sectional properties vary in the X (axial) direction and integration is more difficult. Although some cases can be handled by the analytical procedures already established, in general it is much more satisfactory to adopt numerical procedures to carry out the integrations. The computer is an ideal tool for this purpose.

We begin by looking at the process of evaluating end rotations of non-prismatic beams since these have importance later in arriving at beam stiffnesses. It is convenient to use the concept of flexibility coefficients introduced in section 4.2 and to carry out the integrations using Simpson's rule. In Fig. 6.27(a) we see a non-prismatic beam in which EI_z is a function of x along the axis.

Figure 6.27 End flexibilities for non-prismatic beam

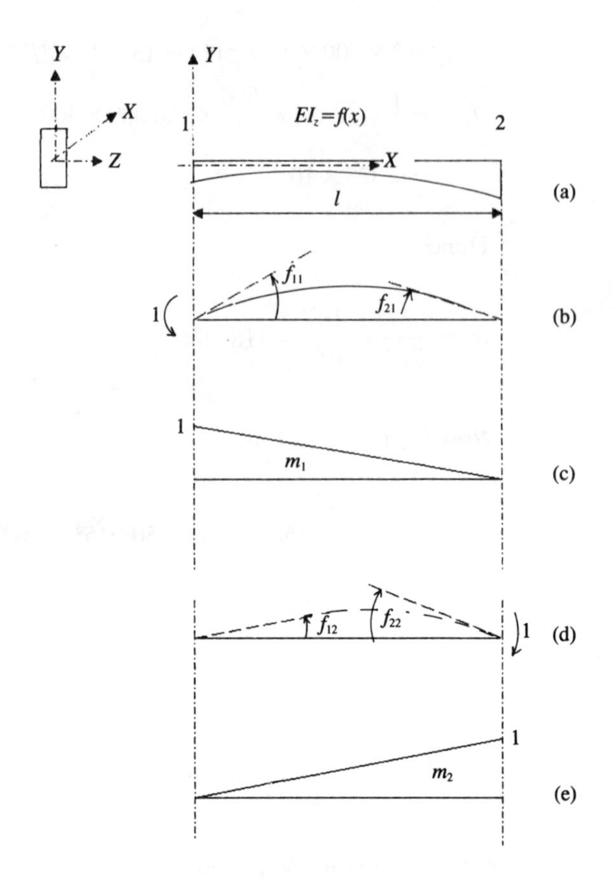

We assume that our results will not be significantly affected by the fact that the centroidal axis of the beam may no longer be straight. We take the beam to be simply-supported, i.e. subject to purely vertical restraint at ends 1 and 2, and apply a unit moment at end 1 and then at end 2 as in diagrams (b) and (d). The bending moment M_z distributions resulting from this are given in (c) and (e), and we label them m_1 and m_2 respectively. (The reader might like to refer to section 4.2, where we evaluated end flexibilities for a uniform member.)

Now

$$\left.\begin{aligned} f_{11} &= \int_0^l \frac{m_1^2}{EI}\,dx \\ f_{22} &= \int_0^l \frac{m_2^2}{EI}\,dx \\ f_{12} = f_{21} &= \int_0^l \frac{m_1 m_2}{EI}\,dx \end{aligned}\right\} \qquad [6.145]$$

where $EI = EI_z = f(x)$. To carry out the integrations using Simpson's rule, we divide the beam into n segments of equal length h, where n is even. Values of $m_r m_s/EI = \phi$ are calculated at the $n+1$ stations along the beam; r and s are given values 1 and 2 according to eqns [6.145]. Simpson's rule then gives us

$$f_{rs} = h/3[\phi_1 + 4\phi_2 + 2\phi_3 + 4\phi_4 + \dots + 2\phi_{n-1} + 4\phi_n + \phi_{n+1}]$$
$$= h/3[\text{sum of end ordinates} + 4(\text{even ordinates}) + 2(\text{other odd ordinates})] \quad [6.146]$$

Example 6.14

The beam shown in Fig. 6.28 spans 12 m and has a central section

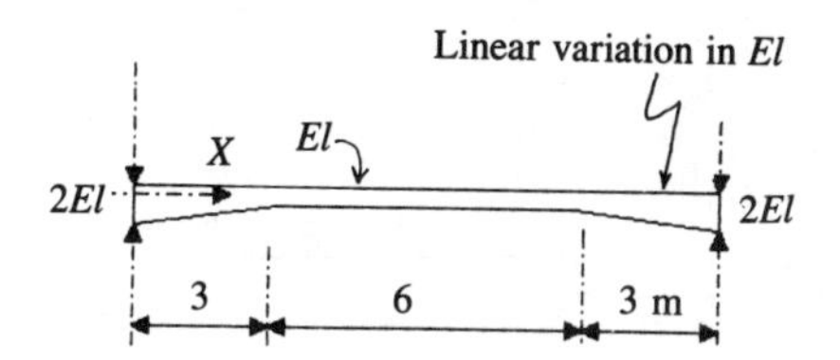

Figure 6.28 Example 6.14

6 m long with uniform flexural rigidity EI. The end sections of the beam have a linear variation in EI to $2EI$ at the ends. We wish to determine the end flexibilities for the beam in terms of EI. The solution is set out in Table 6.3. Thus

Table 6.3 Solution for example 6.14

x (m)	0	1.5	3	4.5	6	7.5	9	10.5	12
EI	$2EI$	$1.5EI$	EI	EI	EI	EI	EI	$1.5EI$	EI
m_1	1.0	0.875	0.75	0.625	0.5	0.375	0.25	0.125	0
m_2	0	0.125	0.25	0.375	0.5	0.625	0.75	0.875	1.0
m_1^2	1.0	0.7656	0.5625	0.3906	0.25	0.1406	0.0625	0.0156	0
m_1m_2	0	0.1094	0.1875	0.2344	0.25	0.2344	0.1875	0.1094	0
m_1^2/EI	0.5	0.5104	0.5625	0.3906	0.25	0.1406	0.0625	0.0104	0
m_1m_2/EI	0	0.0729	0.1875	0.2344	0.25	0.2344	0.1875	0.0729	0

$$EIf_{11} = \frac{1.5}{3}[(0.5+0) + 4(0.5104 + 0.3906 + 0.1406 + 0.0104) + 2(0.5625 + 0.25 + 0.0625)]$$
$$= 3.229 = EIf_{22}$$

$$EIf_{12} = \frac{1.5}{3}[(0+0) + 4(0.0729 + 0.2344 + 0.2344 + 0.0729) + 2(0.1875 + 0.25 + 0.1875)]$$
$$= 1.854 = EIf_{21}$$

Example 6.15

Here we seek the deflection at the tip of the rectangular cross-section cantilever beam shown in Fig. 6.29. The depth tapers uniformly from 0.3 m to 0.15 m and the width is constant at 0.1 m. We can use the flexibility coefficient method to obtain a displacement in a beam by placing unit load at the position where the displacement is required. This happens to be at the applied load position. The shape of the bending moment diagram is therefore the same for both the applied load and the unit load, with the moment at the fixed end being $10 \times 5 = 50$ kN m for the applied load and $1 \times 5 = 5$ for the unit load. Using eqns [4.18] and [4.19],

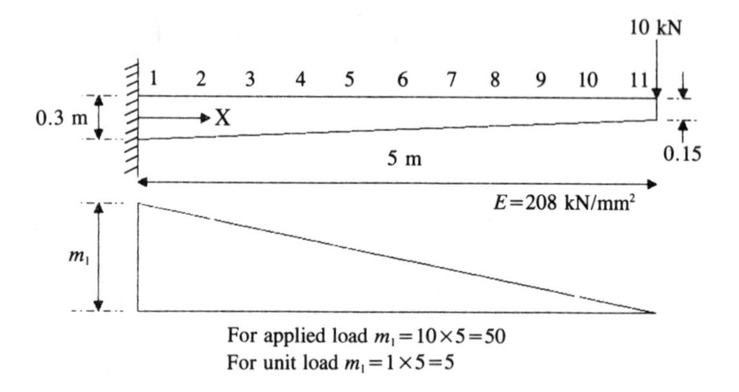

Figure 6.29 Example 6.15

$$y_i = \frac{\partial U}{\partial F_i} = \int \frac{M}{EI}\left(\frac{\partial M}{\partial F_i}\right) dx$$

where M is due to the applied load and $\partial M/\partial F_i$ is the bending moment distribution due to unit load. Thus we can construct Table 6.4, where we use $I = bd^3/12$ to obtain I and divide the beam into ten equal segments.

Table 6.4 Solution for example 6.15

x	0	0.5	1.0	1.5	2.0	2.5	3.0	3.5	4.0	4.5	5.0
$I \times 10^6$	225	192.9	164.0	138.2	115.2	94.92	77.18	61.79	48.60	37.43	28.13
EI (kN m^2)	46 800	40 123	34 112	28 746	23 962	19 743	16 053	12 852	10 109	7785	5851
M (kN μ)	50	45	40	35	30	25	20	15	10	5	0
$\partial M/\partial F_i$	5	4.5	4.0	3.5	3.0	2.5	2.0	1.5	1.0	0.5	0
$\frac{M}{EI}\frac{\partial M}{\partial F_i} \times 10^3$	5.342	5.047	4.690	4.261	3.756	3.166	2.492	1.751	0.989	0.321	0

Hence

$$y_{10} = \frac{0.5}{3 \times 10^3}[(5.342 + 0) + 4(5.047 + 4.261 + 3.166 + 1.751 + 0.321) + 2(4.690 + 3.756 + 2.492 + 0.989)]$$

$$= 14.6 \text{ mm}$$

6.9.1 Organization and accuracy of computation

One of the advantages of the numerical method just described is that its organization is ideally suited to the computer. With its arrays of numbers and its repetitive nature, it is easily programmed. For polynomials up to degree three, Simpson's rule gives an 'exact' result. In other cases the accuracy will depend on the number of ordinates taken. In most structures, a division of the member into ten equal segments will be satisfactory. If, however, there is a discontinuity in the *EI* variation at an even-numbered station, this will reduce the accuracy. The effect can easily be overcome by halving the intervals, thus causing the discontinuity to fall at an odd-numbered station.

Exercises

6.1 For the thin-walled cross-section shown in Fig. 6.30, determine (a) the location of the centroid; (b) I_z (c) I_y.

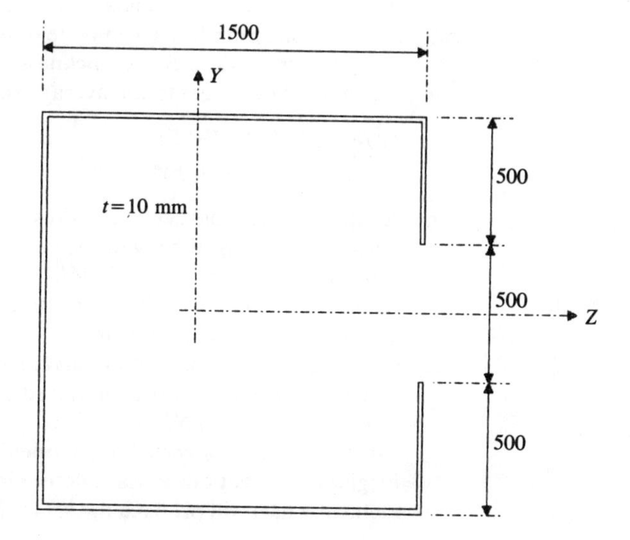

Figure 6.30 Exercise 6.1

6.2 If the thin-walled section shown in Fig. 6.31 is subjected to an applied moment $M_z = 14$ kN m, determine the distribution of direct stress over the cross-section.

Figure 6.31 Exercise 6.2

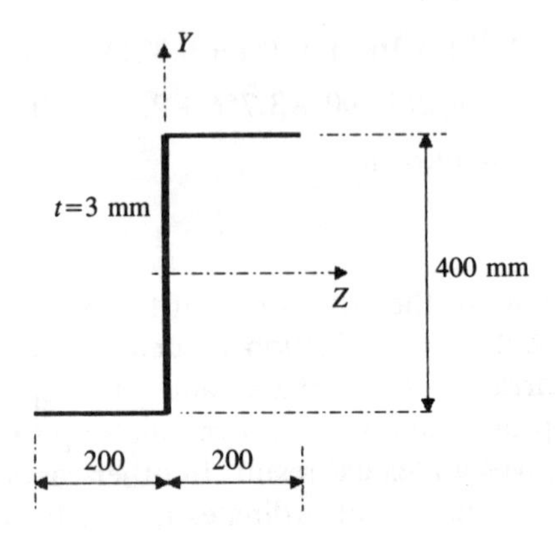

6.3 A cantilever beam AB, 3 m long, is fixed at A and simply-supported at B by a beam CD of the same material and cross-section. The beam CD is 4 m long and is arranged at right angles to the cantilever in plan. The beam CD is simply-supported at its ends C and D and the point of support of the cantilever is at the centre of CD. A vertical load W is carried by the beams at B. The section modulus of both beams is 300 cm^3. If the maximum bending stress in the beams is 150 N/mm^2, what is the maximum allowable value of the load W?

6.4 A beam of uniform flexural rigidity $EI = 20 \times 10^3$ kN m^2 and length 8 m is simply-supported at its ends and carries a uniformly distributed load of 10 kN/m, 4 m long. The load is centrally disposed on the beam. Determine
(a) the deflection at the centre of the beam
(b) the deflection at the ends of the load.

6.5 An I-section girder has a web plate of depth d and thickness t and flange plates of width B and thickness T. Show that the ratio of the maximum shear stress to the average shear stress is

$$\frac{td^3/8 + (dBT/2)(d+T)}{td^3/12 + (BT/2)(d+T)^2}$$

6.6 A solid circular cross-section, radius r, is subjected to a transverse shear force V coincident with the vertical diameter. Show that the maximum shear stress is $(4/3)(V/\pi r^2)$.

6.7 A square section hollow box has a side of length 1500 mm and a uniform wall thickness of 15 mm.
(a) Determine the shear flow, shear stress and angle of twist per unit length when the box is subjected to a pure twisting moment of 1000 kN m. $G = 80$ kN/mm^2.
(b) If the box is converted to an 'open' section by introducing a longitudinal slit at one corner, determine the value of the twisting moment which will produce the same shear stress as in (a).

6.8 The cross-section of a beam is the solid section triangle shown in Fig. 6.32. Show that, due to transverse shear S_y,

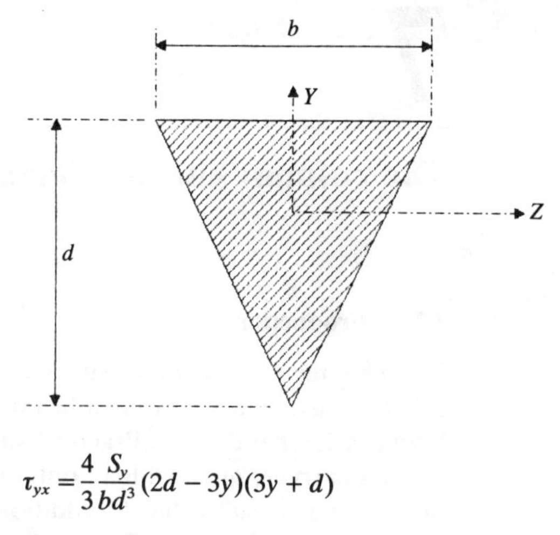

Figure 6.32 Exercise 6.8

$$\tau_{yx} = \frac{4}{3}\frac{S_y}{bd^3}(2d - 3y)(3y + d)$$

6.9 Determine the location of the shear centre of the thin-walled cross-section shown in Fig. 6.33.

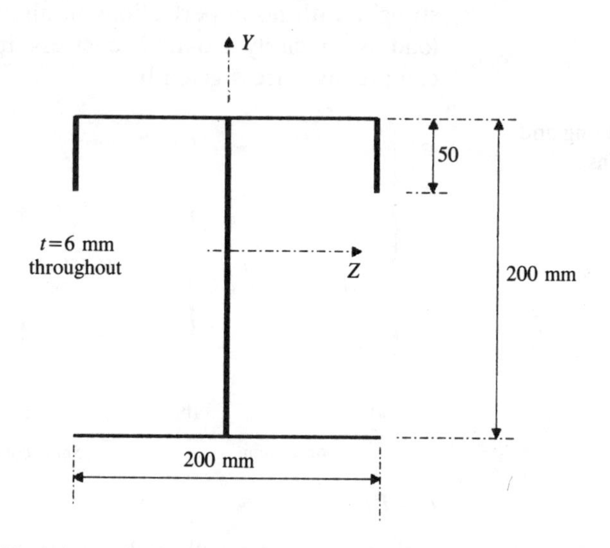

Figure 6.33 Exercise 6.9

6.10 A beam AB has a length l and a rectangular cross-section of width b. The depth of the beam varies uniformly from d_0 at end A to $2d_0$ at end B. Determine (a) the end flexibilities (b) the end stiffnesses.

CHAPTER

7

The column and the beam-column

7.1 Introduction

This chapter is concerned with structural members in which the effective axial load is significant (Timoshenko, Gere 1961; Horne, Merchant 1965). Practical situations vary from the simple column carrying an axial load only, to more complex cases where the member is subjected to additional actions involving bending and possibly twisting. In Fig. 7.1(a) we see a column carrying an axial load P. We assume 'ideal' circumstances: the column is straight with no imperfections in alignment or material, and the load is precisely axial. The stress in the column is a simple compressive stress given by

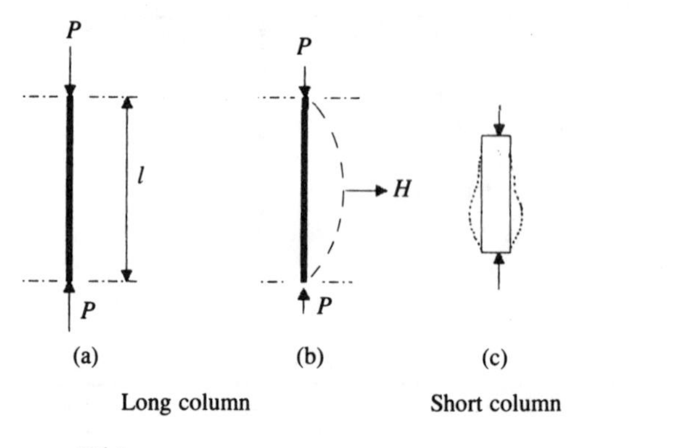

Figure 7.1 Long and short columns

$$\sigma = P/A \qquad [7.1]$$

and this situation will apply as the load P is increased until the material of the column begins to yield. However, in practice we shall not have the ideal conditions and failure of the column may take place by instability. Suppose that, at some particular value of the load P, we 'disturb' the column by a small lateral force H as in Fig. 7.1(b) and then remove H. Three possible situations can arise:

1. The column may return to the straight condition and is said to be stable.
2. The deflections may increase and the column is said to be unstable.

3. The column may remain in the slightly bent condition and is said to be in neutral equilibrium.

The last situation is of great importance in structural mechanics, and the value of the load in this condition is the 'critical' or 'Euler' load of the column.

It is convenient to classify columns as 'short' or 'long' depending on whether behaviour is controlled by stability (long column) or by a simple stress situation described by eqn [7.1] (short column: Fig. 7.1(c)).

The term 'squash' load is used for short columns when there is no tendency to instability. The squash load is

$$P(\text{squash}) = \sigma_y A \qquad [7.2]$$

where σ_y is the yield stress of the material and A is the cross-sectional area. The squash load is an important benchmark in the structural design of steel columns.

7.2 The short column

The characteristic of the short column is, as we have seen, the absence of instability tendencies. The stress due to an applied axial load is given by eqn [7.1]. In many cases, the effect of the axial load will be accompanied by bending in the column. This will happen when bending moments are applied through loading being eccentric to the column axis or inclined to the axis. In Fig. 7.2(a) the applied load P is inclined to the axis of the column.

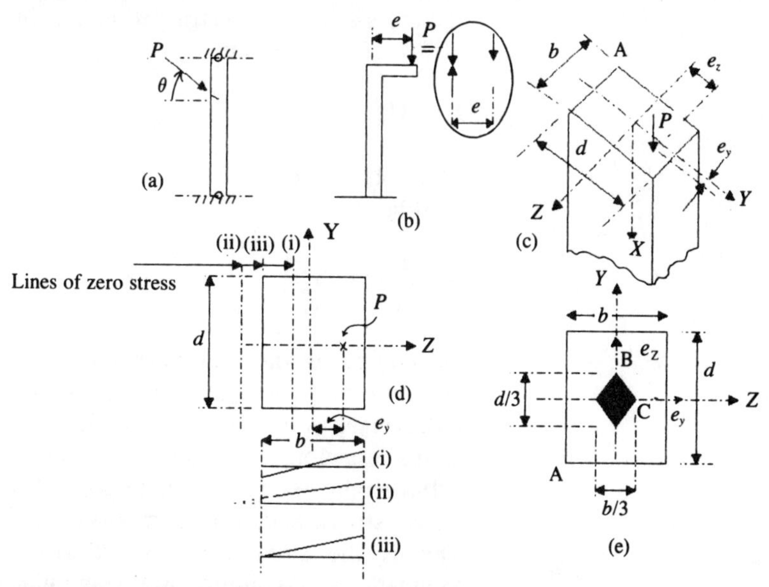

Figure 7.2 Combined axial load and bending in short columns

The vertical component $P \sin \theta$ acts as an axial load on the column, and the horizontal component $P \cos \theta$ will cause bending. In diagram (b) the load P is parallel to the longitudinal axis but eccentric by distance e. The resultant effect is an axial load P and a couple Pe. In diagram (c) the load P is eccentric to both Z and Y axes and thus moments are applied about $Z(Pe_z)$ and about $Y(Pe_y)$.

In general, bending stresses in a short column follow eqn [6.26] for asymmetrical sections and eqn [6.27] for symmetrical sections. To find the total stress at a point we simply add the stress due to the axial load (P/A) to the bending stresses. For example in Fig. 7.2(d) bending is entirely about the Y axis, and we can use eqn [6.27] with the term added for the axial load. Thus, assuming tension positive,

$$\sigma_x = -\frac{P}{A} - \frac{M_y z}{I_y}$$

$$= -\frac{P}{A} - \frac{Pe_y z}{I_y}$$

If we temporarily reverse the sign convention when we are dealing with columns and designate compressive stress as positive, then

$$\sigma_x = \frac{P}{A} + \frac{Pe_y z}{I_y} \qquad [7.3]$$

Equation [7.3] is independent of y, so the stress is constant for z constant. We can therefore envisage the stress distribution over the cross-section as a straight line, as in diagrams (i), (ii) and (iii) of (d). If we now put

$$I_y = Ak_y^2 \qquad [7.4]$$

where k_y is the 'radius of gyration' about the Y axis, then eqn [7.3] becomes

$$\sigma_x = \frac{P}{A}\left(1 + \frac{e_y z}{k_y^2}\right) \qquad [7.5]$$

Depending on the relative magnitudes of the two terms in eqn [7.3] we can have a reversal of compression to tension as in (i), no reversal as in (ii), or the limiting case of (iii) where we have zero stress along an edge of the column.

In certain structures it is important to avoid the production of tensile stress, and so it is important to identify the limiting case. This is the situation in brick and masonry piers and with foundations. Assuming a rectangular cross-section as in Fig. 7.2(d),

$$I_y = db^3/12$$

Hence

$$k_y^2 = b^2/12$$

and eqn [7.5] becomes

$$\sigma_x = \frac{P}{A}\left(1 + \frac{12e_y z}{b^2}\right)$$

For zero stress at $z = -b/2$,

$$\sigma_x = 0 = 1 - \frac{6e_y}{b}$$

$$e_y = b/6$$

Thus, providing $e_y \not> b/6$, we can say that no tension is caused by P at eccentricity e_y.

Consider now the load P eccentric to both Y and Z axes as in Fig. 7.2(c):

$$\sigma_x = \frac{P}{A} + \frac{Pe_y z}{I_y} + \frac{Pe_z y}{I_z} \qquad [7.6]$$

$$= \frac{P}{A}\left(1 + \frac{e_y z}{k_y^2} + \frac{e_z y}{k_z^2}\right) \qquad [7.7]$$

The condition for zero stress is therefore

$$0 = 1 + \frac{e_y z}{k_y^2} + \frac{e_z y}{k_z^2} \qquad [7.8]$$

We can use eqn [7.8] to determine limits on the eccentricities of the load in each quadrant of the cross-section of the column. If the load is in the first quadrant then the condition for zero tension in the diagonally opposite corner at point A ($z = -b/2$, $y = -d/2$) is

$$0 = 1 + \frac{e_y}{k_y^2}\left(-\frac{b}{2}\right) + \frac{e_z}{k_z^2}\left(-\frac{d}{2}\right)$$

and since $k_y^2 = b^2/12$ and $k_z^2 = d^2/12$,

$$0 = 1 - \frac{6e_y}{b} - \frac{6e_z}{d} \qquad [7.9]$$

Equation [7.9] gives us line BC in Fig. 7.2(e), and we can obtain similar lines in the other three quadrants. Together they constitute an area within which the load point must lie to avoid the development of tensile stress. The shaded area is called the 'core' of the section. This is clearly a geometrical property of any cross-section.

Example 7.1

A short column has the cross-section shown in Fig. 7.3. Vertical loads of 9000 kN can be applied at A and/or A′. Determine the maximum compressive and tensile stresses in the cross-section.

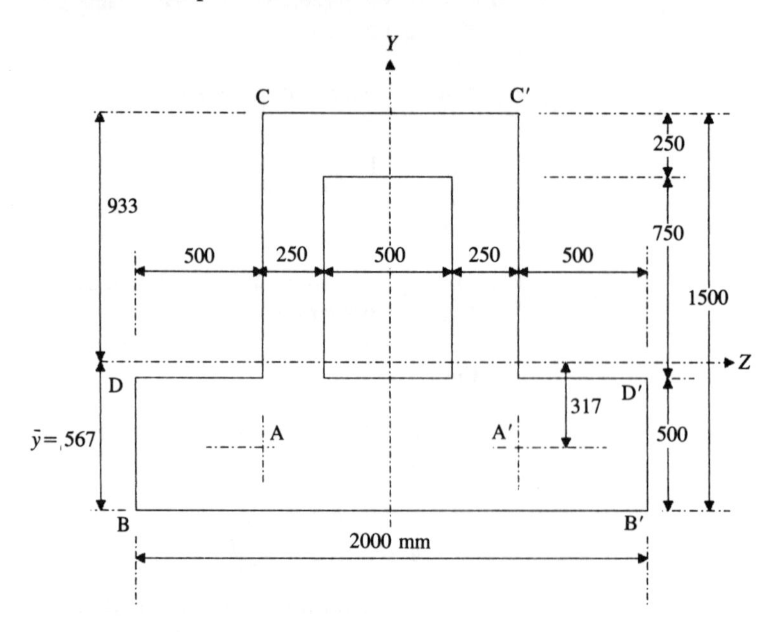

Figure 7.3 Example 7.1

$$A = 2000 \times 500 + 1000 \times 1000 - 750 \times 500$$
$$= 1\,625\,000 \text{ mm}^2$$
$$A\bar{y} = 2000 \times 500 \times 250 + 1000 \times 1000 \times 1000 - 500 \times 750 \times 875$$
$$\bar{y} = 567$$

Thus the eccentricity of both loads about the Z axis is

$$e_z = -567 + 250 = -317 \text{ mm}$$
$$I_z = 2000 \times \frac{567^3}{3} - 1500 \times \frac{67^3}{3} + 1000 \times \frac{933^3}{3} - 500 \times \frac{683^3}{3}$$
$$= 33.9 \times 10^{10} \text{ mm}^4$$
$$I_y = 500 \times \frac{2000^3}{12} + 1000 \times \frac{1000^3}{12} - 750 \times \frac{500^3}{12}$$
$$= 40.89 \times 10^{10} \text{ mm}^4$$

The compressive stress at any point with coordinates (z, y) is, from eqn [7.6],

$$\sigma = \frac{P}{A} + \frac{Pe_y z}{I_y} + \frac{Pe_z y}{I_z}$$

We need to consider the stresses produced by both loads acting together and by a single load acting, since the latter situation might produce the maximum tension in the column.

Both loads acting

We have $e_z = -317$ and $e_y = 0$. The maximum compression will occur along the face BB′ and will be

$$\frac{2 \times 9000 \times 10^3}{1625 \times 10^3} + \frac{2 \times 9000 \times 10^3(-317)(-567)}{33.9 \times 10^{10}}$$
$$= 11.08 + 9.54 = 20.6\,\text{N/mm}^2$$

The maximum tension under this loading will be along face CC′ ($y = 933$) and will be

$$\frac{2 \times 9000 \times 10^3}{1625 \times 10^3} + \frac{2 \times 9000 \times 10^3(-317)(933)}{33.9 \times 10^{10}}$$
$$= 11.08 - 15.7 = -4.6\,\text{N/mm}^2$$

Load at A′ only

We have $e_z = -317$ and $e_y = 500$. The maximum compressive stress will be at B′ and is

$$5.54 + \frac{9000 \times 10^3(-317)(-567)}{33.9 \times 10^{10}} + \frac{9000 \times 10^3(500)(1000)}{40.89 \times 10^{10}}$$
$$= 5.54 + 4.77 + 11.01 = 21.3\,\text{N/mm}^2$$

The maximum tensile stress will be at C or D. We calculate both:

$$\sigma_C = 5.54 + \frac{9000 \times 10^3(-317)(933)}{33.9 \times 10^{10}} + \frac{9000 \times 10^3(500)(-500)}{40.89 \times 10^{10}}$$
$$= 5.54 - 7.85 - 5.50 = -7.8\,\text{N/mm}^2$$

$$\sigma_D = 5.54 + \frac{9000 \times 10^3(-317)(-67)}{33.9 \times 10^{10}} + \frac{9000 \times 10^3(500)(-1000)}{40.89 \times 10^{10}}$$
$$= -4.9\,\text{N/mm}^2$$

The maximum stresses are therefore as follows: maximum compression = 21.3 N/mm^2 (at B′), and maximum tension = 7.8 N/mm^2 (at C), both occurring with a single load acting at A′.

7.3 The long column

We now consider columns in which there is a tendency to instability as described in section 7.1. We assume the column

(Fig. 7.4(a)) to be in a state of neutral equilibrium following the

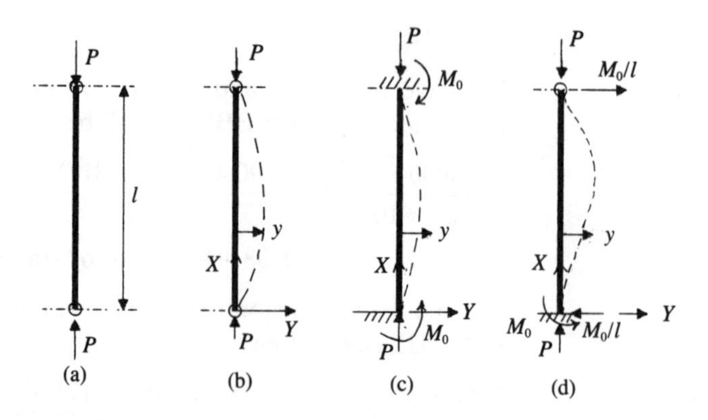

Figure 7.4 Euler buckling of columns

removal of the lateral disturbing force; thus the bending moment is

$$M = Py = -EI\frac{d^2y}{dx}$$

or

$$\frac{d^2y}{dx^2} + \frac{Py}{EI} = 0 \qquad [7.10]$$

where y is the lateral deflection of the column in the X–Y plane. Putting

$$P/EI = n^2 \qquad [7.11]$$

then

$$\frac{d^2y}{dx^2} + n^2y = 0 \qquad [7.12]$$

the solution of which is

$$y = A\cos nx + B\sin nx \qquad [7.13]$$

The constants A and B are determined from the end conditions. At $x = 0$, $y = 0$; hence $A = 0$. At $x = l$, $y = 0$; hence $B\sin nl = 0$, or

$$B\sin l\sqrt{(P/EI)} = 0 \qquad [7.14]$$

Now $B \neq 0$, otherwise y would be zero everywhere. Hence

$$\sin l\sqrt{(P/EI)} = 0 \qquad [7.15]$$

There are an infinite number of solutions to eqn [7.15], i.e.

$$l\sqrt{(P/EI)} = \pi, 2\pi, 3\pi, \ldots \qquad [7.16]$$

or

$$P = \frac{\pi^2 EI}{l^2}; \quad \frac{(2\pi)^2 EI}{l^2}; \quad \frac{(3\pi)^2 EI}{l^2}; \ldots \qquad [7.17]$$

Clearly, from physical considerations, the critical value of P is the lowest one; this is the Euler load P_E. Thus

$$P_E = \frac{\pi^2 EI}{l^2} \qquad [7.18]$$

and

$$y = B \sin nx$$

$$= B \sin \frac{\pi x}{l} \qquad [7.19]$$

Equation [7.19] represents a half-sine wave from $x = 0$ to $x = l$ (Fig. 7.4(b)). The coefficient B is undetermined since the column is in *any* bent half-sine wave providing y is small.

We have tacitly assumed that the column will bend in the plane of minimum flexural rigidity and that this is a plane of symmetry. In these circumstances there will be no twisting of the column. However, with unsymmetrical sections, instability may be by combined bending and twisting. Putting

$$I = Ak^2 \qquad [7.20]$$

in eqn [7.18] we get $P_E = \pi^2 EAk^2/l^2$ or

$$\frac{P_E}{A} = \sigma_E = \frac{\pi^2 E}{(l/k)^2} \qquad [7.21]$$

The term l/k is the 'slenderness ratio' and is an important indicator of the instability tendency of columns.

Consider now the column with direction-fixed ends as in Fig. 7.4(c). We have

$$M = Py - M_0$$

$$EI\frac{d^2y}{dx^2} = -M = M_0 - Py$$

$$\frac{d^2y}{dx^2} + n^2 y = \frac{M_0}{EI} \qquad [7.22]$$

The solution of eqn [7.22] is

$$y = A \cos nx + B \sin nx + \frac{M_0}{P} \qquad [7.23]$$

At $x = 0$, $y = 0$; hence $A = -M_0/P$. At $x = 0$, $dy/dx = 0$; hence $B = 0$, and

$$y = -\frac{M_0}{P}\cos nx + \frac{M_0}{P}$$

At $x = l$, $y = 0$; hence

$$0 = \frac{M_0}{P}(1 - \cos nl)$$

i.e.

$$\cos nl = 1$$

The solution of this is

$$nl = 2\pi, 4\pi, 6\pi, \ldots$$

Hence

$$P = n^2 EI$$
$$= \frac{(2\pi)^2 EI}{l^2}; \qquad \frac{(4\pi)^2 EI}{l^2}; \ldots$$

The Euler load is the smallest of these, i.e.

$$P_E = \frac{4\pi^2 EI}{l^2} \qquad [7.24]$$

The column deflections are

$$y = A \cos nx + \frac{M_0}{P}$$

i.e.

$$y = \frac{M_0}{P}\left(1 - \cos\frac{2\pi x}{l}\right) \qquad [7.25]$$

Again the deflection is indeterminate since we cannot evaluate M_0.

Consider now the column with one end direction-fixed and the other pinned, as in Fig. 7.4(d). We have

$$M = Py - M_0 + \frac{M_0 x}{l}$$

(note the horizontal shear force M_0/l at each end of the column). Thus

$$EI\frac{d^2y}{dx^2} = -Py + M_0 - M_0\frac{x}{l}$$

$$\frac{d^2y}{dx^2} + n^2 y = \frac{M_0}{P} n^2\left(1 - \frac{x}{l}\right) \qquad [7.26]$$

The solution is

$$y = A \cos nx + B \sin nx + \frac{M_0}{P}\left(1 - \frac{x}{l}\right) \qquad [7.27]$$

At $x = 0$, $dy/dx = 0$; hence $B = M_0/nPl$. At $x = 0$, $y = 0$; hence

$A = -M_0/P$. Hence

$$y = \frac{M_0}{nPl} \sin nx - \frac{M_0}{P} \cos nx + \frac{M_0}{P}\left(1 - \frac{x}{l}\right) \qquad [7.28]$$

Also, at $x = l$, $y = 0$; hence

$$0 = \frac{M_0}{nPl} \sin nl - \frac{M_0}{P} \cos nl$$

or

$$\tan nl = nl \qquad [7.29]$$

The smallest non-trivial solution to eqn [7.29] is $nl = 4.4934$, giving

$$\sqrt{\left(\frac{P}{EI}\right)} = \frac{4.4934}{l}$$

or

$$P_E = \frac{20.2EI}{l^2} \qquad [7.30]$$

Alternatively,

$$P_E = \frac{\pi^2 EI}{(0.7l)^2} \qquad [7.31]$$

in which some rounding-off has been carried out to give a compact result incorporating π^2.

Comparing eqns [7.18], [7.24] and [7.31] we see that they all take the form

$$P_E = \frac{\pi^2 EI}{\lambda^2} \qquad [7.32]$$

where

$$\lambda = \begin{cases} l & \text{for pinned ends} \\ l/2 & \text{for direction-fixed ends} \\ 0.7l & \text{for one end fixed and one pinned} \end{cases}$$

λ is called the 'effective length' of the column and will be seen to be the length of a half-sine wave on the deflected shape of the buckled column. We could now deduce the Euler load for a column fixed at the lower end and free at the upper end. The half-sine wave will have a length $2l$ for such a column; hence $\lambda = 2l$ and

$$P_E = \frac{\pi^2 EI}{(2l)^2} = \frac{\pi^2 EI}{4l^2}$$

7.4 The column with initial curvature

The initial assumption we made in section 7.1 that the column was perfectly straight will never be satisfied in practice and all practical columns will have some lack of straightness. Furthermore there will be other imperfections in manufacture and unintended eccentricities of loading which will reduce the reliability of the Euler theory. The effect of practical imperfections is more marked with lower slendernesses l/k, and the Euler theory becomes more reliable when l/k is large. Perry's column theory attempts to reflect this practical situation and is based on the following assumptions:

1. The imperfections in material, manufacture and loading can be represented by assuming the column to have an initial curvature which is taken to be a sinusoidal curve (Fig. 7.5).

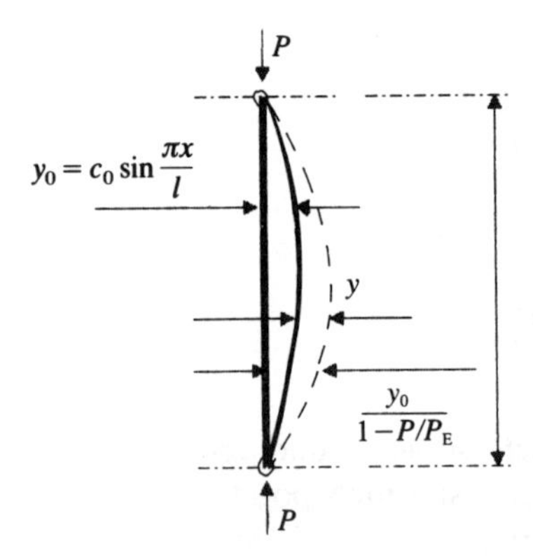

Figure 7.5 Perry's theory for initial curvature

2. The column behaves elastically.

The unloaded shape of the column is taken to be

$$y_0 = c_0 \sin \frac{\pi x}{l} \qquad [7.33]$$

Hence

$$M = P(y + y_0)$$

and

$$EI\frac{d^2y}{dx^2} = -P(y + y_0) = -P\left(y + c_0 \sin \frac{\pi x}{l}\right)$$

or

$$\frac{dy^2}{dx^2} + n^2\left(y + c_0 \sin \frac{\pi x}{l}\right) = 0 \qquad [7.34]$$

The solution is

$$y = A\cos nx + B\sin nx + \frac{n^2 c_0 \sin \pi x/l}{\pi^2/l^2 - n^2}$$

At $x = 0$, $y = 0$; hence $A = 0$. At $x = l$, $y = 0$; hence $B = 0$. Hence

$$y = \frac{n^2 c_0 \sin \pi x/l}{\pi^2/l^2 - n^2}$$

$$= \frac{P}{P_E - P} c_0 \sin\frac{\pi x}{l}$$

The total departure from straightness at load P is therefore

$$y_0 + y = y_0\left[\frac{1}{1 - P/P_E}\right] \qquad [7.35]$$

The term $1/(1 - P/P_E)$ in eqn [7.35] is a 'magnification' factor. It determines the magnification of the lateral deflection when the load is applied.

7.5 A design formula for columns

We shall now develop the Perry formula used as a basis for the design of columns in BS 5950 (Appendix C). The maximum bending moment will be at $x = l/2$ where, from eqn [7.35], and [7.33],

$$y_0 + y = \frac{c_0}{1 - P/P_E}$$

Hence

$$M_{max} = \frac{Pc_0}{1 - P/P_E} \qquad [7.36]$$

The design formula is expressed in stresses rather than loads, so we put

$$\sigma_c = \frac{P}{A} = \frac{\text{design load}}{\text{cross-sectional area}}$$

Also,

$$\sigma_E = \frac{P_E}{A}$$

The maximum compressive stress is, from eqn [7.36],

$$\sigma_{max} = \frac{P}{A} + \frac{Pc_0}{1 - P/P_E}\frac{d}{I} \qquad [7.37]$$

where I is the second moment of area for bending about the weak axis, and d is the distance from the centroidal axis to the extreme fibre. Putting $I = Ak^2$ and $\eta = c_0 d/k^2$ in eqn [7.37], we obtain

$$\sigma_{max} = \sigma_c\left(1 + \frac{\eta\sigma_E}{\sigma_E - \sigma_c}\right) \qquad [7.38]$$

Putting $\sigma_{max} = \sigma_y$, the yield stress, eqn [7.38] becomes

$$(\sigma_E - \sigma_c)(\sigma_y - \sigma_c) = \eta\sigma_E\sigma_c \qquad [7.39]$$

This is the equation quoted in Appendix C of BS 5950. Equation [7.39] is a quadratic in σ_c; the smallest solution is

$$\sigma_c = \frac{\sigma_y + (\eta + 1)\sigma_E}{2} - \sqrt{\left\{\left[\frac{\sigma_y + (\eta + 1)\sigma_E}{2}\right]^2 - \sigma_y\sigma_E\right\}} \qquad [7.40]$$

which can also be expressed as

$$\sigma_c = \frac{\sigma_E\sigma_y}{\phi + (\phi^2 - \sigma_E\sigma_y)^{1/2}} \qquad [7.41]$$

where

$$\phi = \frac{\sigma_y + (\eta + 1)\sigma_E}{2} \qquad [7.42]$$

Equation [7.40] or the alternative [7.41] is the Perry formula for the intensity of loading ($\sigma_c = P/A$) which will cause the maximum stress in the column to reach yield.

Example 7.2

A column is pin-ended for bending about both principal axes. The column is 3500 mm long and the following data apply:

$I_z = 5383\ \text{cm}^4$

$I_y = 2041\ \text{cm}^4$

$A = 75.8\ \text{cm}^2$

$\sigma_y = 240\ \text{N/mm}^2$

$E = 207\ \text{kN/mm}^2$

Using the Perry formula with $\eta = 0.003l/k$, determine the maximum load for the column.

I_y is critical since $I_y \leqslant I_z$. Therefore

$$k^2 = I/A = \frac{2041 \times 10^2}{75.8} = 2692.6$$

Hence

$$\sigma_E = \frac{\pi^2 E}{(l/k)^2} = \frac{\pi^2 \times 207 \times 10^3 \times 2692.6}{3500^2} = 449 \text{ N/mm}^2$$

$$\eta = \frac{0.003 \times 3500}{\sqrt{2692.6}} = 0.20\,235$$

Therefore

$$\sigma_c = \frac{240 + 1.20\,235 \times 449}{2} - \sqrt{\left[\left(\frac{240 + 1.20\,235 \times 449}{2}\right)^2 - 240 \times 449\right]}$$

$$= 179.5 \text{ N/mm}^2$$

7.6 Variable section columns

The analysis of variable section columns is best carried out using numerical procedures on the computer. We shall investigate a method of computing the elastic critical load P_E for a non-prismatic column using the standard finite difference approximation for curvature and a matrix iterative method to determine P_E. The reader will recognize this as a standard technique for the solution of eigenvalue problems.

Suppose we have a pin-ended column as in Fig. 7.6(a). We divide the column length into equal segments of length h and number the n internal nodes only. The governing differential equation of bending is, as we have already seen, eqn [7.10], here given as

Figure 7.6 Example 7.3: variable section columns

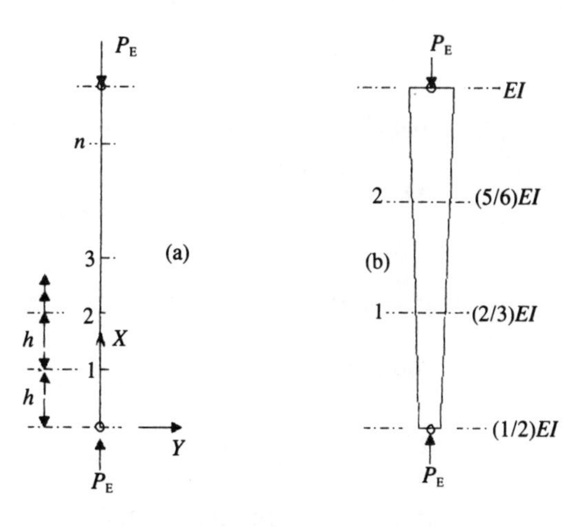

$$EI\frac{d^2y}{dx^2} = -P_E y \tag{7.43}$$

where now $EI = f(x)$ along the column.

The finite difference approximation to d^2y/dx^2 is

$$\left(\frac{d^2y}{dx^2}\right)_i = \frac{1}{h^2}(y_{i-1} - 2y_i + y_{i+1}) \qquad [7.44]$$

Substituting eqn [7.44] in eqn [7.43], at a typical node i,

$$(-y_{i-1} + 2y_i - y_{i+1}) = P_E y_i h^2/(EI)_i \qquad [7.45]$$

We can now write an equation like [7.45] at each internal node in the column. The value of y at the ends is zero, so these do not appear in the equations we write. Thus

$$\begin{bmatrix} 2 & -1 & & & & & \\ -1 & 2 & -1 & & & 0 & \\ & -1 & 2 & -1 & & & \\ & \cdot & \cdot & \cdot & & & \\ & & \cdot & \cdot & \cdot & & \\ & & & \cdot & \cdot & \cdot & \\ & 0 & & & \cdot & \cdot & -1 \\ & & & & & -1 & 2 \end{bmatrix} \begin{bmatrix} y_1 \\ y_2 \\ y_3 \\ \cdot \\ \cdot \\ \cdot \\ y_{n-1} \\ y_n \end{bmatrix}$$

$$= h^2 P_E \begin{bmatrix} 1/(EI)_1 & & & & 0 \\ & 1/(EI)_2 & & & \\ & & 1/(EI)_3 & & \\ & & & \cdot & \\ & & & & \cdot \\ & 0 & & & \cdot \\ & & & & 1/(EI)_n \end{bmatrix} \begin{bmatrix} y_1 \\ y_2 \\ y_3 \\ \cdot \\ \cdot \\ \cdot \\ y_n \end{bmatrix} \qquad [7.46]$$

Which we write in the compact form

$$\boldsymbol{Ay} = h^2 P_E \boldsymbol{Cy}$$

Putting

$$\gamma^2 = h^2 P_E \qquad [7.47]$$

and premultiplying both sides of eqn [7.46] by $\boldsymbol{A}^{-1}$, we have

$$\boldsymbol{A}^{-1}\boldsymbol{Cy} = \frac{1}{\gamma^2}\boldsymbol{y} \qquad [7.48]$$

Equation [7.48] should be recognized as a standard eigenvalue problem.

We commence the iterative solution by assuming an initial eigenvector $\boldsymbol{y}_0$; then

$$A^{-1}Cy_0 \simeq \frac{1}{\gamma^2} y_0$$

Putting $A^{-1}Cy_0 = y_1$, then

$$y_1 \simeq \frac{1}{\gamma_1^2} y_0$$

giving $\gamma_1^2 \simeq y_0/y_1$ or, generally,

$$\gamma_i^2 = y_{i-1}/y_i \qquad [7.49]$$

Now we cannot form a ratio of two column matrices as required by eqn [7.49], so we form the individual ratios of corresponding elements and adopt the largest as representative. (Some methods express the ratio using the sums of the elements in the column matrices.) The iterative process then continues:

$$A^{-1}Cy_1 \simeq \frac{1}{\gamma^2} y_1 = y_2$$

So in general,

$$A^{-1}Cy_i \simeq \frac{1}{\gamma^2} y_i = y_{i+1} \qquad [7.50]$$

The numerical procedure described can be shown to lead to the largest value of $1/\gamma^2$ and hence to the smallest value of P_E, which is of course the critical Euler load.

Example 7.3

We shall compute the critical load of the column shown in Fig. 7.6(b) using the iterative procedure. We take only two internal points in order to keep the matrices small and easy to follow. On the computer, we could use many more internal points of subdivision if necessary. The flexural rigidity of the column varies linearly from $EI/2$ at the bottom to EI at the top. Now

$$A = \begin{bmatrix} 2 & -1 \\ -1 & 2 \end{bmatrix}; \qquad A^{-1} = \frac{1}{3}\begin{bmatrix} 2 & 1 \\ 1 & 2 \end{bmatrix}$$

$$C = \frac{1}{EI}\begin{bmatrix} 3/2 & 0 \\ 0 & 6/5 \end{bmatrix}; \qquad A^{-1}C = \frac{1}{10EI}\begin{bmatrix} 10 & 4 \\ 5 & 8 \end{bmatrix}$$

For the initial eigenvector, we take

$$y_0 = \begin{bmatrix} 0.5 \\ 0.4 \end{bmatrix}$$

Then

$$y_1 = A^{-1}Cy_0 = \frac{1}{EI}\begin{bmatrix}0.66\\0.57\end{bmatrix}$$

Hence the first approximation to γ^2 is

$$\gamma^2 \simeq \frac{y_0}{y_1} = \frac{0.5}{0.66/EI} = 0.76EI$$

where $h = l/3$. Therefore

$$P_E = \gamma^2/h^2 = 0.76EI \times 9/l^2 = 6.84EI/l^2$$

(Note: $\pi^2 EI/l^2 = 9.87EI/l^2$, so we are going in the right direction since clearly $P_E < \pi^2 EI/l^2$.)

The second approximation is

$$y_2 = A^{-1}Cy_1$$
$$= \frac{1}{(EI)^2}\begin{bmatrix}0.888\\0.786\end{bmatrix}$$

and

$$\gamma^2 \simeq \frac{y_1}{y_2} = \frac{0.66EI}{0.888} = 0.743EI$$

which gives $P_E = 6.69EI/l^2$. A third iteration gives $P_E = 6.65EI/l^2$.

Although the process has converged satisfactorily, it has done so to an approximate result. The reader might like to repeat this calculation with a finer division, say six equal segments with five internal points of subdivision. The result is $P_E = 7.1EI/l^2$, which is a somewhat improved estimate (see exercise 7.3).

7.7 The beam-column and the stability functions

The beam stiffnesses we examined in section 4.5 need modification if a significant axial load is present in the member. The effect of a compressive axial load in a 'beam-column' is to reduce the stiffnesses in general. The theoretical development produces the stability functions (Livesley 1964). The theory is well known, so we shall give only a brief introduction to it using the notation of Horne and Merchant (1965).

Referring to Fig. 7.7, a moment M_1 is applied at end 1 to a prismatic member which is fixed at end 2. A rotation θ_1 is produced at end 1. The member carries an axial load P. Now

Figure 7.7 Prismatic member stiffnesses in terms of stability functions

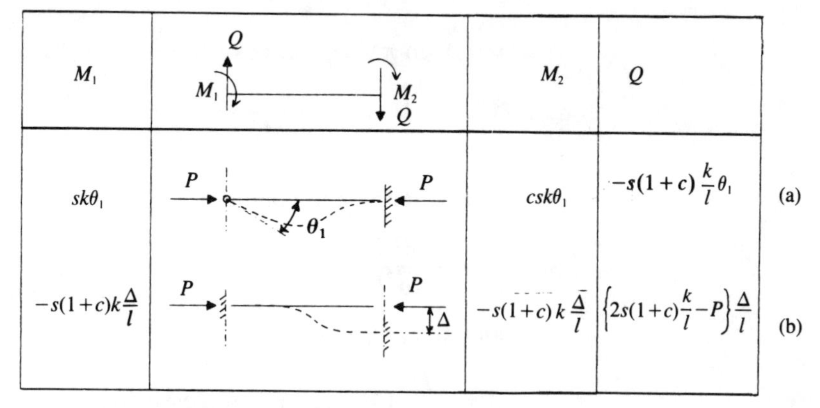

$$s=\frac{\alpha(1-2\alpha\cot 2\alpha)}{\tan\alpha-\alpha} \qquad c=\frac{2\alpha-\sin 2\alpha}{\sin 2\alpha-2\alpha\cos 2\alpha}$$

$$\alpha=\frac{\pi}{2}\sqrt{\varrho} \qquad \varrho=P/P_E=\ Pl/\pi^2k$$

$$k=EI/l$$

$$M=-M_1-Qx+Py=-EI\frac{d^2y}{dx^2}$$

Therefore

$$EI\frac{d^2y}{dx^2}=M_1+Qx-Py \qquad [7.51]$$

We put

$$\rho=P/P_E \qquad [7.52]$$

where $P_E=\pi^2EI/l^2$ is the Euler load for a pin-ended member, and also put

$$EI/l=k \qquad [7.53]$$

Also from Fig. 7.7,

$$Q=-\frac{(M_1+M_2)}{l} \qquad [7.54]$$

Hence the governing differential equation is

$$\frac{d^2y}{dx^2}+\frac{\pi^2}{l^2}\rho y=-\frac{1}{kl}\left[(M_1+M_2)\frac{x}{l}-M_1\right] \qquad [7.55]$$

The solution is

$$y = A \sin \pi\sqrt{\rho}\frac{x}{l} + B \cos \pi\sqrt{\rho}\frac{x}{l} - \frac{l}{\pi^2 \rho k}\left[(M_1 + M_2)\frac{x}{l} - M_1\right] \quad [7.56]$$

Applying the end conditions: at $x = 0$, $y = 0$,

$$B = -\frac{M_1 l}{4\alpha^2 k}$$

and at $x = l$, $y = 0$,

$$A = \frac{l}{4\alpha^2 k}(M_1 \cot 2\alpha + M_2 \operatorname{cosec} 2\alpha)$$

where

$$\alpha = \frac{\pi}{2}\sqrt{\rho} \quad [7.57]$$

Now at $x = l$, $\mathrm{d}y/\mathrm{d}x = 0$; hence, differentiating eqn [7.56] and substituting we obtain

$$\frac{M_2}{M_1} = \frac{2\alpha - \sin 2\alpha}{\sin 2\alpha - 2\alpha \cos 2\alpha} = c \quad [7.58]$$

Now

$$\frac{\mathrm{d}y}{\mathrm{d}x} = -\theta_1 \quad \text{at} \quad x = 0 \quad [7.59]$$

Applying condition [7.59] and substituting for M_2 in eqn [7.58] we obtain

$$M_1 = \frac{\alpha k(1 - 2\alpha \cot 2\alpha)\theta_1}{\tan \alpha - \alpha} \quad [7.60]$$

which we can conveniently put in the form

$$M_1 = sk\theta_1 \quad [7.61]$$

where

$$s = \frac{(1 - 2\alpha \cot 2\alpha)\alpha}{\tan \alpha - \alpha} \quad [7.62]$$

Substituting for M_1 in eqn [7.58] we find

$$M_2 = csk\theta_1 \quad [7.63]$$

Also the shear force Q is, from eqn [7.54],

$$Q = -s(1 + c)\frac{k}{l}\theta_1 \quad [7.64]$$

Consider now the purely transverse displacement Δ in Fig.

7.7(b). We can deduce the expressions for M_1, M_2 and Q for this case from the previous case if we apply rotations $-\Delta/l$ to each end of the beam; then we have the same geometrical conditions as in diagram (b). Thus

$$\left.\begin{aligned} M_1 &= -sk\frac{\Delta}{l} - sck\frac{\Delta}{l} = -s(1+c)\frac{k\Delta}{l} \\ M_2 &= -s(1+c)k\frac{\Delta}{l} \end{aligned}\right\} \qquad [7.65]$$

The transverse shear Q is obtained by taking moments about either end,

$$Q = [2s(1+c)k/l - P]\Delta/l \qquad [7.66]$$

Table 7.1 gives values of the stability functions s and c over a limited range of values of $\rho(=P/P_{\text{E}})$. For a more complete tabulation the reader is referred to other sources (Horne, Merchant 1965; Majid 1978).

Table 7.1 s and c stability functions for axial compression ($\rho = P/(\pi^2 EI/l^2)$)

ρ	s	c	ρ	s	c
0.0	4.0000	0.5000	1.0	2.467	1.000
0.1	3.8667	0.5260	1.1	2.283	1.111
0.2	3.7297	0.5550	1.2	2.090	1.249
0.3	3.5889	0.5875	1.3	1.889	1.424
0.4	3.4439	0.6242	1.4	1.678	1.656
0.5	3.2945	0.6659	1.5	1.457	1.973
0.6	3.1403	0.7136	1.6	1.224	2.435
0.7	2.9809	0.7687	1.7	0.978	3.166
0.8	2.8159	0.8330	1.8	0.717	4.497
0.9	2.6450	0.9090	1.9	0.439	7.661

Example 7.4

For the column shown in Fig. 7.8(a) the critical (Euler) load as a pin-ended column is $\pi^2 EI/l^2$. If a moment M_1 is applied at end 1, the effect of this is to reduce the bending stiffness of the column. We can express the relationship between the end moment and the corresponding rotation θ_1 by temporarily fixing the column axis at end 1 as at (b) and then applying a moment at end 2 to produce the rotation θ_2. Then we apply a moment at 1 with 2 fixed as at (c). This is an application of the principle of superposition. Thus we can obtain diagram (a) by adding diagrams (b) and (c):

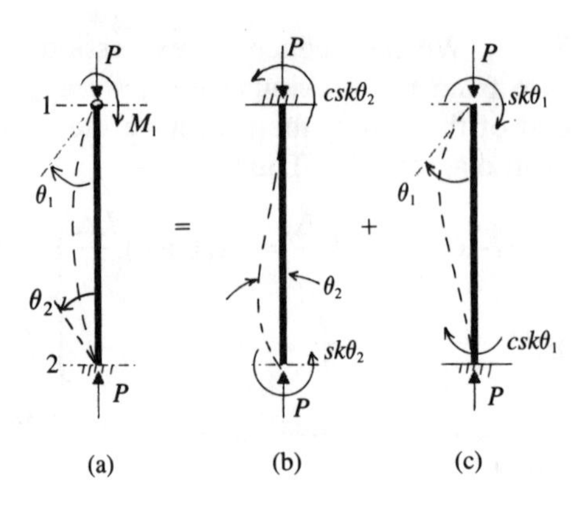

Figure 7.8 Example 7.4

$$M_1 = sk\theta_1 - csk\theta_2$$
$$M_2 = csk\theta_1 - sk\theta_2 = 0$$

Hence

$$\theta_2 = c\theta_1$$
$$M_1 = sk\theta_1 - c^2 sk\theta_1$$
$$= sk(1 - c^2)\theta_1$$

Now if the value of the axial load is increased (i.e. ρ increasing) then $s(1 - c^2)$ will reduce until $s(1 - c^2) = 0$. This occurs at $\rho = 1$, where

$$s(1 - c^2) = 2.4674(1 - 1) = 0$$

Thus the stiffness of the column vanishes at $\rho = 1$, and we have the critical load for the pin-ended column

$$P = P_E = \pi^2 EI/l^2$$

We can apply the same reasoning to the column in diagram (c), where the stiffness will vanish when $M_1 = sk\theta_1 = 0$, i.e. when $s = 0$. This situation is outside the scope of Table 7.1 but occurs at $\rho = 2.05$, and hence the critical load is

$$P = 2.05\pi^2 EI/l^2 \simeq 20.2EI/l^2$$

Example 7.5

In Fig. 7.9 a column AB, pinned at the bottom, is supported at the top by a beam AC, fixed at C. *EI* is the same for the beam and column. Determine the critical value of the load *P*.

The critical load will be reached when the rotational stiffness of joint A vanishes. The column then takes up a buckled form as shown, this being accompanied by bending of the beam. Now the relationship between moment and rotation of A is

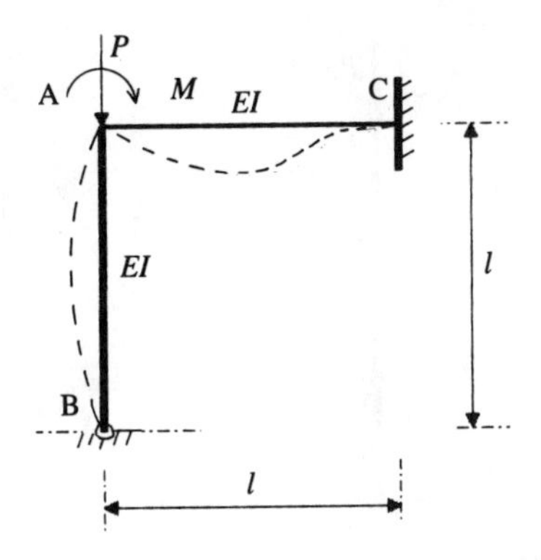

Figure 7.9 Example 7.5

$$M = [sk(1 - c^2) + 4EI/l]\theta = 0$$

since no moment is applied. Hence

Table 7.2 Solution for example 7.5

ρ	s	c	$s(1 - c^2) + 4$
1.0	2.467	1.00	4
1.5	1.457	1.973	−0.214
1.4	1.678	1.656	1.08
1.45	1.567	1.814	0.410

$$s(1 - c^2) + 4 = 0$$

This can be solved by trial and error as in Table 7.2. We see that $\rho = 1.47$ will be a good approximation. Hence

$$P_{crit} = 1.47\pi^2 EI/l^2 = 14.5EI/l^2$$

7.8 An approximate theory for laterally loaded columns

The theoretical analysis of the beam-column can be complex, as we have seen. In certain circumstances a good approximation can be obtained if the deflected form can be taken as sinusoidal. This is what is known as the 'Perry approximation'.

Suppose, for the beam-column shown in Fig. 7.10, we can take the deflected shape as

$$y = c \sin \frac{\pi x}{l} \qquad [7.67]$$

and

$$M_z = M = EI \frac{d^2 y}{dx^2} = -M' - Py \qquad [7.68]$$

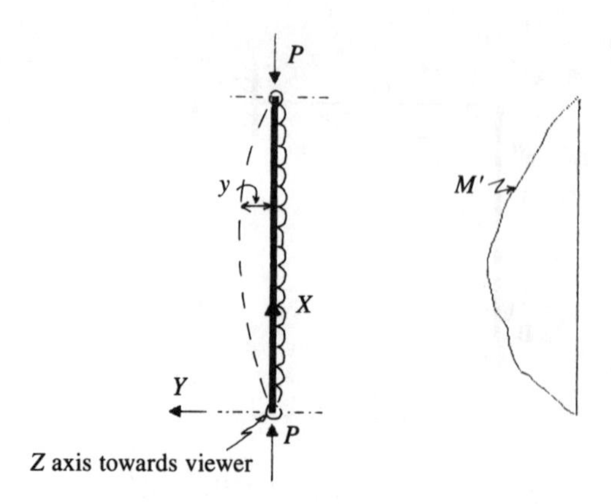

Figure 7.10 Perry's approximation for beam-columns

where M is the bending moment at x along the beam, and M' is the bending moment at x due to lateral loads alone. Now

$$\frac{d^2y}{dx^2} = -\frac{\pi^2}{l^2} c \sin\frac{\pi x}{l}$$

Hence in eqn [7.68],

$$EI\frac{\pi^2}{l^2} c \sin\frac{\pi x}{l} = M' + Py$$

or

$$P_E y = M' + Py \qquad [7.69]$$

Hence

$$y = \frac{M'}{P_E - P} \qquad [7.70]$$

and therefore

$$M = M' + \frac{PM'}{P_E - P}$$

$$= \frac{P_E}{P_E - P} M' \qquad [7.71]$$

The factor $P_E/(P_E - P)$ is the 'magnification factor'; it is to be applied to the bending moment due to lateral load alone to obtain the bending moment due to both lateral and axial loads.

Example 7.6

A pin-ended column of length l carries a load $P = P_E/2$ and a lateral load W at the mid-point. Using the Perry approximation, show that the maximum deflection due to the lateral load is increased by a factor of about 2.4 when the axial load is applied. Find the value of the maximum bending moment.

Now

$$y = \frac{M'}{P_E - P} = \frac{Wl/4}{\pi^2 EI/2l^2} = 2.43\frac{Wl^3}{48EI}$$

and

$$M_{max} = \frac{Wl}{4}\frac{P_E}{P_E - P}$$

$$= \frac{Wl}{4}\frac{1}{1 - P/P_E}$$

$$= \frac{Wl}{2}$$

Exercises

7.1 A pin-ended column of length l carries an axial load P and a single concentrated lateral load W at mid-height. Show that the maximum bending moment in the column is

$$\frac{W}{2n}\tan\frac{nl}{2}$$

where $n = \sqrt{(P/EI)}$.

7.2 A steel column is 4.2 m long and carries an axial load of 3200 kN. The end conditions of support are such that about the Z (major) axis the effective length factor is 1.0 and about the Y (minor) axis the effective length factor is 0.7.

$$I_z = 50\,832 \text{ cm}^4$$

$$I_y = 16\,230 \text{ cm}^4$$

$$A = 252.3 \text{ cm}^2$$

$$\sigma_y = 250 \text{ N/mm}^2$$

$$E = 207 \text{ kN/mm}^2$$

Using the Perry formula with $\eta = 0.003l/k$, determine the load factor.

7.3 Refine the analysis in example 7.3 using five internal points of subdivision to show that the critical load of the tapered column is approximately $7.1EI/l^2$.

7.4 For the structure shown in Fig. 7.11, show that the critical value of the load P is approximately $0.75\pi^2 EI/l^2$.

Figure 7.11 Exercise 7.4

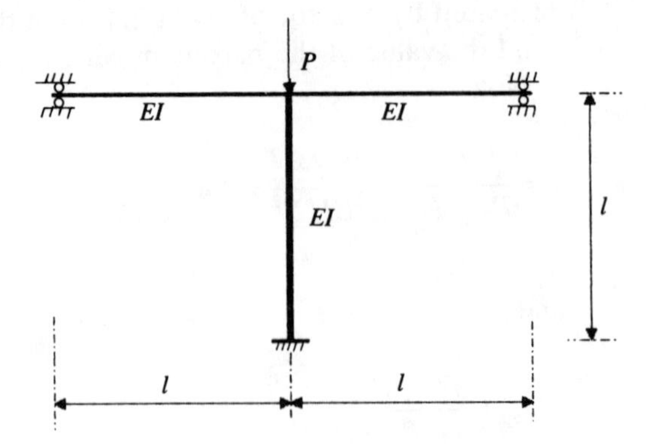

7.5 A column carries an axial load $P = 0.2P_E$ and a uniformly distributed lateral load w per unit length. Using the Perry approximation for laterally loaded columns, show that the maximum bending moment in the column is increased by 25 per cent compared with that due to the lateral load alone.

CHAPTER

8

The stiffness method

8.1 Introduction

In Chapter 4 we introduced some fundamental concepts, including those of flexibility and stiffness, and we went on to develop matrix formulations of methods of analysis based on these principles. This was an introductory development in order to establish some of the fundamental ideas related to these methods. The treatment of the flexibility method in section 4.4 will not be extended here since this method is severely limited in scope when compared with the stiffness method. The latter method is, however, of supreme importance in modern analysis and deserves a much fuller treatment than we were able to give it in the earlier chapter. It would be useful for the reader, at this stage, to review section 4.5 and to look in particular at eqns [4.49]–[4.57], since we shall be meeting these again in what we are to do.

The stiffness method was extensively developed during the 1950s for skeletal structures once the computer was available to solve the large sets of simultaneous equilibrium equations it produces. The 1960s saw the extension of the method to encompass structural continua such as plates and shells, and the name 'finite element analysis' was applied to this development. Such was the impetus for finite element analysis that it was accompanied by a fundamental review and restatement of the principles of the stiffness method, putting it on a much more comprehensive footing. The title 'finite element analysis', in the structural engineering context, embraces all of what had been known as the 'stiffness method'. The purpose of this chapter is to explore the 'stiffness' method using the concept of finite elements.

We saw in section 1.3 that structures can be 'kinematically indeterminate', and we saw how the degrees of kinematical indeterminacy can be obtained. Attention is focused on the nodal displacements in coordinate directions, and the degree of kinematical indeterminacy is equal to the number of independent nodal displacements. These are often referred to as the 'degrees of freedom' since each is kinematically independent of the others. There is usually little difficulty in deciding the kinematical indeterminacy of a structure, and often we proceed to a structural analysis without giving much thought to this aspect. In general the indeterminacy will be the total number of nodal displacements omitting the constrained displacements at the supports. This was how we arrived at eqn [1.11]:

$$n_k = {}^{3}_{6}(n-1) - c + r$$

In practice, we can usually count the degrees of freedom purely by inspection and can sometimes omit displacements in directions which have little or no consequence for us. For example, if we apply eqn [1.11] to the cantilever of Fig. 8.1 we get

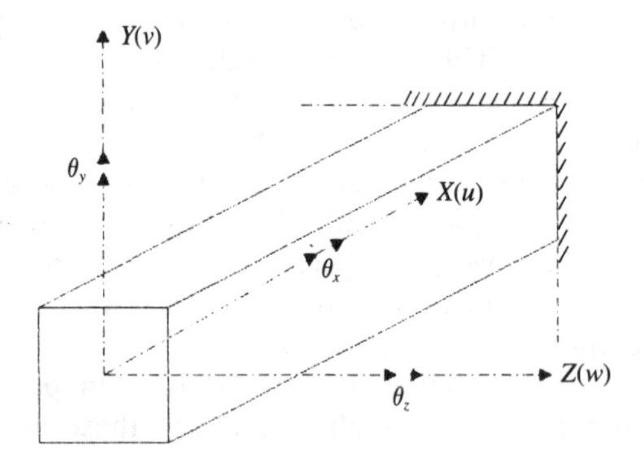

Figure 8.1 Nodal degrees of freedom of a cantilever

$$n_k = 6(2-1) - 0 + 0 = 6$$

which we can see as the six independent displacements u, v, w, θ_x, θ_y, θ_z. Although these displacements are independent of one another, they are all uniquely determined once a specific load has been applied. If there is no torsion on the beam then θ_x is either zero or of no interest to us. In addition, the displacement u in the axial direction will be negligible in most cases in the absence of significant axial load; so we are left with v, w, θ_y and θ_z. If, furthermore, we are dealing with loading and deformation in the Y–X plane only, then $w = 0$ and $\theta_y = 0$; so we are left with v and θ_z. If this is the final number of kinematical indeterminacies for our problem then we express the equilibrium of the structure in these two coordinate directions to obtain values of v and θ_z. A knowledge of these displacements constitutes the solution to the structural analysis. We can then calculate other quantities such as stress resultants which are functions of v and θ_z.

Although we have focused attention on displacements developing at the nodes, and this is all that is required in skeletal structures, with structural continua we must also examine what is happening along the boundaries of the elements (nodal lines and element faces). The usual procedure is to specify the 'shape' of the deformation and constrain the nodal lines and interfaces between elements to conform to this displacement function. An 'exact' solution will be obtained if the correct continuity of displacement and, where necessary, displacement derivatives (slopes) is achieved. Frequently a degree of approximation will be introduced due to the particular displacement function used. It will be appreciated that this approach allows us to continue to express equilibrium at the nodes only, just as we do with skeletal structures, and so we can use methods of computation suitable in general for the stiffness method.

The complete method breaks down into a number of stages:

1. Discretize the structure. With skeletal structures the members themselves form the 'discrete' elements, although we frequently introduce additional nodes at strategic points within members and thereby increase the number of elements. In continuous structures, discretization can involve a variety of shapes of elements to suit the geometry, and can be carried out at a 'coarse' level with comparatively few elements or at a 'refined' level with a fine division into many elements. Graded nets of elements are also possible. Some computer systems carry out the discretization more or less automatically and generate an 'optimum' mesh for the problem.
2. Establish the force–displacement relationships for all the elements. This involves computing the stiffness matrix for each element in local coordinates, i.e. the nodal coordinates local and 'natural' to the element. This will usually involve some transformation so that the local stiffnesses are transformed into the 'global' system for the structure as a whole.
3. The structural elements are then 'assembled' to form the actual structure to be analysed. Mathematically this means the transfer of the element stiffnesses into a stiffness matrix $\boldsymbol{K}$ for the complete structure. This stiffness matrix contains the coefficients in the equations representing nodal equilibrium in terms of nodal displacements $\boldsymbol{r}$.
4. The applied load matrix $\boldsymbol{R}$ forming the right-hand sides of the equations is then computed from the actual applied loads by expressing them as equivalent nodal loads as necessary.
5. The nodal force equilibrium equations

   $$\boldsymbol{K}\boldsymbol{r} = \boldsymbol{R} \qquad [8.1]$$

   are then solved to obtain $\boldsymbol{r}$.
6. The displacements $\boldsymbol{r}$ are then substituted in element force–displacement relationships to produce element forces.

We shall now examine each of these stages and develop the

theory involved. We start with the fundamental matter of displacement shape functions.

8.2 Shape functions for element displacements

003 We start by considering a very simple element with a single coordinate and a linear distribution of displacement, as in Fig. 8.2(a). The single coordinate is in the X (axial) direction and the

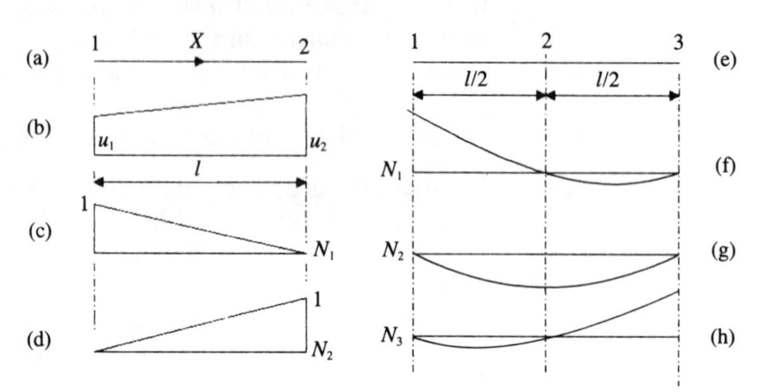

Figure 8.2 Linear and quadratic shape functions for single-coordinate element

displacement of any point distant x from end 1 is u. As we have assumed a linear distribution of displacement u, the distribution is as at diagram (b). Now we can represent this in a general way as

$$u = \alpha_1 + \alpha_2 x$$

or, with the matrix $\boldsymbol{d}$ containing internal displacements,

$$\boldsymbol{d} = [1 \quad x]\begin{bmatrix}\alpha_1\\\alpha_2\end{bmatrix} = \boldsymbol{L\alpha} \qquad [8.2]$$

Now the nodal displacements are u_1 and u_2, i.e.

$$\begin{bmatrix}u_1\\u_2\end{bmatrix} = \begin{bmatrix}1 & 0\\1 & l\end{bmatrix}\begin{bmatrix}\alpha_1\\\alpha_2\end{bmatrix}$$

or

$$\boldsymbol{s} = \boldsymbol{A\alpha} \qquad [8.3]$$

$$\boldsymbol{\alpha} = \boldsymbol{A}^{-1}\boldsymbol{s} \qquad [8.4]$$

Hence

$$\boldsymbol{d} = \boldsymbol{LA}^{-1}\boldsymbol{s} \qquad [8.5]$$

We can invert $\boldsymbol{A}$ to obtain

$$\boldsymbol{A}^{-1} = \begin{bmatrix}1 & 0\\-1/l & 1/l\end{bmatrix} \qquad [8.6]$$

Hence

$$\boldsymbol{\alpha} = \begin{bmatrix} 1 & 0 \\ -1/l & 1/l \end{bmatrix} \begin{bmatrix} u_1 \\ u_2 \end{bmatrix}$$

i.e.

$$\alpha_1 = u_1$$

$$\alpha_2 = -\frac{1}{l}u_1 + \frac{1}{l}u_2 = \frac{1}{l}(u_2 - u_1)$$

Substituting in eqn [8.2],

$$u = u_1 + \frac{1}{l}(u_2 - u_1)x \qquad [8.7]$$

which is clearly correct.

The purpose of this little exercise has of course been to produce the general matrix eqns [8.2] and [8.3] for use with bigger problems. To reach eqn [8.7] we have inverted matrix $\boldsymbol{A}$, and this can be avoided if we change the form of eqns [8.2] and [8.7] as follows. Put

$$u = N_1 u_1 + N_2 u_2 \qquad [8.8]$$

Now if we study the form of eqn [8.8] we see that N_1 represents the internal displacement u if $u_2 = 0$ and $u_1 = 1$; similarly, N_2 represents u if $u_1 = 0$ and $u_2 = 1$. N_1 and N_2 are called 'shape' functions or interpolation functions. A shape function associated with a particular nodal displacement gives the displacement field over the element when unit value is given to that nodal displacement and all other nodal displacements are zero.

Comparing eqns [8.7] and [8.8] we see that

$$N_1 = 1 - x/l \qquad [8.9]$$

$$N_2 = x/l \qquad [8.10]$$

Shape functions [8.9] and [8.10] are shown at (c) and (d) in Fig. 8.2. We can generalize eqn [8.8] by writing it as

$$\boldsymbol{d} = \boldsymbol{N}\boldsymbol{s} \qquad [8.11]$$

By comparing eqns [8.5] and [8.11] we can relate the shape functions to the original polynomial [8.2] as

$$\boldsymbol{N} = \boldsymbol{L}\boldsymbol{A}^{-1} \qquad [8.12]$$

Finding eqn [8.8] reduces to the problem of obtaining the shape functions $\boldsymbol{N}$. This is an important step in finite element analysis.

We now look at a slightly more difficult problem. We retain our single-coordinate element but now wish to represent the internal displacement u by a quadratic function. We write

$$u = \alpha_1 + \alpha_2 x + \alpha_3 x^2 \qquad [8.13]$$

$$\boldsymbol{d} = [1 \quad x \quad x^2]\begin{bmatrix} \alpha_1 \\ \alpha_2 \\ \alpha_3 \end{bmatrix} = \boldsymbol{L\alpha} \qquad [8.14]$$

We now need three nodes corresponding to the three coefficients in the quadratic functions [8.13]. We take node 2 at the centre of the element as in Fig. 8.2(e). Thus, substituting nodal values of x in eqn [8.14],

$$\begin{bmatrix} u_1 \\ u_2 \\ u_3 \end{bmatrix} = \begin{bmatrix} 1 & 0 & 0 \\ 1 & l/2 & l^2/4 \\ 1 & l & l^2 \end{bmatrix}\begin{bmatrix} \alpha_1 \\ \alpha_2 \\ \alpha_3 \end{bmatrix}$$

or

$$\boldsymbol{s} = \boldsymbol{A\alpha} \qquad [8.15]$$

We can invert $\boldsymbol{A}$ and obtain

$$\boldsymbol{A}^{-1} = \begin{bmatrix} 1 & 0 & 0 \\ -3/l & 4/l & -1/l \\ 2/l^2 & -4/l^2 & 2/l^2 \end{bmatrix}$$

We can now obtain the shape functions from eqn [8.12]:

$$[N_1, N_2, N_3] = \left[\left(1 - \frac{3x}{l} + \frac{2x^2}{l^2}\right), \left(\frac{4x}{l} - \frac{4x^2}{l^2}\right), \left(\frac{-x}{l} + \frac{2x^2}{l^2}\right)\right] \qquad [8.16]$$

These shape functions are shown in Fig. 8.2 at (f), (g) and (h). The reader should note that a property of the shape functions is that their sum is equal to unity.

Now it is unnecessary to invert matrix A in order to obtain the shape functions; they can be obtained using standard mathematical interpolation methods. If we are to represent nodal displacements u, v, w only and not derivatives, we can use Lagrange polynomials. If we need to represent derivatives (slopes) also, we can use Hermitian polynomials (Desai 1979; Cheung, Yeo 1979).

8.2.1 Beam element

The beam is of course one of the most frequently used finite elements. In Fig. 8.3 we have a uniform beam 1–2 of length l. If we neglect axial displacements (X direction), then we need to identify four degrees of freedom: v_1, θ_1, v_2 and θ_2. The polynomial displacement function must therefore contain four

Figure 8.3 Beam element

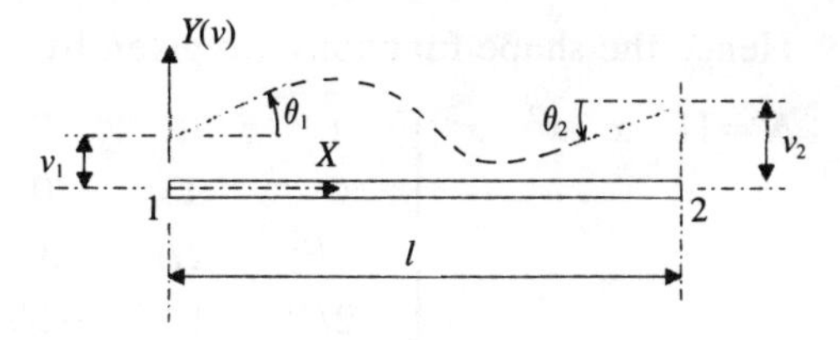

constants, so we use

$$v = \alpha_1 + \alpha_2 x + \alpha_3 x^2 + \alpha_4 x^3 = [1 \quad x \quad x^2 \quad x^3]\begin{bmatrix} \alpha_1 \\ \alpha_2 \\ \alpha_3 \\ \alpha_4 \end{bmatrix} \qquad [8.17]$$

Then

$$\frac{dv}{dx} = \alpha_2 + 2\alpha_3 x + 3\alpha_4 x^2$$

$$\frac{d^2v}{dx^2} = 2\alpha_3 + 6\alpha_4 x$$

$$\frac{d^3v}{dx^3} = 6\alpha_4$$

$$\frac{d^4v}{dx^4} = 0$$

If shear strains are neglected then $\theta_1 = (dv/dx)_1$ and $\theta_2 = (dv/dx)_2$, so the nodal displacements are

$$\begin{bmatrix} v_1 \\ \theta_1 \\ v_2 \\ \theta_2 \end{bmatrix} = \begin{bmatrix} 1 & 0 & 0 & 0 \\ 0 & 1 & 0 & 0 \\ 1 & l & l^2 & l^3 \\ 0 & 1 & 2l & 3l^2 \end{bmatrix}\begin{bmatrix} \alpha_1 \\ \alpha_2 \\ \alpha_3 \\ \alpha_4 \end{bmatrix}$$

or

$$\boldsymbol{s} = \boldsymbol{A\alpha}$$

Now

$$v = \boldsymbol{L}\boldsymbol{A}^{-1}\boldsymbol{s} = \boldsymbol{N}\boldsymbol{s} \qquad [8.18]$$

and

$$\boldsymbol{A}^{-1} = \begin{bmatrix} 1 & 0 & 0 & 0 \\ 0 & 1 & 0 & 0 \\ -3/l^2 & -2/l & 3/l^2 & -1/l \\ 2/l^3 & 1/l^2 & -2/l^3 & 1/l^2 \end{bmatrix}$$

Hence the shape functions are given by

$$N = [1 \quad x \quad x^2 \quad x^3]\begin{bmatrix} 1 & 0 & 0 & 0 \\ 0 & 1 & 0 & 0 \\ -3/l^2 & -2/l & 3/l^2 & -1/l \\ 2/l^3 & 1/l^2 & -2/l^3 & 1/l^2 \end{bmatrix}$$

$$N = \left[\left(1 - \frac{3x^2}{l^2} + \frac{2x^3}{l^3}\right), x\left(1 - \frac{2x}{l} + \frac{x^2}{l^2}\right), \left(\frac{3x^2}{l^2} - \frac{2x^3}{l^3}\right), x\left(\frac{x^2}{l^2} - \frac{x}{l}\right)\right] \qquad [8.19]$$

8.2.2 *Isoparametric element*

Before we proceed to use shape functions in deriving element stiffness matrices, we give a brief introduction to a special type of element which is of great importance in analysis – the 'isoparametric' element.

Consider again the linear, single-coordinate element of Fig. 8.2(a). We show this again in Fig. 8.4 with a global origin O_G and a local origin O_L, now taken at the mid-point of the element. The global coordinate is x and the local coordinate is ξ as shown. Now we already have from eqns [8.8], [8.9] and [8.10], for the displacement u,

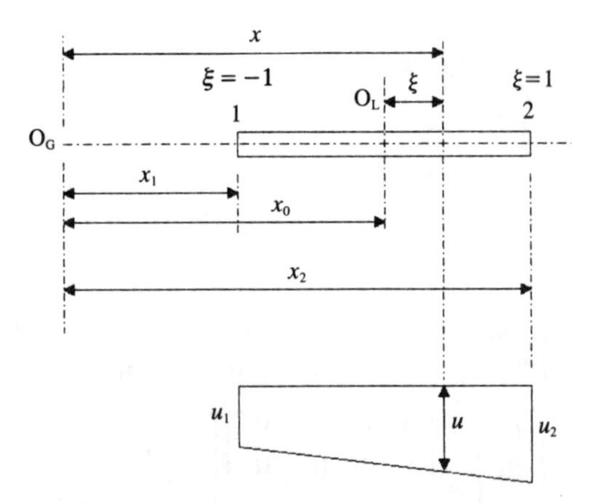

Figure 8.4 Isoparametric element

$$u = N_1u_1 + N_2u_2$$
$$= \frac{1}{2}(1 - \xi)u_1 + \frac{1}{2}(1 + \xi)u_2 \qquad [8.20]$$

Now we can express the global coordinate x as

$$x = x_1 + (x_2 - x_1)\frac{1 + \xi}{2}$$
$$= \frac{1}{2}(1 - \xi)x_1 + \frac{1}{2}(1 + \xi)x_2 \qquad [8.21]$$

Comparing eqns [8.20] and [8.21] we see that the same functions are used to describe both the element internal displacement (u) and the element geometry (x). Such an element is called 'isoparametric' (Desai 1979; Cheung, Yeo 1979).

8.3 Element stiffness matrix

The mathematical representation of the deformation shapes of the structural element having been established by eqn [8.11], we can now proceed to determine the stiffness properties of the element described in the stiffness matrix. The preliminary theoretical work was done in section 4.5 and we produced eqn [4.49]:

$$\boldsymbol{k} = \int \boldsymbol{B}^{\mathrm{T}}\boldsymbol{D}\boldsymbol{B}\,\mathrm{d}vol$$

It will be useful to review the matrices in this equation before proceeding:

$\boldsymbol{k}$ stiffness matrix
$\boldsymbol{B}$ matrix of strains due to unit nodal displacements, the actual strains in the element being represented as $\boldsymbol{\varepsilon} = \boldsymbol{Bs}$, where
$\boldsymbol{s}$ matrix of nodal displacements
$\boldsymbol{D}$ elasticity matrix relating stress $\boldsymbol{\sigma}$ and strain $\boldsymbol{\varepsilon}$ as $\boldsymbol{\sigma} = \boldsymbol{D\varepsilon} = \boldsymbol{DBs}$

Whilst it is not usually a difficult matter to set up the matrices required for the integration in eqn [4.49], the integration may be more difficult. In certain circumstances the integration can be carried out in 'closed form', but in many cases numerical integration is required. In any event a numerical approach will be required in the computer. The most generally used method for numerical integration in finite element work is the well-known Gauss-Legendre quadrature (Cheung, Yeo 1979). In one dimension, the Gauss quadrature formula gives

$$\int_{-1}^{+1} f(\xi)\,\mathrm{d}x = \sum_{i=1}^{n} \omega_i f(\xi)_i \qquad [8.22]$$

where

$f(\xi)$ is the function to be integrated
n is the number of Gauss points used in the range of integration
ω_i are the weighting coefficients
$f(\xi)_i$ is the value of $f(\xi)$ at abscissa ξ_i.

Values of weighting coefficients and abscissae have been chosen to optimize the result and can be found in several standard texts (e.g. Kopal 1961). A brief extract is given in Table 8.1 for $n = 2$, 3 and 4.

Table 8.1 Abscissae and weighting coefficients for Gauss quadrature

n	$\pm\xi_i$	ω_i
2	0.57 735	1.00 00
3	0.00 00	0.88 889
	0.77 460	0.55 556
4	0.33 998	0.65 215
	0.86 114	0.34 785

Any interval $a \leqslant x \leqslant b$ can be transformed to the interval $-1 \leqslant \xi \leqslant +1$ by the change of variable

$$x = \frac{1}{2}(a+b) + (b-a)\frac{\xi}{2} \qquad [8.23]$$

and

$$dx = \frac{1}{2}(b-a)\, d\xi \qquad [8.24]$$

The Gauss method gives an exact evaluation of the integral for polynomials of degree $2n-1$ or less. For example $n=2$ will give an exact integration of a cubic expression. The corresponding Gauss quadrature formula in two dimensions is

$$\int_{-1}^{+1}\int_{-1}^{+1} f(\xi, \eta)\, d\xi\, d\eta = \sum_{j=1}^{m}\sum_{i=1}^{n} \omega_i \omega_j f(\xi_i, \eta_j) \qquad [8.25]$$

The situation for a rectangular area of integration is shown in Fig. 8.5. For $n=2$ the integral will be

$$\begin{aligned}\int_{-1}^{+1}\int_{-1}^{+1} f(\xi, \eta)\, d\xi\, d\eta &= \omega_{1,1}^2 F_{1,1} + \omega_{1,2}\omega_{2,1}F_{1,2} \\ &\quad + \omega_{2,1}\omega_{1,2}F_{2,1} + \omega_{2,2}^2 F_{2,2} \\ &= F_{1,1} + F_{1,2} + F_{2,1} + F_{2,2}\end{aligned}$$

since all ω are unity for $n=2$. The notation is $F_{ij} = f(\xi, \eta)_{i,j}$.

We now apply eqn [4.49] to the problem of a simple beam element for which we have already evaluated shape functions as in eqn [8.19]. First we have to express the 'strain' in the element. In the case of the simple beam the 'strain' is a scalar quantity equal to the curvature; thus

$$\text{'}\varepsilon\text{'} = -d^2 v/dx^2 \qquad [8.26]$$

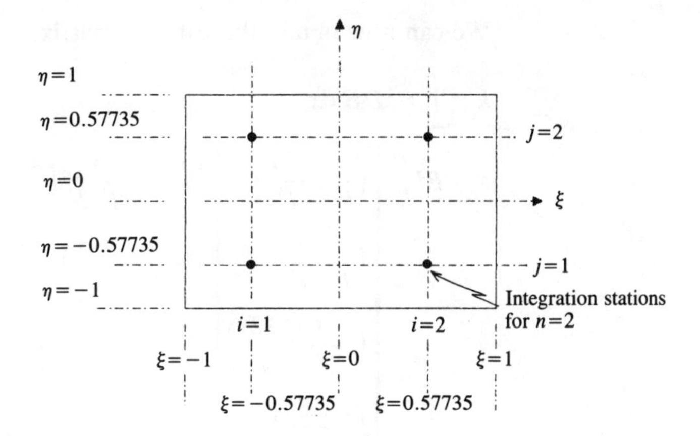

Figure 8.5 Gauss quadrature in two dimensions

The matrix of internal displacements $\boldsymbol{d}$ contains only one item, namely v the transverse displacement. So we have from eqn [8.11]

$$v = \boldsymbol{Ns}$$

$$= \left[\left(1 - \frac{3x^2}{l^2} + \frac{2x^3}{l^3}\right),\ x\left(1 - \frac{2x}{l} + \frac{x^2}{l^2}\right),\ \left(\frac{3x^2}{l^2} - \frac{2x^3}{l^3}\right),\ x\left(\frac{x^2}{l^2} - \frac{x}{l}\right)\right] \begin{bmatrix} v_1 \\ \theta_1 \\ v_2 \\ \theta_2 \end{bmatrix} \quad [8.27]$$

$$\boldsymbol{\varepsilon} = \left[\left(\frac{6}{l^2} - \frac{12x}{l^3}\right), \left(\frac{4}{l} - \frac{6x}{l^2}\right), \left(-\frac{6}{l^2} + \frac{12x}{l^3}\right), \left(\frac{2}{l} - \frac{6x}{l^2}\right)\right] \begin{bmatrix} v_1 \\ \theta_1 \\ v_2 \\ \theta_2 \end{bmatrix} \quad [8.28]$$

$$= \boldsymbol{Bs}$$

from eqn [4.46]. Now from eqn [4.47],

'stress' = bending moment
$= \boldsymbol{D\varepsilon} = EI\, \mathrm{d}^2v/\mathrm{d}x^2$ for the beam

Hence

$$\boldsymbol{D} = EI$$

We can now obtain the stiffness matrix:

$$\boldsymbol{k} = \int_0^l \boldsymbol{B}^{\mathrm{T}}\boldsymbol{D}\boldsymbol{B}\,\mathrm{d}x$$

$$\boldsymbol{k} = EI\int_0^l \begin{bmatrix} \left(\frac{6}{l^2} - \frac{12x}{l^3}\right) \\ \left(\frac{4}{l} - \frac{6x}{l^2}\right) \\ \left(-\frac{6}{l^2} + \frac{12x}{l^3}\right) \\ \left(\frac{2}{l} - \frac{6x}{l^2}\right) \end{bmatrix} \left[\left(\frac{6}{l^2} - \frac{12x}{l^3}\right), \left(\frac{4}{l} - \frac{6x}{l^2}\right), \left(-\frac{6}{l^2} + \frac{12x}{l^3}\right), \left(\frac{2}{l} - \frac{6x}{l^2}\right)\right] \mathrm{d}x \qquad [8.29]$$

$$= \begin{bmatrix} k_{11} & k_{12} & k_{13} & k_{14} \\ k_{21} & k_{22} & k_{23} & k_{24} \\ k_{31} & k_{32} & k_{33} & k_{34} \\ k_{41} & k_{42} & k_{43} & k_{44} \end{bmatrix} \qquad [8.30]$$

In this case the integrals can be evaluated explicitly. However, we shall evaluate a typical stiffness, say k_{23}, using the Gauss method:

$$k_{23} = EI\int_0^l \left(\frac{4}{l} - \frac{6x}{l^2}\right)\left(-\frac{6}{l^2} + \frac{12x}{l^3}\right) \mathrm{d}x$$

Put

$$x = \frac{1}{2}(0 + l) + (l - 0)\frac{\xi}{2} = \frac{l}{2} + \frac{l\xi}{2} = \frac{l}{2}(1 + \xi)$$

and

$$\mathrm{d}x = \frac{l}{2}\mathrm{d}\xi$$

from eqns [8.23] and [8.24]. Then

$$k_{23} = EI\frac{l}{2}\int_{-1}^{+1} \left(\frac{6}{l^3}\xi - \frac{18}{l^3}\xi^2\right) \mathrm{d}\xi = \frac{3EI}{l^2}\int_{-1}^{+1} (\xi - 3\xi^2)\,\mathrm{d}\xi$$

Using the two-point Gauss formula,

$$k_{23} = \frac{3EI}{l^2}[0.57735 - 3(0.57735)^2 + -0.57735$$

$$- 3(-0.57725)^2]$$

$$= -\frac{6EI}{l^2}$$

In this case, of course, the integrals could be evaluated explicitly;

however, we have taken the opportunity to employ the Gauss formula. The remaining integrals can easily be carried out to produce the following results:

$$\boldsymbol{k} = \frac{EI}{l^3}\begin{bmatrix} 12 & 6l & -12 & 6l \\ 6l & 4l^2 & -6l & 2l^2 \\ -12 & -6l & 12 & -6l \\ 6l & 2l^2 & -6l & 4l^2 \end{bmatrix} \qquad [8.31]$$

8.3.1 Stiffnesses by inversion of flexibilities

In section 4.2 we introduced the concepts of flexibility and stiffness and showed (eqn [4.5]) that one was the inverse of the other. In example 4.4 we obtained the stiffness matrix we needed by inverting the corresponding flexibility matrix. This is frequently the quickest way to arrive at a stiffness matrix for a beam-type element, so we shall consider this now as an alternative to the finite element type of approach.

In Fig. 8.6(a) we have a simple beam 1–2. We apply a unit moment (note that the positive sense of this moment is clockwise about an axis towards the viewer) at end 1, producing flexibilities (rotations) f_{11} and f_{21} as at (b). The bending moment diagram is

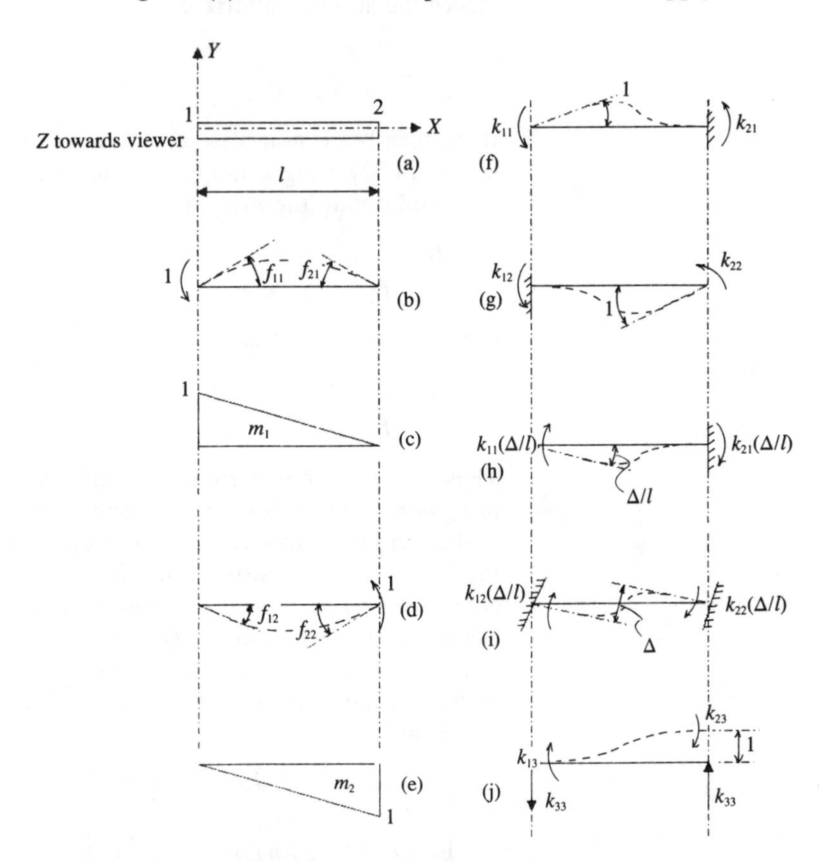

Figure 8.6 Beam stiffnesses from inverse flexibility

given at (c) and is drawn on the tension side of the member. We carry out a similar operation at end 2 and obtain diagrams (d) and (e). Now the flexibilities are,

$$\left.\begin{aligned} f_{11} &= \int_0^l \frac{m_1^2}{EI}\,\mathrm{d}x \\ f_{12} = f_{21} &= \int_0^l \frac{m_1 m_2}{EI}\,\mathrm{d}x \\ f_{22} &= \int_0^l \frac{m_2^2}{EI}\,\mathrm{d}x \end{aligned}\right\} \qquad [8.32]$$

If our member is of uniform flexural rigidity EI then the flexibilities are easily evaluated using the table of product integrals (Appendix 4), and we obtain

$$\boldsymbol{f} = \frac{l}{6EI}\begin{bmatrix} 2 & -1 \\ -1 & 2 \end{bmatrix} \qquad [8.33]$$

Hence the stiffness matrix is

$$\boldsymbol{k} = \boldsymbol{f}^{-1} = \frac{2EI}{l}\begin{bmatrix} 2 & 1 \\ 1 & 2 \end{bmatrix} \qquad [8.34]$$

If the member is non-prismatic then we can evaluate the integrals in eqn [8.32] using a numerical method, for example Simpson's rule, and obtain the integrals in

$$\boldsymbol{f} = \begin{bmatrix} f_{11} & f_{12} \\ f_{21} & f_{22} \end{bmatrix} \qquad [8.35]$$

of which the inverse is

$$\boldsymbol{k} = \boldsymbol{f}^{-1} = \frac{1}{f_{11}f_{22} - f_{12}^2}\begin{bmatrix} f_{22} & -f_{12} \\ -f_{21} & f_{11} \end{bmatrix} \qquad [8.36]$$

Thus we can compute rotational stiffnesses for a prismatic beam using eqn [8.34] or for a non-prismatic beam using eqn [8.36].

We can now introduce a translational degree of freedom using the principle of superposition. If we wish to obtain stiffnesses corresponding to a pure translation Δ as in Fig. 8.6(j), then we can produce a rotation (Δ/l) at end 1 as in (h), and a rotation (Δ/l) at end 2 as in (i), and add the results. We then have all the stiffnesses we need for the beam and can set up the stiffness matrix as

$$\boldsymbol{k} = \begin{bmatrix} 4EI/l & 2EI/l & -6EI/l^2 \\ 2EI/l & 4EI/l & -6EI/l^2 \\ -6EI/l^2 & -6EI/l^2 & 12EI/l^3 \end{bmatrix} \qquad [8.37]$$

This is the same as eqn [8.31] with one transverse degree of freedom eliminated and some rearrangement.

The method just described is very suitable for determining stiffnesses of non-prismatic members using the numerical integration procedures described in section 6.9.

8.4 Consistent nodal loads

The displacements and stiffnesses we have considered so far are all in the coordinate directions at the nodes of the element. It follows that the applied loads must also be at the nodes and in directions corresponding to the coordinates we have defined. Frequently loads are applied between nodes and it is necessary to determine the equivalent or 'consistent' effects at the nodes. We can develop an expression for the consistent nodal loads matrix by equating the work done by the actual applied loads to that done by the equivalent nodal loads. Thus in Fig. 8.7(a) the work

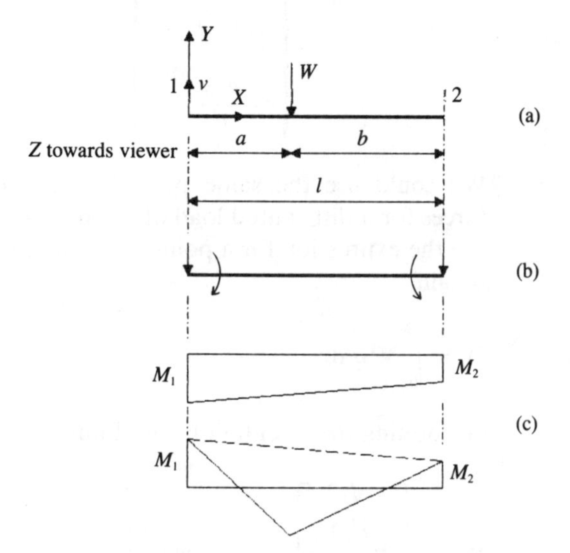

Figure 8.7 (a)(b) Consistent nodal loads (c) fixed-end effects

done by typical load W is Wv_a, where v_a is the transverse deflection of the beam at $x = a$. From eqn [8.18] we have

$$v = \mathbf{N}\mathbf{s}$$

for the whole beam; hence at $x = a$

$$v_a = \mathbf{N}_a\mathbf{s} \qquad [8.38]$$

where we obtain $\mathbf{N}_a$ by substituting $x = a$ in the shape functions $\mathbf{N}$. The work done by the load W is therefore $W\mathbf{N}_a\mathbf{s}$. If the equivalent nodal loads are

$$\mathbf{W}_e = \begin{bmatrix} V_1 \\ M_1 \\ V_2 \\ M_2 \end{bmatrix} \qquad [8.39]$$

then the work done by these loads is $\boldsymbol{W}_e^T\boldsymbol{s}$. Hence

$$\boldsymbol{W}_e^T\boldsymbol{s} = W\boldsymbol{N}_a\boldsymbol{s}$$

The equivalent nodal loads are therefore

$$\boldsymbol{W}_e = W\boldsymbol{N}_a^T \qquad [8.40]$$

Taking the shape functions $\boldsymbol{N}$ from eqn [8.19] and substituting $x = a$, we obtain

$$\boldsymbol{W}_e = \begin{bmatrix} V_1 \\ M_1 \\ V_2 \\ M_2 \end{bmatrix} = W \begin{bmatrix} \frac{b^2}{l^3}(3a+b) \\ \frac{ab^2}{l^2} \\ \frac{a^2}{l^3}(a+3b) \\ -\frac{a^2b}{l^2} \end{bmatrix} \qquad [8.41]$$

We could use the same procedure to obtain equivalent nodal forces for a distributed load of intensity q. Alternatively we could use the expression for a point load and integrate. In any event we obtain

$$\boldsymbol{W}_e = \int_0^l \boldsymbol{N}^T q \, \mathrm{d}x \qquad [8.42]$$

Performing the integration we obtain

$$\boldsymbol{W}_e = q \begin{bmatrix} l/2 \\ l^2/12 \\ l/2 \\ -l^2/12 \end{bmatrix} \qquad [8.43]$$

8.5 Evaluation of stresses and stress resultants

We have yet to consider how the structure stiffness matrix is assembled and how the solution of the equilibrium equations is carried out; we shall do this shortly. However, it will be useful at this stage to consider the equations leading to the computation of stresses and stress resultants once the nodal displacements have been obtained. For any particular element, we can determine the internal stresses $\boldsymbol{\sigma}$ from eqns [4.47] and [4.48]:

$$\boldsymbol{\sigma} = \boldsymbol{D}\boldsymbol{\varepsilon} = \boldsymbol{D}\boldsymbol{B}\boldsymbol{s}$$

The interpretation of this general equation in the case of the beam element was, from eqn [8.26],

$$`\boldsymbol{\sigma}' = M$$

$$`\boldsymbol{D}' = EI$$

$$`\boldsymbol{\varepsilon}' = -\mathrm{d}^2 v/\mathrm{d}x^2$$

Thus we can write

$$M = -EI\frac{\mathrm{d}^2 v}{\mathrm{d}x^2} = -\frac{\mathrm{d}^2 \boldsymbol{N}}{\mathrm{d}x^2}\boldsymbol{s} \qquad [8.44]$$

So we can express the bending moment in the beam in terms of the nodal displacements $\boldsymbol{s}$. Thus, using eqn [8.28],

$$M = EI\left\{\left(\frac{6}{l^2} - \frac{12x}{l^3}\right)v_1 + \left(\frac{4}{l} - \frac{6x}{l^2}\right)\theta_1 + \left(-\frac{6}{l^2} + \frac{12x}{l^3}\right)v_2 + \left(\frac{2}{l} - \frac{6x}{l^2}\right)\theta_2\right\} \qquad [8.45]$$

Putting $x = 0$ we obtain

$$M_1 = EI\left\{\frac{4\theta_1}{l} + \frac{2\theta_2}{l} - \frac{6}{l^2}(v_2 - v_1)\right\}$$

and putting $x = l$ we obtain

$$M_2 = -EI\left\{\frac{4\theta_2}{l} + \frac{2\theta_1}{l} - \frac{6}{l^2}(v_2 - v_1)\right\}$$

These are the moment stress resultants at ends 1 and 2. Between 1 and 2 the bending moment is obtained from eqn [8.45], which is linear in x.

It should be emphasised that the evaluation of stress resultants from a set of displacements caused by equivalent nodal loads does not give the total bending moment in the beam. We need to add a distribution of bending moments due to a set of 'fixed-end' effects. The signs allocated to the fixed-end forces are opposite to those for the equivalent nodal forces. Figure 8.7(b) and (c) illustrate this point.

8.6 Influence of shear strains

The influence of shear is to increase deformations, and so this effect is more readily seen when we examine flexibilities. Generally, the effect of shear on beam deflections is negligible unless the beam has a small span/depth ratio. In practice there are circumstances in which it is necessary to make an allowance for shear deformation, so we shall now outline a suitable method. In section 8.3 we examined a method of obtaining element stiffnesses by inversion of flexibilities. If we repeat that analysis including shear strains, eqns [8.32] become

$$\left.\begin{aligned} f_{11} &= \int_0^l \frac{m_1^2}{EI}\,dx + f\int_0^l \frac{v_1^2}{GA}\,dx \\ f_{12} = f_{21} &= \int_0^l \frac{m_1 m_2}{EI}\,dx + f\int_0^l \frac{v_1 v_2}{GA}\,dx \\ f_{22} &= \int_0^l \frac{m_2^2}{EI}\,dx + f\int_0^l \frac{v_2^2}{GA}\,dx \end{aligned}\right\} \qquad [8.46]$$

where v_1 is the shear force distribution due to unit moment at end 1, v_2 is the shear force distribution due to unit moment at end 2, and f is a dimensionless form factor, dependent only on the shape of the cross-section:

$$f = \frac{A}{I^2}\int \frac{Q^2}{b^2}\,dA \qquad [8.47]$$

The integral in eqn [8.47] is evaluated over the cross-section, where A is the cross-sectional area, b is the width, and Q is the first moment of area of the part of the cross-section above the level considered about the centroidal axis. Values of the form factor f for some common structural cross-sectional shapes are given in Table 8.2.

Table 8.2 Form factors f for shear deflections in beams

Cross-sectional shape	f
Rectangle	1.2
Circle	10/9
Hollow circle	2
I-section or hollow rectangle	$\simeq A/A_{\text{web}}$

Now the v_1 and v_2 distributions are as shown in Fig. 8.8, so we

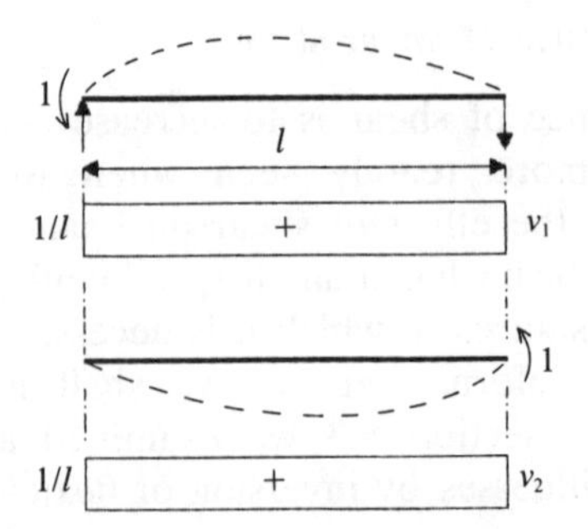

Figure 8.8 End rotations of beam due to shear

can evaluate eqn [8.46] for a prismatic beam:

$$f_{11} = \frac{l}{3EI} + \frac{f}{lGA} = f_{22} \qquad [8.48]$$

$$f_{12} = f_{21} = -\frac{l}{6EI} + \frac{f}{lGA} \qquad [8.49]$$

Now

$$\boldsymbol{f} = \begin{bmatrix} f_{11} & f_{12} \\ f_{21} & f_{22} \end{bmatrix}$$

$$\boldsymbol{k} = \boldsymbol{f}^{-1} = \frac{1}{f_{11}f_{22} - f_{12}^2} \begin{bmatrix} f_{22} & -f_{12} \\ -f_{21} & f_{11} \end{bmatrix} \qquad [8.50]$$

Hence substituting eqns [8.48] and [8.49] in eqn [8.50]:

$$\boldsymbol{k} = \frac{2EI}{l(1+2g)} \begin{bmatrix} 2+g & 1-g \\ 1-g & 2+g \end{bmatrix} \qquad [8.51]$$

where

$$g = \frac{6fEI}{l^2GA} \qquad [8.52]$$

In a similar way the effect of shearing strains on equivalent nodal forces can be obtained. It can be shown (Jenkins 1969) that the equivalent end moments are unaffected by shear for symmetrical loading, and thus standard values based on bending alone can be used.

Now we can expand the stiffness matrix of eqn [8.51] to include a transverse displacement, just as we did to obtain eqn [8.37]. We get

$$\boldsymbol{k} = \frac{2EI}{l(1+2g)} \begin{bmatrix} 2+g & 1-g & -3/l \\ 1-g & 2+g & -3/l \\ -3/l & -3/l & 6/l^2 \end{bmatrix} \qquad [8.53]$$

8.7 Transformation of stiffness from local to global coordinates

The element stiffness matrices we have met so far in this chapter are constructed in 'local' coordinates in that they relate to the natural coordinate axes for the element in question. It would be possible to use these stiffness matrices as they are and produce a stiffness matrix for the complete structure in one matrix operation according to eqn [4.56]:

$$\boldsymbol{K} = \boldsymbol{A}^{\mathrm{T}}\boldsymbol{k}\boldsymbol{A}$$

However, in practice this is impracticable for all but the smallest structures because of the consequent demands on computational time and the space in the computer. In practice the transformation is done element by element and then the structure stiffness $\boldsymbol{K}$ is produced by simply moving numbers from each $\boldsymbol{k}$ into $\boldsymbol{K}$. We shall look at this process in the next section, but here we establish the method of transforming element stiffness $\boldsymbol{k}$ in local coordinates to element stiffness in global coordinates. We take the opportunity to develop the beam element stiffnesses in the complete three-dimensional form using the two-dimensional results we obtained in eqn [8.31].

We order the nodal forces and corresponding nodal displacements (Fig. 8.9) at node 1 as follows:

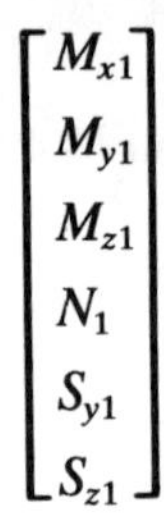

$$\begin{bmatrix} M_{x1} \\ M_{y1} \\ M_{z1} \\ N_1 \\ S_{y1} \\ S_{z1} \end{bmatrix}$$

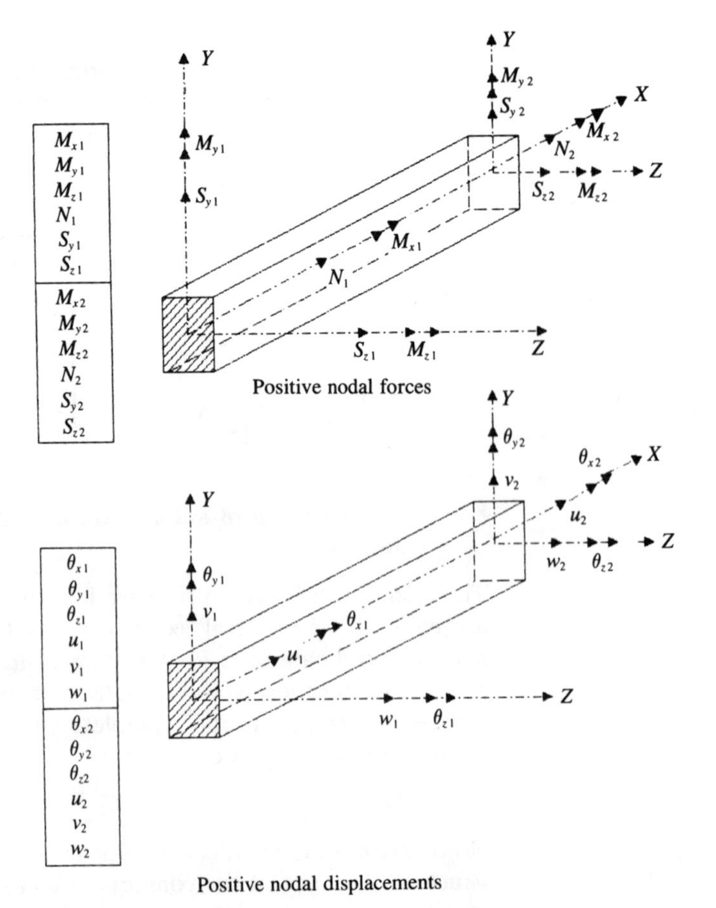

Figure 8.9 Nodal forces and displacements for three-dimensional beam element

and similarly at node 2. We can then immediately insert stiffnesses from eqn [8.31] with $I = I_z$ as follows:

$$12EI_z/l^3 \rightarrow k_{5,5}, \quad k_{11,11}$$

$$6EI_z/l^2 \rightarrow k_{3,5}, \quad k_{5,3}, \quad k_{9,5}, \quad k_{5,9}$$

$$4EI_z/l \rightarrow k_{3,3}, \quad k_{9,9}$$

$$2EI_z/l \rightarrow k_{3,9}, \quad k_{9,3}$$

The remainder of the matrix can be constructed for the terms involving EI_y, EA and GJ, and the complete matrix is shown as eqn [8.54]. It may help the reader, in the construction of this matrix, to imagine that a typical ith column of $\boldsymbol{k}$ contains the nodal forces consistent with unit imposed value of the ith displacement, all other nodal displacements being zero.

$$\begin{bmatrix} M_{x1} \\ M_{y1} \\ M_{z1} \\ N_1 \\ S_{y1} \\ S_{z1} \\ \hline M_{x2} \\ M_{y2} \\ M_{z2} \\ N_2 \\ S_{y2} \\ S_{z2} \end{bmatrix} = \left[\begin{array}{cccccc|cccccc}
\frac{GJ}{l} & 0 & 0 & 0 & 0 & 0 & -\frac{GJ}{l} & 0 & 0 & 0 & 0 & 0 \\
0 & \frac{4EI_y}{l} & 0 & 0 & 0 & -\frac{6EI_y}{l^2} & 0 & \frac{2EI_y}{l} & 0 & 0 & 0 & \frac{6EI_y}{l^2} \\
0 & 0 & \frac{4EI_z}{l} & 0 & \frac{6EI_z}{l^2} & 0 & 0 & 0 & \frac{2EI_z}{l} & 0 & -\frac{6EI_z}{l^2} & 0 \\
0 & 0 & 0 & \frac{EA}{l} & 0 & 0 & 0 & 0 & 0 & -\frac{EA}{l} & 0 & 0 \\
0 & 0 & \frac{6EI_z}{l^2} & 0 & \frac{12EI_z}{l^3} & 0 & 0 & 0 & \frac{6EI_z}{l^2} & 0 & -\frac{12EI_z}{l^3} & 0 \\
0 & -\frac{6EI_z}{l^2} & 0 & 0 & 0 & \frac{12EI_y}{l^3} & 0 & -\frac{6EI_y}{l^2} & 0 & 0 & 0 & -\frac{12EI_y}{l^3} \\
\hline
-\frac{GJ}{l} & 0 & 0 & 0 & 0 & 0 & \frac{GJ}{l} & 0 & 0 & 0 & 0 & 0 \\
0 & \frac{2EI_y}{l} & 0 & 0 & 0 & -\frac{6EI_y}{l^2} & 0 & \frac{4EI_y}{l} & 0 & 0 & 0 & \frac{6EI_y}{l^2} \\
0 & 0 & \frac{2EI_z}{l} & 0 & \frac{6EI_z}{l^2} & 0 & 0 & 0 & \frac{4EI_z}{l} & 0 & -\frac{6EI_z}{l^2} & 0 \\
0 & 0 & 0 & -\frac{EA}{l} & 0 & 0 & 0 & 0 & 0 & \frac{EA}{l} & 0 & 0 \\
0 & 0 & -\frac{6EI_z}{l^2} & 0 & -\frac{12EI_z}{l^3} & 0 & 0 & 0 & -\frac{6EI_z}{l^2} & 0 & \frac{12EI_z}{l^3} & 0 \\
0 & \frac{6EI_y}{l^2} & 0 & 0 & 0 & -\frac{12EI_y}{l^3} & 0 & \frac{6EI_y}{l^2} & 0 & 0 & 0 & \frac{12EI_y}{l^3}
\end{array}\right] \begin{bmatrix} \theta_{x1} \\ \theta_{y1} \\ \theta_{z1} \\ u_1 \\ v_1 \\ w_1 \\ \hline \theta_{x2} \\ \theta_{y2} \\ \theta_{z2} \\ u_2 \\ v_2 \\ w_2 \end{bmatrix} \quad [8.54]$$

Equation [8.54] is written more compactly in partitioned form:

$$\begin{bmatrix} \boldsymbol{S}_1 \\ \boldsymbol{S}_2 \end{bmatrix} = \begin{bmatrix} \boldsymbol{k}_{11} & \boldsymbol{k}_{12} \\ \boldsymbol{k}_{21} & \boldsymbol{k}_{22} \end{bmatrix} \begin{bmatrix} \boldsymbol{s}_1 \\ \boldsymbol{s}_2 \end{bmatrix} \quad [8.55]$$

To transform these stiffnesses to global coordinates we imagine a new set of coordinates axes $\bar{X}$, $\bar{Y}$ and $\bar{Z}$ obtained by rotating the local axes X, Y and Z through angles $\cos^{-1} \bar{X}OX$, $\cos^{-1} \bar{Y}OY$ and $\cos^{-1} \bar{Z}OZ$ respectively. The relationship between nodal forces in the local and global coordinate systems is, from resolution of forces,

$$\begin{bmatrix} \bar{M}_{x1} \\ \bar{M}_{y1} \\ \bar{M}_{z1} \\ \hline \bar{N}_1 \\ \bar{S}_{y1} \\ \bar{S}_{z1} \end{bmatrix} = \left[\begin{array}{ccc|ccc} \lambda_{\bar{x}x} & \lambda_{\bar{x}y} & \lambda_{\bar{x}z} & 0 & 0 & 0 \\ \lambda_{\bar{y}x} & \lambda_{\bar{y}y} & \lambda_{\bar{y}z} & 0 & 0 & 0 \\ \lambda_{\bar{z}x} & \lambda_{\bar{z}y} & \lambda_{\bar{z}z} & 0 & 0 & 0 \\ \hline 0 & 0 & 0 & \lambda_{\bar{x}x} & \lambda_{\bar{x}y} & \lambda_{\bar{x}z} \\ 0 & 0 & 0 & \lambda_{\bar{y}x} & \lambda_{\bar{y}y} & \lambda_{\bar{y}z} \\ 0 & 0 & 0 & \lambda_{\bar{z}x} & \lambda_{\bar{z}y} & \lambda_{\bar{z}z} \end{array}\right] \begin{bmatrix} M_{x1} \\ M_{y1} \\ M_{z1} \\ \hline N_1 \\ S_{y1} \\ S_{z1} \end{bmatrix} \qquad [8.56]$$

and similarly at end 2, where

$\lambda_{\bar{x}x} = \cos \bar{X}OX$ etc.

Now eqn [8.56] can be written

$$\bar{\boldsymbol{S}}_1 = \boldsymbol{\lambda} \boldsymbol{S}_1 \qquad [8.57]$$

where $\boldsymbol{\lambda}$ is a matrix of direction cosines. Now we know from the principle of contragredience (section 4.4.3) that forces and displacements transform contragrediently. Hence

$$\boldsymbol{s}_1 = \boldsymbol{\lambda}^{\mathrm{T}} \bar{\boldsymbol{s}}_1 \qquad [8.58]$$

or $\bar{\boldsymbol{s}}_1 = (\boldsymbol{\lambda}^{\mathrm{T}})^{-1} \boldsymbol{s}_1$ and $\bar{\boldsymbol{s}}_2 = (\boldsymbol{\lambda}^{\mathrm{T}})^{-1} \boldsymbol{s}_2$. Now it can be shown that $(\boldsymbol{\lambda}^{\mathrm{T}})^{-1} = \boldsymbol{\lambda}$, this being an 'orthogonal' transformation. Hence

$$\bar{\boldsymbol{s}} = \boldsymbol{\lambda} \boldsymbol{s} \qquad [8.59]$$

and from eqn [8.55],

$$\boldsymbol{S}_1 = \boldsymbol{k}_{11} \boldsymbol{s}_1 + \boldsymbol{k}_{12} \boldsymbol{s}_2 \qquad [8.60]$$

Hence

$$\boldsymbol{\lambda} \boldsymbol{S}_1 = \boldsymbol{\lambda} \boldsymbol{k}_{11} \boldsymbol{s}_1 + \boldsymbol{\lambda} \boldsymbol{k}_{12} \boldsymbol{s}_2 \qquad [8.61]$$

Substituting in eqn [8.61] from eqns [8.57] and [8.58],

$$\bar{\boldsymbol{S}}_1 = \boldsymbol{\lambda} \boldsymbol{k}_{11} \boldsymbol{\lambda}^{\mathrm{T}} \bar{\boldsymbol{s}}_1 + \boldsymbol{\lambda} \boldsymbol{k}_{12} \boldsymbol{\lambda}^{\mathrm{T}} \bar{\boldsymbol{s}}_2 \qquad [8.62]$$

and similarly,

$$\bar{\boldsymbol{S}}_2 = \boldsymbol{\lambda} \boldsymbol{k}_{21} \boldsymbol{\lambda}^{\mathrm{T}} \bar{\boldsymbol{s}}_1 + \boldsymbol{\lambda} \boldsymbol{k}_{22} \boldsymbol{\lambda}^{\mathrm{T}} \bar{\boldsymbol{s}}_2 \qquad [8.63]$$

In partitioned form,

$$\begin{bmatrix} \bar{\boldsymbol{S}}_1 \\ \bar{\boldsymbol{S}}_2 \end{bmatrix} = \begin{bmatrix} \boldsymbol{\lambda} \boldsymbol{k}_{11} \boldsymbol{\lambda}^{\mathrm{T}} & \boldsymbol{\lambda} \boldsymbol{k}_{12} \boldsymbol{\lambda}^{\mathrm{T}} \\ \boldsymbol{\lambda} \boldsymbol{k}_{21} \boldsymbol{\lambda}^{\mathrm{T}} & \boldsymbol{\lambda} \boldsymbol{k}_{22} \boldsymbol{\lambda}^{\mathrm{T}} \end{bmatrix} \begin{bmatrix} \bar{\boldsymbol{s}}_1 \\ \bar{\boldsymbol{s}}_2 \end{bmatrix} \qquad [8.64]$$

which is the transformed version of eqn [8.55].

We can now use this three-dimensional transformation to produce stiffness matrices for particular beam-type elements. For example, the stiffness matrix for a member in a plane frame (M_z, N, S_y non-zero) can be obtained simply by deleting rows 1, 2, 6,

7, 8 and 12 and the corresponding columns of the 12×12 matrix in eqn [8.54]. The local to global transformation is then carried out with corresponding deletions in the transformation matrix $\boldsymbol{\lambda}$. On this basis we have, for a member lying in the X–Y plane,

$$\begin{bmatrix} M_{z1} \\ N_1 \\ S_{y1} \\ \hdashline M_{z2} \\ N_2 \\ S_{y2} \end{bmatrix} = \left[\begin{array}{ccc:ccc} \frac{4EI_z}{l} & 0 & \frac{6EI_z}{l^2} & \frac{2EI_z}{l} & 0 & -\frac{6EI_z}{l^2} \\ 0 & \frac{EA}{l} & 0 & 0 & -\frac{EA}{l} & 0 \\ \frac{6EI_z}{l^2} & 0 & \frac{12EI_z}{l^3} & \frac{6EI_z}{l^2} & 0 & -\frac{12EI_z}{l^3} \\ \hdashline \frac{2EI_z}{l} & 0 & \frac{6EI_z}{l^2} & \frac{4EI_z}{l} & 0 & -\frac{6EI_z}{l^2} \\ 0 & -\frac{EA}{l} & 0 & 0 & \frac{EA}{l} & 0 \\ -\frac{6EI_z}{l^2} & 0 & -\frac{12EI_z}{l^3} & -\frac{6EI_z}{l^2} & 0 & \frac{12EI_z}{l^3} \end{array}\right] \begin{bmatrix} \theta_{z1} \\ u_1 \\ v_1 \\ \hdashline \theta_{z2} \\ u_2 \\ v_2 \end{bmatrix} \quad [8.65]$$

If the axis of the member makes an angle θ with the X axis, then

$$\boldsymbol{\lambda} = \left[\begin{array}{ccc:ccc} \cos\theta & -\sin\theta & 0 & 0 & 0 & 0 \\ \sin\theta & \cos\theta & 0 & 0 & 0 & 0 \\ 0 & 0 & 1 & 0 & 0 & 0 \\ \hdashline 0 & 0 & 0 & \cos\theta & -\sin\theta & 0 \\ 0 & 0 & 0 & \sin\theta & \cos\theta & 0 \\ 0 & 0 & 0 & 0 & 0 & 1 \end{array}\right]$$

For a plane truss element, we can take the matrix from example 4.3 and augment it with zeros in the rows corresponding to P_{y1} and P_{y2}, as in Fig. 8.10:

$$\boldsymbol{k} = \frac{EA}{l}\begin{bmatrix} 1 & 0 & -1 & 0 \\ 0 & 0 & 0 & 0 \\ -1 & 0 & 1 & 0 \\ 0 & 0 & 0 & 0 \end{bmatrix} \quad [8.66]$$

Now

$$\boldsymbol{\lambda} = \begin{bmatrix} \lambda_{\bar{x}x} & \lambda_{\bar{x}y} & 0 & 0 \\ \lambda_{\bar{y}x} & \lambda_{\bar{y}y} & 0 & 0 \\ 0 & 0 & \lambda_{\bar{x}x} & \lambda_{\bar{x}y} \\ 0 & 0 & \lambda_{\bar{y}x} & \lambda_{\bar{y}y} \end{bmatrix} = \begin{bmatrix} \cos\theta & -\sin\theta & 0 & 0 \\ \sin\theta & \cos\theta & 0 & 0 \\ 0 & 0 & \cos\theta & -\sin\theta \\ 0 & 0 & \sin\theta & \cos\theta \end{bmatrix} \quad [8.67]$$

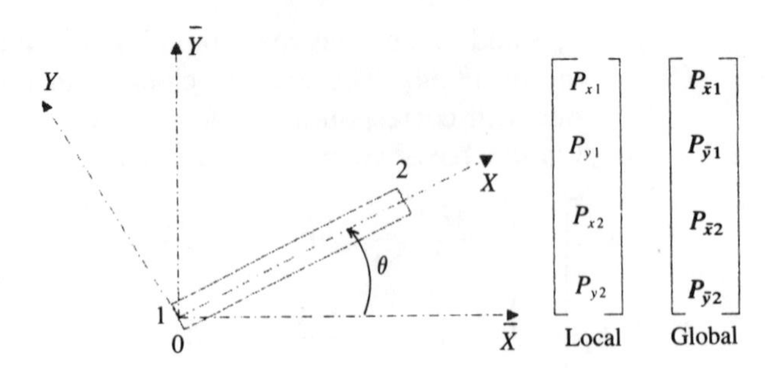

Figure 8.10 Plane truss element

Thus

$$\bar{k} = \lambda k \lambda^{\mathrm{T}}$$

$$= \frac{EA}{l}\begin{bmatrix} \cos^2\theta & \cos\theta\sin\theta & -\cos^2\theta & -\cos\theta\sin\theta \\ \cos\theta\sin\theta & \sin^2\theta & -\cos\theta\sin\theta & -\sin^2\theta \\ -\cos^2\theta & -\cos\theta\sin\theta & \cos^2\theta & \cos\theta\sin\theta \\ -\cos\theta\sin\theta & -\sin^2\theta & \cos\theta\sin\theta & \sin^2\theta \end{bmatrix} \quad [8.68]$$

8.8 Assembly of structure stiffness matrix

In this section we consider methods of assembling the stiffness matrix of the complete structure, $\boldsymbol{K}$, from the stiffness properties of the discrete elements. Effectively this is the process of setting up the matrix of coefficients of the equilibrium equations of the structure. The number of such equations will be equal to the number of degrees of freedom of the structure. In section 4.5.3 we derived a matrix formulation which produced $\boldsymbol{K}$ from the stiffnesses of the unassembled structure $\boldsymbol{k}$ using a transformation matrix $\boldsymbol{A}$ and obtained eqn [4.56]:

$$\boldsymbol{K} = \boldsymbol{A}^{\mathrm{T}}\boldsymbol{k}\boldsymbol{A}$$

Now this method is of limited practical use in view of the storage demands made by it. In practice the matrix $\boldsymbol{K}$ is generally formed by direct methods, and it is this approach which we shall now consider.

In section 4.5.3 we illustrated the matrix stiffness method in example 4.4 using a five-span continuous beam (Fig. 4.5). We reconsider this beam and take a more direct approach to the equilibrium equations. We shall then review what we have done in order to generalize the method in a form suitable for computers. The five-span beam is shown in Fig. 8.11(a). We assume that the applied loads have already been transformed into equivalent nodal forces R_1 to R_6, and initially we shall take the six rotations of the beam axis at the supports as being the degrees of freedom. Now if we imagine all degrees of freedom constrained (made zero) except for r_1, to which we give unit

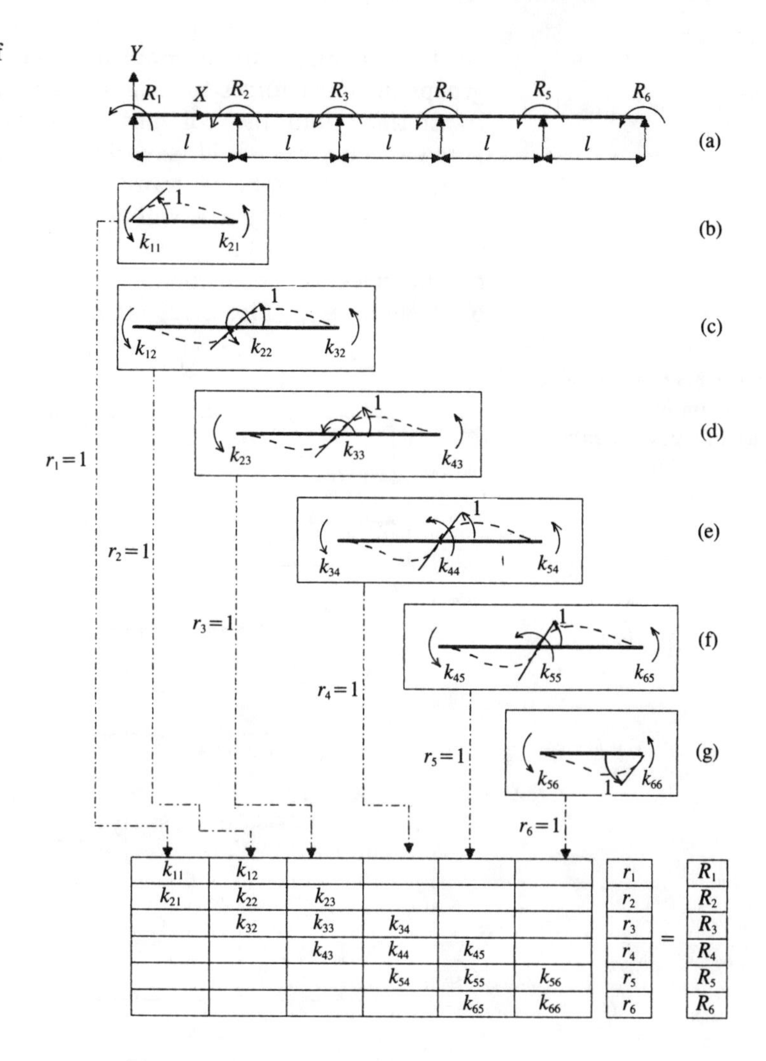

Figure 8.11 Assembly of ***K*** for continuous beam

value, then the forces required to be applied to the beam at the coordinates 1 to 6 are as shown in Fig. 8.11(b). Such forces are confined to the coordinates r_1 and r_2 in this case because the freedoms 2 to 6 are constrained. Similarly, in diagrams (c) to (g) we apply unit value to r_2 to r_6 in turn.

Now we express equilibrium of the beam in the direction of each of the coordinates r_1 to r_6. First, at r_1 the applied load R_1 must be in equilibrium with the internal moments. The moment at node 1 is $k_{11}r_1$ from diagram (b) and $k_{12}r_2$ from (c). There are no contributions from (d) to (g). Hence

$$k_{11}r_1 + k_{12}r_2 = R_1$$

Similarly at node 2,

$$k_{21}r_1 + k_{22}r_2 + k_{23}r_3 = R_2$$

As the equations are assembled it will be apparent that the groups of nodal forces in diagrams (b), (c), (d) etc. are being transformed into the columns of $\boldsymbol{K}$ as shown in Fig. 8.11. Alternatively, we could say that the group of forces at a particular node is being transferred into the rows of $\boldsymbol{K}$. The symmetry of the matrix $\boldsymbol{K}$ will now be very apparent.

Before we leave this particular structure we make the point that the five-span beam could be analysed with only four degrees of freedom as in Fig. 8.12, since the element stiffnesses for a

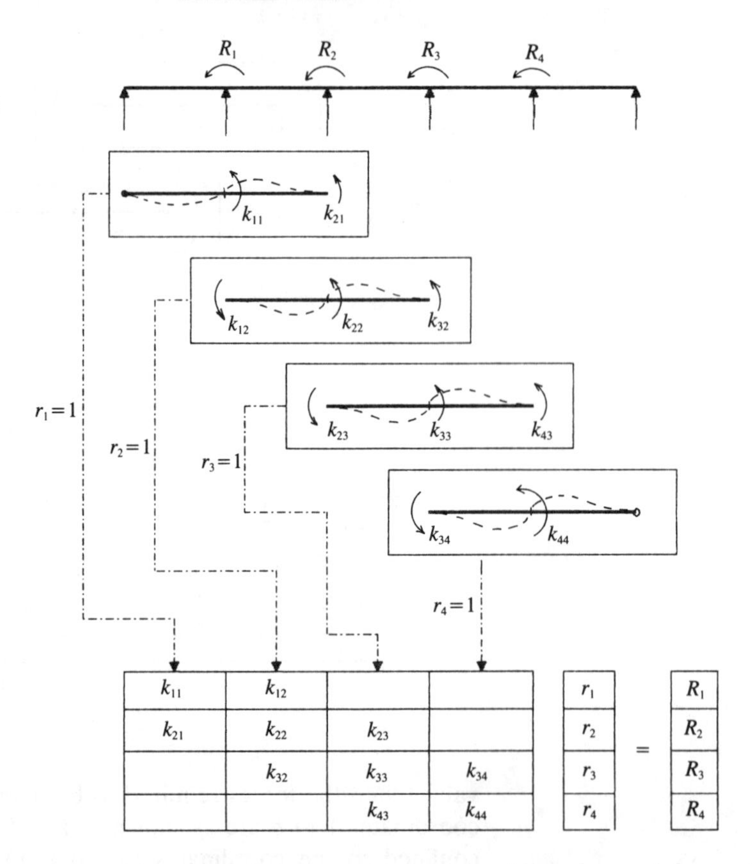

Figure 8.12 Assembly of $\boldsymbol{K}$ for continuous beam with minimum number of degrees of freedom

beam with a hinged end are known, enabling k_{11} and k_{44} to be written down. Thus it is unnecessary to know the rotations at nodes 1 and 6 to analyse the forces in the beam. The matrices for six and four degrees of freedom are given in section 4.5.3.

Now what we have just done needs a little more organization to make it suitable for computer programming. Suppose we have a member with ends 1 and 2 (Fig. 8.13) and that the stiffness matrix $\boldsymbol{k}$ is known in global coordinates, any necessary transformations having been done according to the procedures of section 8.7. Suppose also that the nodes of the structure have all been numbered 1, 2, . . . , N, including any nodes partially or completely constrained by external reactions. Further, suppose

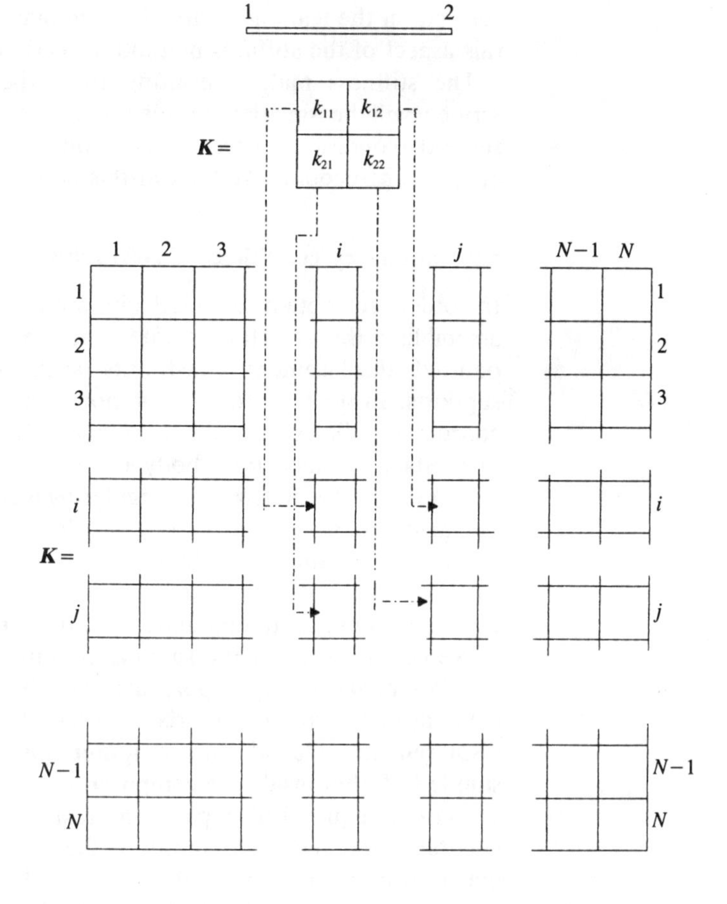

Figure 8.13 Transfer of element stiffness into structure stiffness

that end 1 of the element coincides with node i and end 2 with node j of the assembled structure. The stiffness transfers are as shown in Fig. 8.13, each stiffness being added on to what is already there; $\boldsymbol{K}$ has initially been set up as a null matrix. The transfers are

$$k_{11} \rightarrow K_{ii}$$

$$k_{12} \rightarrow K_{ij}$$

$$k_{22} \rightarrow K_{jj}$$

$$k_{21} \rightarrow K_{ji}$$

Now the individual stiffnesses in $\boldsymbol{k}$ may be single quantities or, more likely in practice, they may be submatrices. The same rules apply for the transfers but of course the number and ordering of the degrees of freedom at an element node must correspond with that at the structure nodes. Having commenced the assembly of $\boldsymbol{K}$ with an $N \times N$ null matrix, the parts of the matrix into which no stiffnesses are added will remain null; so in Figs 8.11 and 8.12 we see that the non-zero parts of the matrices lie on a band centred on the leading diagonal of the matrix. We shall touch on this aspect of the stiffness method in section 8.10.

The stiffness matrix resulting from the procedure just described will be singular because we have not yet observed the support conditions which exert constraints on some of the degrees of freedom. We look at this now.

8.9 Boundary conditions of constraint

In order to obtain a relatively straightforward process for assembling the structure stiffness matrix we have temporarily omitted displacement constraints such as those existing at supports. In other words, all the nodes are free to displace in all coordinate directions under applied loads. Clearly this is unsatisfactory since rigid body movement can take place; as a consequence the matrix $\boldsymbol{K}$ is singular as it stands. The matrix will be made non-singular when we have introduced sufficient constraints to eliminate all rigid body movements. Usually the total number of constraints will be rather more than the minimum required to eliminate rigid body movements.

We could represent the support conditions, assuming that the 'displacements' at the supports are zero, by deleting the rows and columns of the stiffness matrix corresponding to the zero support displacements. Outside the computer this would appear to be a simple task, but inside the computer a considerable programming exercise is required to 'repack' the resulting matrix. The method generally adopted is to multiply the diagonal element of the matrix $\boldsymbol{K}$ corresponding to the specified displacements, say r_i, by a large number, say 10^{20}. If the specified displacement is Δ_i, then the ith element of the load vector is replaced by $k_{ii} \times 10^{20} \times \Delta_i$. The result will be that all stiffnesses in the ith equation will be 'swamped' by the large number, and we shall get

$$r_i = \frac{R_i}{k_{ii} \times 10^{20}} = \frac{k_{ii} \times \Delta_i \times 10^{20}}{k_{ii} \times 10^{20}} = \Delta_i$$

Usually, of course, a displacement at a constrained support will be zero. In these circumstances we do not need to interfere with the load vector, and we shall obtain effectively

$$r_i = \frac{R_i}{k_{ii} \times 10^{20}} = 0$$

8.10 Solution of the equilibrium equations

This subject attracted a good deal of attention in the early phase of the development of computerized methods. Generally the method employed in computer programs is the Gauss elimination method, frequently with some refinement intended to speed the process, to give more accurate results or to reduce storage demands in the computer. The stiffness method is generally a very 'well-behaved' method numerically, and usually gives no difficulty with arithmetical rounding errors even with quite large sets of equations (Taniguchi, Shiraishi, Soga 1985). The structural equilibrium equations are naturally 'well conditioned', the diagonal elements usually dominating the matrix. The physical reasons for this will be evident if one considers that the force required to produce a displacement (diagonal element) will generally be larger than the corresponding forces (off-diagonal elements) to constrain displacements at other coordinates. The Gauss method is too well known to need explanation; however, there are some features of the stiffness matrix which deserve attention from the computational point of view.

The stiffness matrices in Figs 8.11 and 8.12 reveal a typical 'banded' form in that all the non-zero elements are concentrated on and close to the leading diagonal. This is characteristic of the stiffness method and can be explained by physical reasoning. The forces generated at nodes due to the imposition of a unit displacement at a node, all other nodal displacements constrained, constitute one column of the stiffness matrix. Because of the nodal constraints, the effect will not spread very far from the node where the displacement is imposed. This effect can be seen in Figs 8.11 and 8.12 very clearly. The banding is even more marked in very large sets of equations.

The second feature of the structural stiffness matrix is its symmetry. This is an indirect result of Maxwell's reciprocal theorem. Considerable computational advantage can be taken of this feature in programming the elimination method. The equations can be represented in the computer by storing one-half of the band only, as in Fig. 8.14. Some control can be exercised

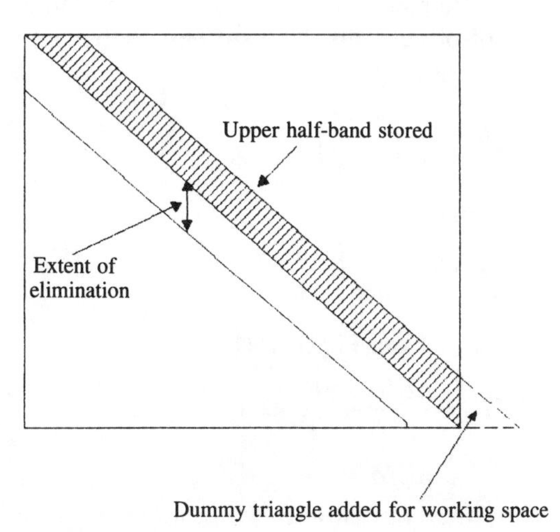

Figure 8.14 Storage of band matrix in elimination method

by the analyst over the width of the band. It can be shown that the half-band width is related to the maximum difference between the numbered degrees of freedom in the direction of numbering of the nodes. Thus, for a 'rectangular' array of numbered nodes, the narrowest band will develop if the nodes are numbered across the shorter dimension of the rectangle. With the very large sets of equations produced by the finite element method, other special methods have to be used to economize on storage and obtain rapid and accurate results. Transfer of data, usually element stiffness matrices, from a 'back-up' store (say disc files) to the working store only when needed can reduce storage demands very substantially but can slow the process down unacceptably. Most commercial packages for finite element analysis will use a 'front solver' type of equation solving process, in which considerable saving results from the recognition of the 'sparse' nature of the matrices involved (Cheung, Yeo 1979).

8.11 Support displacement, prestrain and thermal loading

In the structural equilibrium equations $\boldsymbol{Kr} = \boldsymbol{R}$, the matrix $\boldsymbol{R}$ contains the nodal loads to be applied to the structure. The matrix $-\boldsymbol{R}$ contains the constraining forces applied to the structure, and such constraining forces may arise from effects other than normal loading. The effect of support displacement was considered in the previous section, and the procedure is straightforward where the value of a nodal displacement is known and the stiffness matrix contains that particular degree of freedom. An alternative procedure will now be outlined for cases where the coordinate corresponding to the specific support displacement is not included in the structure stiffness matrix.

In Fig. 8.15(a) the support B is subjected to a vertical settlement δ and we wish to determine the resulting nodal displacements. The stiffness matrix in nodal coordinates r_1 and r_2 is easily shown to be,

$$\boldsymbol{K} = \frac{EI}{l}\begin{bmatrix} 8 & 2 \\ 2 & 8 \end{bmatrix} \qquad [8.69]$$

Then

$$\boldsymbol{K}^{-1} = \frac{l}{30EI}\begin{bmatrix} 4 & -1 \\ -1 & 4 \end{bmatrix} \qquad [8.70]$$

Now from Fig. 8.15(b)

$$R_1 = \frac{6EI\delta}{l^2} - \frac{6EI\delta}{l^2} = 0$$

$$R_2 = \frac{6EI\delta}{l^2}$$

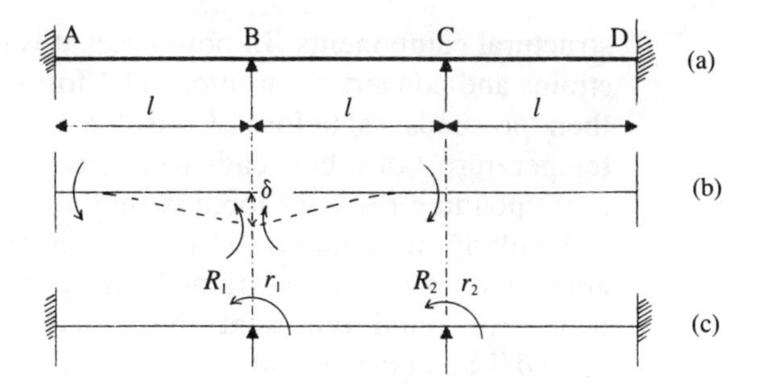

Figure 8.15 Support displacement, prestrain and thermal effects

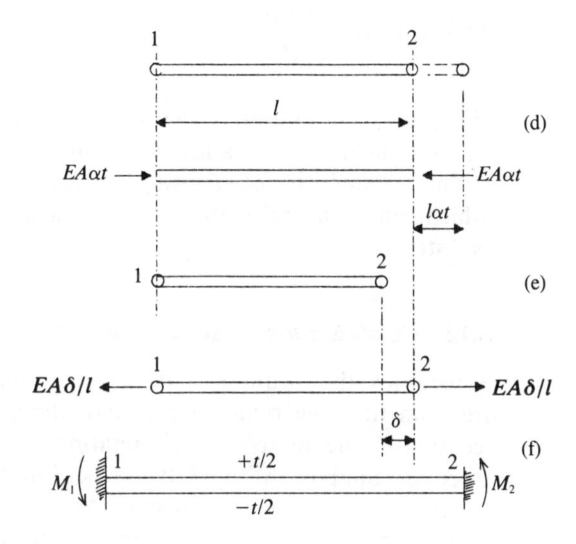

Hence

$$\boldsymbol{r} = \boldsymbol{K}^{-1}\boldsymbol{R} = \begin{bmatrix} -\delta/5l \\ 4\delta/5l \end{bmatrix}$$

The reader might now like to show that the resulting bending moments are

$$M_A = 28EI\delta/5l^2$$

$$M_B = 26EI\delta/5l^2$$

Sometimes a support yield is dependent on the value of the reactive force. In these cases the simplest way is to replace the particular coordinate in the support by a spring of known stiffness and include the spring as an element in the structure. Most commercial packages for structural analysis contain provision for the inclusion of springs.

We consider now the existence of initial strains. These may be due to temperature variation or to strains caused by lack of fit in structural components. In both cases it is possible to evaluate the strains and convert them into nodal force restraints. The analysis then proceeds as before. Consider Fig. 8.15(d). A change in temperature t of a bar leads to a 'free' extension of $l\alpha t$, and the corresponding restraint force is therefore $\pm EA\alpha t$ at nodes 1 and 2. Similarly, if member 1–2 in diagram (e) were too short by an amount δ, it could be strained into position by a force $EA\delta/l$; hence we could represent this situation by restraining forces $\pm EA\delta/l$ at nodes 1 and 2. If a beam 1–2 as in diagram (f) is subjected to a temperature change $-t/2$ to $+t/2$ across the depth d of the beam, then the restraining moments will be

$$M_1 = -M_2 = -\frac{EI\alpha t}{d} \tag{8.71}$$

We now go on to look at two further topics which are sometimes of value in the stiffness method. One is the condensation of the stiffness matrix by eliminating degrees of freedom. The other, which can be useful with very large structures, is the method of 'substructures'.

8.12 Condensation of stiffness matrix

Sometimes the stiffness matrix may refer to more degrees of freedom than we really require. In these circumstances we can 'condense' the matrix by eliminating the unwanted degrees of freedom. Suppose we have two sets of nodal coordinates: a set $\boldsymbol{r}_1$ in which we are not interested, and a set $\boldsymbol{r}_2$ in which we are interested. We assemble the stiffness matrix in partitioned form:

$$\begin{bmatrix} \boldsymbol{K}_1 & \boldsymbol{K}_3^{\mathrm{T}} \\ \boldsymbol{K}_3 & \boldsymbol{K}_2 \end{bmatrix} \begin{bmatrix} \boldsymbol{r}_1 \\ \boldsymbol{r}_2 \end{bmatrix} = \begin{bmatrix} \boldsymbol{R}_1 \\ \boldsymbol{R}_2 \end{bmatrix} \tag{8.72}$$

Now we can eliminate $\boldsymbol{r}_1$ by a sequence of matrix operations and obtain

$$\bar{\boldsymbol{K}}\boldsymbol{r}_2 = \bar{\boldsymbol{R}} \tag{8.73}$$

where

$$\bar{\boldsymbol{K}} = (\boldsymbol{K}_2 - \boldsymbol{K}_3\boldsymbol{K}_1^{-1}\boldsymbol{K}_3^{\mathrm{T}}) \tag{8.74}$$

$$\bar{\boldsymbol{R}} = (\boldsymbol{R}_2 - \boldsymbol{K}_3\boldsymbol{K}_1^{-1}\boldsymbol{R}_1) \tag{8.75}$$

$\bar{\boldsymbol{K}}$ is the condensed stiffness matrix and $\bar{\boldsymbol{R}}$ is the condensed load matrix. The process is essentially one of partial elimination of the variables. Since the matrix $\boldsymbol{K}_1$ requires inversion, the process will not be very efficient if $\boldsymbol{K}_1$ is large. (See exercises 8.5 and 8.6.)

8.13 Method of substructures

This method was popular when the storage capacity of computers was more severely limited. It may still be useful if a large structure has to be solved on a small computer. We saw in section 8.10 that the banded nature of the stiffness matrix is a valuable feature of the method in that special methods can be employed to solve the equations using minimal space in the computer. Whilst in some structures it may be possible to number nodes in such a way that the band is uniformly narrow, there are other cases where the nature of the structure produces a local widening of the band, thus committing more storage.

Examples of such structures are shown in Fig. 8.16(a) and (b).

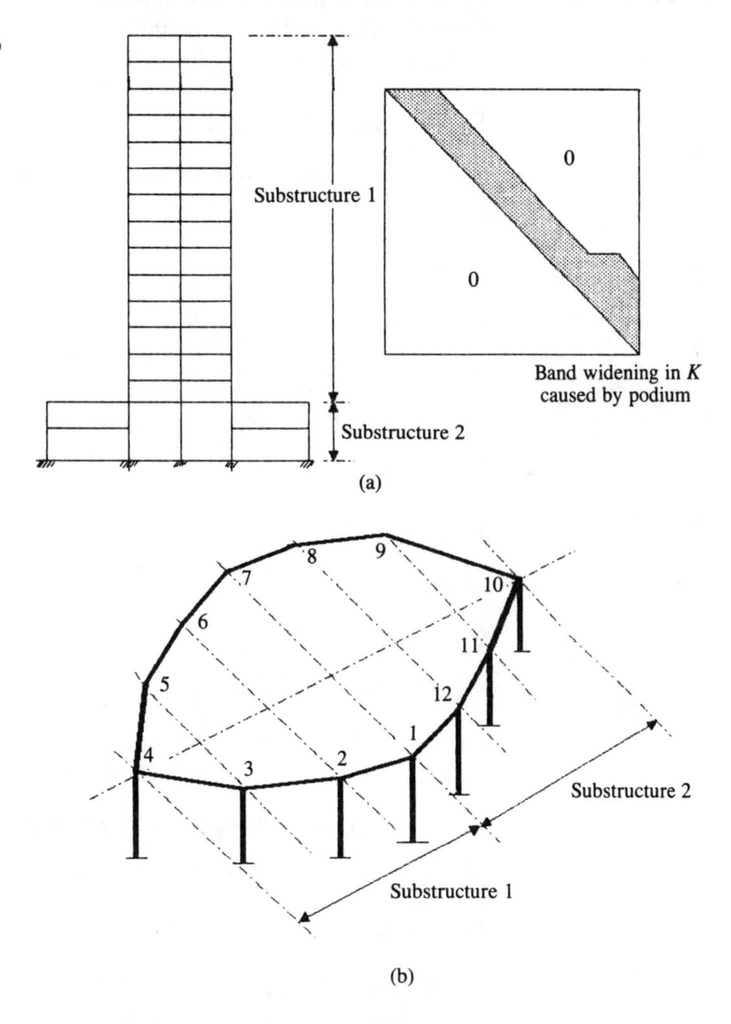

Figure 8.16 Division into substructures

In (a) we have a multi-storey frame with a podium. The widening of the podium will increase the band width. In (b) we have a closed ring where the simple band form is disturbed by the closure of the ring. The procedure is as follows. For substructure 1 we partition the matrices:

$$\begin{bmatrix} \boldsymbol{K}_1 & \boldsymbol{C}_1^{\mathrm{T}} \\ \boldsymbol{C}_1 & \boldsymbol{K}_{1\mathrm{C}} \end{bmatrix} \begin{bmatrix} \boldsymbol{r}_1 \\ \boldsymbol{r}_{\mathrm{C}} \end{bmatrix} = \begin{bmatrix} \boldsymbol{R}_1 \\ \boldsymbol{R}_{1\mathrm{C}} \end{bmatrix} \qquad [8.76]$$

and for substructure 2:

$$\begin{bmatrix} \boldsymbol{K}_{2\mathrm{C}} & \boldsymbol{C}_2 \\ \boldsymbol{C}_2^{\mathrm{T}} & \boldsymbol{K}_2 \end{bmatrix} \begin{bmatrix} \boldsymbol{r}_{\mathrm{C}} \\ \boldsymbol{r}_2 \end{bmatrix} = \begin{bmatrix} \boldsymbol{R}_{2\mathrm{C}} \\ \boldsymbol{R}_2 \end{bmatrix} \qquad [8.77]$$

The subscript C is used for the displacements and forces common to both substructures. The internal forces between the substructures are $\boldsymbol{R}_{1\mathrm{C}}$ and $\boldsymbol{R}_{2\mathrm{C}}$ and the internal components of these are unknown. However, we know that

$$\boldsymbol{R}_{1\mathrm{C}} = -\boldsymbol{R}_{2\mathrm{C}} = \boldsymbol{R}_{\mathrm{C}} \quad \text{(say)} \qquad [8.78]$$

We can now apply the condensation treatment of section 8.12 to give

$$\bar{\boldsymbol{K}}_{1\mathrm{C}} \boldsymbol{r}_{\mathrm{C}} = \bar{\boldsymbol{R}}_{1\mathrm{C}} \qquad [8.79]$$

where

$$\bar{\boldsymbol{K}}_{1\mathrm{C}} = (\boldsymbol{K}_{1\mathrm{C}} - \boldsymbol{C}_1 \boldsymbol{K}_1^{-1} \boldsymbol{C}_1^{\mathrm{T}})$$
$$\bar{\boldsymbol{R}}_{1\mathrm{C}} = (\boldsymbol{R}_{\mathrm{C}} - \boldsymbol{C}_1 \boldsymbol{K}_1^{-1} \boldsymbol{R}_1)$$

and to give

$$\bar{\boldsymbol{K}}_{2\mathrm{C}} \boldsymbol{r}_{\mathrm{C}} = \bar{\boldsymbol{R}}_{2\mathrm{C}} \qquad [8.80]$$

where

$$\bar{\boldsymbol{K}}_{2\mathrm{C}} = (\boldsymbol{K}_{2\mathrm{C}} - \boldsymbol{C}_2 \boldsymbol{K}_2^{-1} \boldsymbol{C}_2^{\mathrm{T}})$$
$$\bar{\boldsymbol{R}}_{2\mathrm{C}} = (-\boldsymbol{R}_{\mathrm{C}} - \boldsymbol{C}_2 \boldsymbol{K}_2^{-1} \boldsymbol{R}_2)$$

Equations [8.79] and [8.80] can now be solved for the internal forces $\boldsymbol{R}_{\mathrm{C}}$ and displacements $\boldsymbol{r}_{\mathrm{C}}$ at the common nodes:

$$[\bar{\boldsymbol{K}}_{1\mathrm{C}} + \bar{\boldsymbol{K}}_{2\mathrm{C}}] \boldsymbol{r}_{\mathrm{C}} = -[\boldsymbol{C}_1 \boldsymbol{K}_1^{-1} \boldsymbol{R}_1 + \boldsymbol{C}_2 \boldsymbol{K}_2^{-1} \boldsymbol{R}_2] \qquad [8.81]$$

$$[\bar{\boldsymbol{K}}_{1\mathrm{C}}^{-1} + \bar{\boldsymbol{K}}_{2\mathrm{C}}^{-1}] \boldsymbol{R}_{\mathrm{C}} = \bar{\boldsymbol{K}}_{1\mathrm{C}}^{-1} (\boldsymbol{C}_1 \boldsymbol{K}_1^{-1} \boldsymbol{R}_1) - \bar{\boldsymbol{K}}_{2\mathrm{C}}^{-1} (\boldsymbol{C}_2 \boldsymbol{K}_2^{-1} \boldsymbol{R}_2) \qquad [8.82]$$

Once $\boldsymbol{r}_{\mathrm{C}}$ is determined, then

$$\boldsymbol{r}_1 = \boldsymbol{K}_1^{-1} [\boldsymbol{R}_1 - \boldsymbol{C}_1^{\mathrm{T}} \boldsymbol{r}_{\mathrm{C}}] \qquad [8.83]$$

$$\boldsymbol{r}_2 = \boldsymbol{K}_2^{-1} [\boldsymbol{R}_2 - \boldsymbol{C}_2^{\mathrm{T}} \boldsymbol{r}_{\mathrm{C}}] \qquad [8.84]$$

A summary of the equations developed in the stiffness method, with equation numbers as references, is contained in Appendix 5.

Exercises

8.1 (a) For a one-dimensional finite element, $\boldsymbol{N}$ is a matrix of shape functions and w is the variable intensity of distributed applied load. Show that the consistent nodal load matrix is given by

$$\boldsymbol{W}_e = \int_0^l w\boldsymbol{N}^T \, dx$$

(b) A one-dimensional, linear, axial force element has a length l and a uniform extensional rigidity EA. The element has a node at each end and a third, internal, node at a point distant $l/3$ from end 1. The applied loading on the element is an axial load varying linearly according to

$$w = w_0 x/l$$

where the origin of x is at node 1. Determine
(i) the shape functions using a parabolic displacement function
(ii) the stiffness matrix of the element
(iii) the consistent nodal load matrix.

8.2 Assemble a stiffness matrix for the two-storey frame with rigid joints shown in Fig. 8.17. Ignore axial strains in the members and consider the nodal displacements in the order 1 to 9 as shown.

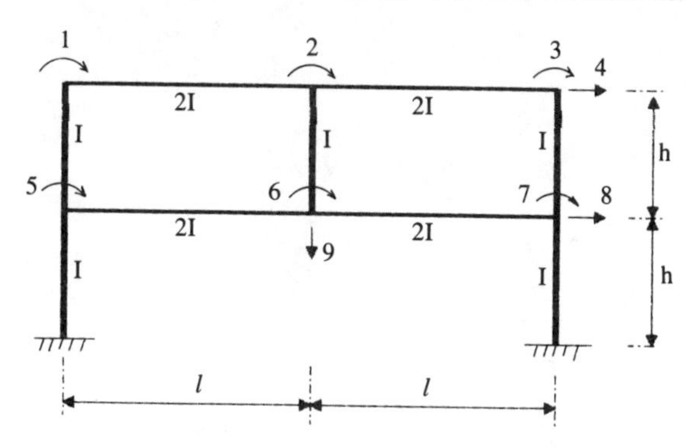

Figure 8.17 Exercise 8.2

8.3 Prepare element stiffness matrices in local coordinates for the members of the frame with inclined columns shown in structure 5F in Part II. Transform the element stiffnesses into global coordinates in the order x, y, θ at the nodes 2 and 3 and hence assemble the structure stiffness matrix. Solve the equilibrium equations and compare the results with those obtained in the Workshop.

8.4 Proceeding from the general 6×6 stiffness matrix for a uniform, plane flexural element 1–2, prepare the corresponding stiffness matrix for the member if end 1 is hinged.

8.5 Proceeding from the general 12×12 stiffness matrix for an element in a space frame, prepare a 6×6 stiffness matrix related to suitable coordinates for an element in a plane grillage, by deleting appropriate rows and columns.

8.6 A beam 1–2 of length $3l$ has a stepped variation in EI. From end 1 to internal node 3 the flexural rigidity is EI, and from node 3 to end 2 it is $2EI$. Length 1–3 is l and length 3–2 is $2l$. Assemble the stiffness matrix for the member condensed to stiffnesses at ends 1 and 2.

8.7 A beam has a solid rectangular cross-section 1.5 m deep by 0.5 m wide. The beam is 5 m long. On the basis of the engineer's theory of bending and including the effects of both shear and bending, determine the end stiffnesses. Consider the validity of these results in view of the depth of this beam. $E/G = 2.5$.

CHAPTER

9

Influence lines and surfaces

9.1 Introduction

Influence lines and surfaces are used to describe effects in structures when applied loads can occupy variable positions, as happens quite frequently and especially in bridge structures. An influence line shows the value of some internal force or other effect as a single unit load traverses the structure. The ordinate to the influence line at some position x is the value of the effect when the unit load is placed at x. In other words the independent variable is the unit load position. Influence lines can be constructed for any effect in a structure, but those more usually encountered are for bending moment, shear force and axial force. Influence lines are sometimes useful for describing displacements in structures such as deflections in beams (sections 9.7 and 9.8). The influence line describes a one-dimensional situation with a single independent variable (x usually). In two-dimensional situations such as plates we need two independent variables to define a load position, which therefore requires an influence 'surface'.

We shall introduce the concept of the influence line from first principles and begin by examining influence lines for bending moment and shear force in simple beams. We shall need to distinguish influence lines for statically determinate and statically indeterminate structures only because the structural analysis is different; the concept of the influence line does not depend on the statical determinacy or otherwise of the structure. We need also to look at influence lines for forces in the members of trusses and study the use of these in handling real live loads.

The concept of the 'envelope' diagram is also useful in structural design, and this is best treated along with influence lines. We shall then look at influence lines for displacements and introduce the topic of influence surfaces.

The computational methods used with influence lines have changed substantially with the introduction of computerized methods. Given sufficient computational power, we can always produce influence diagrams by successively placing a unit load at points in a structure and drawing lines through the ordinates obtained. In the modern computational context, this is often the most economical way to organize the work.

9.2 Influence lines for statically determinate beams

In Fig. 9.1(a) we have a simply-supported beam AB of span l. We choose a representative point C in the beam, distant a from

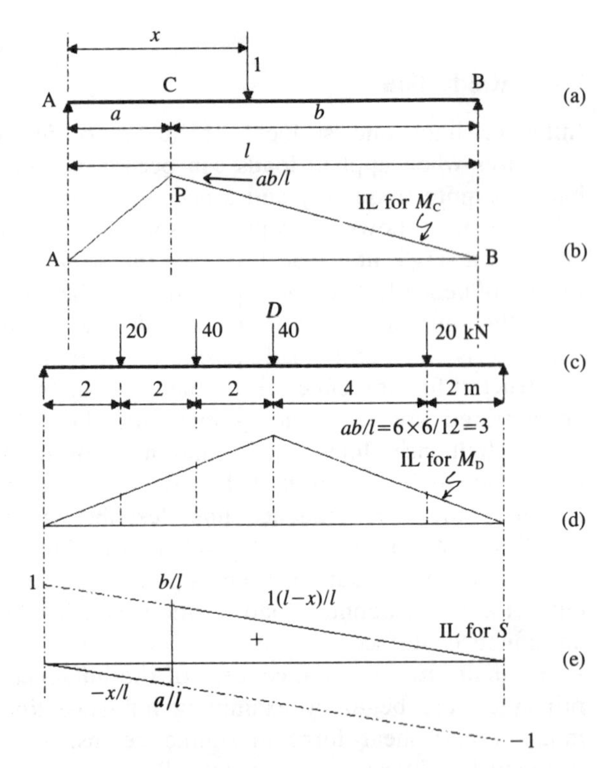

Figure 9.1 Influence lines for simply-supported beam

A and b from B. We are going to construct influence lines for bending moment and shear force at point C, and while we are doing this point C is fixed in position.

Suppose a single unit point load acts at x from A. Now x is the independent variable and $0 \leqslant x \leqslant l$, but suppose for the time being that x lies between C and B. Then

$$M_C = R_A a = 1\frac{l-x}{l}a$$

This is the straight line PB with negative slope in diagram (b); when $x = a$, $M_C = ab/l$. Now suppose that the unit load lies within AC. Then

$$M_C = R_B b = 1\frac{x}{l}b$$

This is the straight line AP with positive slope in diagram (b). At $x = a$, $M_C = ab/l$ as before. Thus the complete influence line for M_C is the triangle APB in (b) with a maximum ordinate of ab/l at point C.

Since the ordinate at any point is equal to the value of M_C when unit load acts at that point, the total effect due to several loads as in diagram (c) is

$$\text{total effect} = \Sigma\, W \times \text{ordinate} \qquad [9.1]$$

For example the influence line for M_D in diagram (c) is shown at (d) and thus, for the loads shown at (c),

$$M_D = \left(20 \times \frac{2}{6} \times 3\right) + \left(40 \times \frac{4}{6} \times 3\right) + (40 \times 3) + \left(20 \times \frac{2}{6} \times 3\right)$$

$$= 240 \text{ kN m}$$

The influence line for shear force at representative point C, as in diagram (a), can be produced in a similar way. Suppose the unit load lies between C and B. Then the shear at C is given by

$$S_C = R_A = 1\frac{l-x}{l}$$

and this is positive if we adopt the sign convention in diagram (e). If the unit load lies in AC, then

$$S_C = -R_B = -1\frac{x}{l}$$

So the complete influence line is as in Fig. 9.1(e). A discontinuity is apparent at C such that S_C is negative for loads on AC and positive for loads on CB. Again eqn [9.1] is used to obtain the total effect due to a number of concentrated loads.

The effect of a distributed load can be seen from Fig. 9.2. If

Figure 9.2 Effect of distributed load

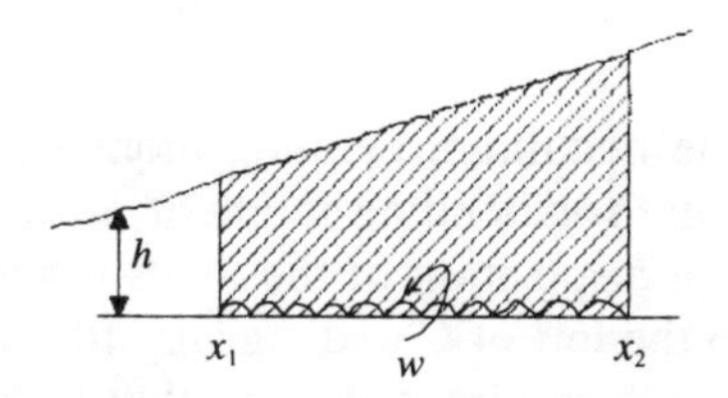

$h = f(x)$ is the ordinate to the influence line, then the effect of a load over elemental length δx is $w\,\delta x h$. Thus the total effect of

load w from $x = x_1$ to $x = x_2$ is

$$\int_{x_1}^{x_2} wh \, \mathrm{d}x$$

which, if w is constant, is

$$\text{effect} = w \times \text{area under influence line covered by load} \qquad [9.2]$$

Statically determinate beams with overhangs or with internal hinges can be handled by similar procedures, statical principles being used to determine reactions and internal forces. The influence lines are generally composed of straight lines; so, by placing the unit load at selected positions in the span, the influence lines can be obtained by joining ordinates by straight lines.

In applying actual live loads using influence lines, it is usually necessary to place the loads in positions to give a maximum value of the effect. Various rules have been devised to assist in doing this. We shall look at one of these now.

Suppose we have a simply-supported beam which has to carry a 'train' of concentrated loads at fixed distances apart, as in Fig. 9.3(a) and (b). We wish to find the placing of the loads to

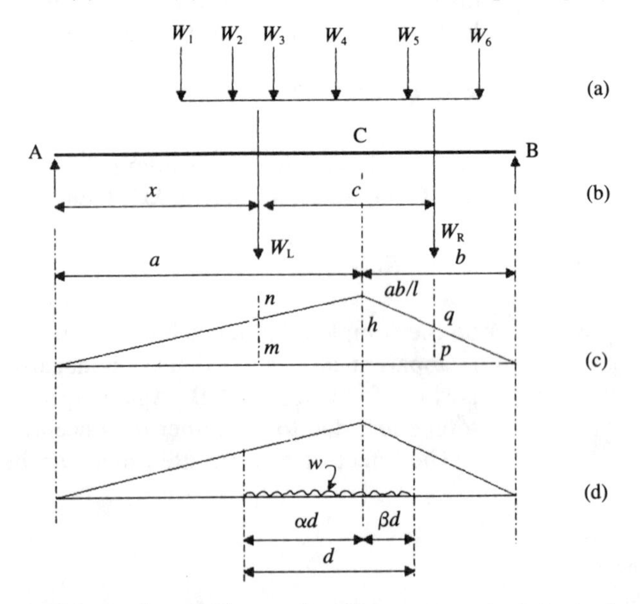

Figure 9.3 Placing loads for maximum bending moment

produce the maximum bending moment at typical point C. The influence line for bending moment at C is given at diagram (c). Suppose the disposition of the loads is such as to involve a group of loads to the left of C and a group to the right. Let the group to the left have a resultant W_L and that to the right a resultant W_R, as in diagram (b). The distance between the lines of action of W_L and W_R will be a constant c, and x will define the position of the loads.

Now

$$M_C = \Sigma\,(W \times \text{ordinate})$$
$$= W_L \times mn + W_R \times pq$$
$$= W_L x \frac{h}{a} + W_R \frac{l - x - c}{b} h$$

$$\frac{dM_C}{dx} = W_L \frac{h}{a} + W_R \frac{h}{b}(-1)$$
$$= h\left(\frac{W_L}{a} - \frac{W_R}{b}\right)$$
$$= 0 \quad \text{for maximum } M_C$$

Thus

$$\frac{W_L}{a} = \frac{W_R}{b} = \frac{\bar{W} - W_L}{l - a}$$

Hence

$$\frac{W_L}{a} = \frac{W_R}{b} = \frac{\bar{W}}{l} \qquad [9.3]$$

where $\bar{W}$ = total load = $\Sigma\, W$. A convenient form of eqn [9.3] for the interpretation of the rule is

$$W_L = \frac{a}{l}\bar{W} \qquad [9.4]$$

Now if we study the typical geometry of a bending moment diagram for a beam carrying concentrated loads, it is clear that a load is always located at the position of maximum bending moment. Thus we must determine which load is the critical one. It will be the one which causes dM_C/dx to change sign. If W_C is the critical load, then if we include W_C in W_L we shall get

$$W_L > \frac{a}{l}\bar{W} \qquad [9.5a]$$

and if we include W_C in W_R we shall get

$$W_L < \frac{a}{l}\bar{W} \qquad [9.5b]$$

All we need to do is to examine likely loads to find the load which satisfies eqns [9.5a] and [9.5b]; we then place this load at C to obtain maximum M_C.

If a uniformly distributed load has a length at least equal to the span, then we simply load the whole span to obtain maximum M_C. If, however, the load has a length, say *d*, shorter than the span, as in Fig. 9.3(d), then we apply the rule [9.4] and get

$$\frac{\alpha dw}{a} = \frac{\beta dw}{b} = \frac{dw}{l} \tag{9.6}$$

It follows as a matter of interest that, with this placing of the distributed load, the ordinates to the influence line at the ends of the load are equal, i.e.

$$h\frac{(a-\alpha d)}{a} = h\frac{(b-\beta d)}{b}$$

$$\alpha\frac{d}{a} = \beta\frac{d}{b}$$

There are other rules which are designed to help us obtain maximum values of effects in beams, but these are generally less helpful than the one described above. A process of judicious trial and error is quite sufficient in most practical circumstances, as we shall see in the following examples.

Example 9.1

We wish to calculate the maximum bending moment that can occur at the centre C of span AB in Fig. 9.4, and also at the

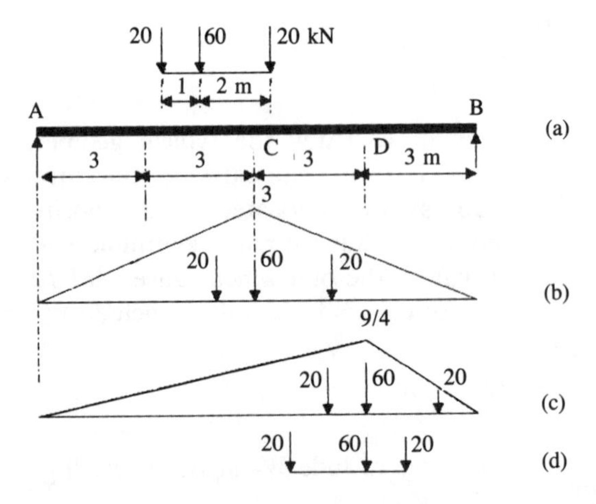

Figure 9.4 Example 9.1

right-hand quarter point D. The loading is shown in diagram (a) and the relevant influence lines in (b) and (c).

Consider the influence line for M_C, using eqn [9.4]:

$$W_L = \frac{a}{l}\bar{W} = \frac{6}{12} \times 100 = 50 \text{ kN}$$

The critical load is the 60 kN load, since $20 + 60 > 50$ and $20 < 50$. Hence for $M_{C\max}$ we place the loads as shown at (b),

and

$$M_{C\,max} = 20 \times \frac{5}{6} \times 3 + 60 \times 3 + 20 \times \frac{4}{6} \times 3$$

$$= 270 \text{ kN m}$$

For the right-hand quarter span point D, as in (c),

$$W_L = \frac{a}{l}\bar{W} = \frac{9}{12} \times 100 = 75 \text{ kN}$$

Again the 60 kN load is critical, since $20 + 60 > 75$ and $20 < 75$. With the loads placed as in diagram (c) we obtain

$$M_D = 20 \times \frac{8}{9} \times \frac{9}{4} + 60 \times \frac{9}{4} + 20 \times \frac{1}{3} \times \frac{9}{4} = 190 \text{ kN m}$$

We should now investigate the effect of reversing the loads so that they lie as in diagram (d). We then get

$$M_D = 20 \times \frac{7}{9} \times \frac{9}{4} + 60 \times \frac{9}{4} + 20 \times \frac{2}{3} \times \frac{9}{4} = 200 \text{ KN m}$$

and this is the maximum moment which can be produced at either left-hand or right-hand quarter span points.

Example 9.2

This example is designed to show how we handle load trains when one or more loads lie off the span when the rule for maximum moment is applied. We wish to obtain the maximum moment at C in the beam of Fig. 9.5(a) due to the loads shown.

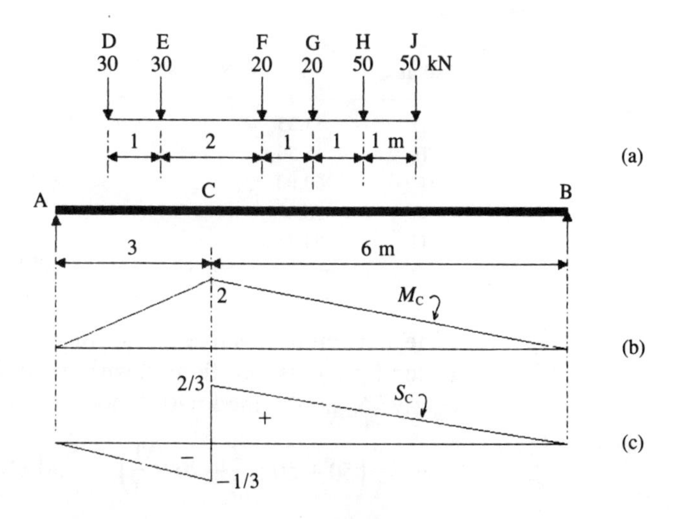

Figure 9.5 Example 9.2

Now

$$W_L = \frac{3}{9} \times 200 = 66.7$$

Hence we place load F at C since $80 > 66.7$ and $60 < 66.7$. However, this puts load D effectively off the span, so we repeat the exercise ignoring load D:

$$W_L = \frac{3}{9} \times 170 = 56.7$$

We now place load G at the section but find that load E has now become ineffective. We apply the rule again, this time ignoring loads D and E:

$$W_L = \frac{3}{9} \times 140 = 46.7$$

So we place load H at C since $90 > 46.7$ and $40 < 46.7$. All the loads considered are now on the span, so we can calculate

$$M_{C\,\max} = 2\left(20 \times \frac{1}{3} + 20 \times \frac{2}{3} + 50 + 50 \times \frac{5}{6}\right) = 223.3 \text{ kN m}$$

For the maximum positive shear force at C there is some uncertainty about the position of the load train. Loads on the negative part of the influence line will reduce the positive shear. We also notice that the 50 kN loads are at the right-hand end of the train and will be more effective if we move the train to the left. Let us examine all possibilities by placing each load D to H at C in turn. The results are given in Table 9.1.

Table 9.1 Solution for example 9.2

Load at C	+ve S_C (kN)
D	53.33
E	45.56
F	60.00
G	58.89
H	54.45

The maximum positive S_C is thus 60 kN and is obtained when the load train is positioned with load F at C. The maximum negative S_C is obtained with load J at section C and is

$$S_C = -\frac{1}{3}\left(50 + 50 \times \frac{2}{3} + 20 \times \frac{1}{3}\right) = -30 \text{ kN m}$$

Example 9.3

The system of beams ABCDEF in Fig. 9.6(a) forms a cantilever and suspended span arrangement. We wish to construct influence

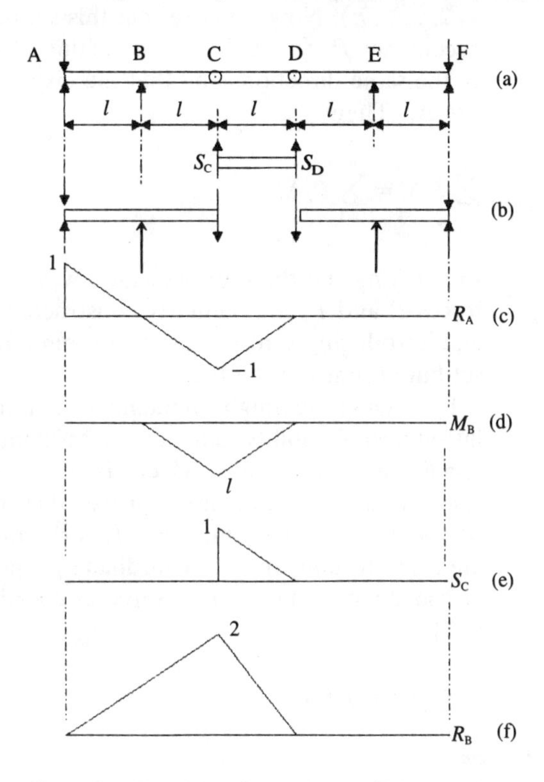

Figure 9.6 Example 9.3

lines for R_A, M_B, S_C and R_B. (The cantilever and suspended span is a 'classic' structure with several built-in advantages, including the fact that it is statically determinate and so is unaffected by differential settlement of supports.) Note the provision for downwards reactions at A and F which will be required under certain loadings.

Now, in approaching the influence lines we should look first at the statics of the suspended span since this is the key to the behaviour of the structure. It helps to separate this span from the rest of the structure as in Fig. 9.6(b) and, since hinges exist at C and D, the only actions transmitted at these positions are vertical shear forces S_C and S_D. Both of these are zero if no load acts on CD. We could summarize the general behaviour as follows. If the unit load lies between A and C, CDEF takes no part in the structural action. Similarly if the unit load lies between D and F, ABCD takes no part. If the load lies between C and D then the complete structure is involved.

We now place the unit load at A and move it step by step along the structure, determining the effects required for the influence lines shown in diagrams (c) to (f) of Fig. 9.6.

9.3 The Maxwell-Betti and Müller-Breslau theorems

Suppose forces P_i $(i = 1, 2, \ldots, n)$ act on an elastic structure producing displacements at corresponding coordinates Δ_i $(i = 1, 2, \ldots, n)$. Now suppose that this set of forces is replaced by a second set P_i' $(i = 1, 2, \ldots, n)$ acting at the same positions and in the same directions and that the displacements produced by P_i' are Δ_i'. Then

$$\sum_{i=1}^{n} P_i \Delta_i' = \sum_{i=1}^{n} P_i' \Delta_i \qquad [9.7]$$

In applying the theorem we can always ensure that both sets of forces P and P' are similarly constructed by including all forces and introducing zero values where some forces are present in one set but absent in the other.

In passing we might remember that in Chapter 4 when we introduced flexibility influence coefficients f_{ij} we observed that $f_{ij} = f_{ji}$, a relationship which is clear from the method of calculation. We can now prove this relationship using the Maxwell–Betti theorem. Since f_{ij} is the displacement at coordinate i due to unit force at coordinate j, and f_{ji} is the displacement at coordinate j due to unit force at coordinate i, then from eqn [9.7]

$$(1)(f_{ij}) = (1)(f_{ji})$$

or

$$f_{ij} = f_{ji} \qquad [9.8]$$

Now the Maxwell-Betti theorem can be developed to provide a means of obtaining influence lines for actions in statically indeterminate structures. The resulting theorem, attributed to Müller-Breslau, states that the ordinates to the influence line for any action in a structure are equal to those of the deflection curve of the structure obtained by removing the action and imposing a unit displacement at its coordinate. This is easily proved using the Maxwell-Betti theorem as follows. In Fig. 9.7(a) we see part of a structure with force R and applied unit

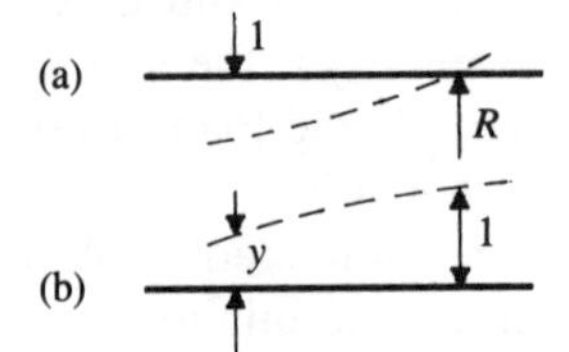

Forces	R	1
Displacements	0	?

Forces	0	0
Displacements	1	y

Figure 9.7 Proof of Müller–Breslau's theorem

load. In (b) the force R and unit load are removed and a unit displacement is imposed in place of R; the structure deflections resulting from this are y. Now, applying eqn [9.7],

$$(R)(1) + (1)(-y) = 0$$

Hence

$$R = y \qquad [9.9]$$

The units in eqn [9.9] will be reconciled if it is observed that the left-hand side has been multiplied by unit displacement and the right-hand side by unit force.

Example 9.4

For the two-span continuous beam shown in Fig. 9.8, we wish to produce expressions for the influence lines for R_B and M_B. For

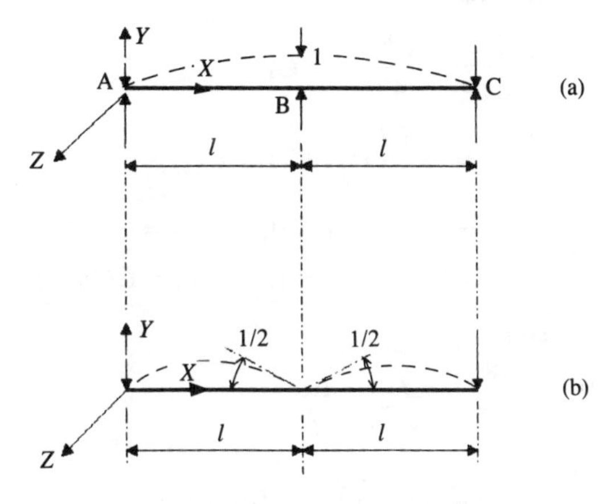

Figure 9.8 Example 9.4

R_B we remove the reactive force and impose a unit vertical displacement as in diagram (a). The force required to produce the unit displacement we know from standard cases to be $48EI/(2l)^3$. So the end reactions are $3EI/l^3$ and the bending moment at x $(0 \leq x \leq l)$ is

$$M_z = \frac{3EI}{l^3}x = -EI\frac{d^2v}{dx^2}$$

Hence

$$EI\frac{dv}{dx} = -\frac{3EI}{2l^3}x^2 + C_1$$

$$= 0 \quad \text{for} \quad x = l; \qquad C_1 = 3EI/2l$$

$$EIv = -\frac{EI}{2l^3}x^3 + \frac{3EI}{2l}x + C_2$$

$$= 0 \quad \text{for} \quad x = 0; \qquad C_2 = 0$$

Hence

$$v = -\frac{x^3}{2l^3} + \frac{3x}{2l}$$

$$= \frac{1}{2}(3k - k^3) \qquad 0 \leqslant k \leqslant 1$$

where $k = x/l$.

Now for M_B we introduce a moment release and then impose unit rotation internally in the beam. The spans are equal, so the slope of each span at B will be 1/2. The moment to produce this we know from standard results to be $(3EI/l)(1/2)$. The end reactions are therefore $-(3/2)(EI/l^2)$. Thus

$$M_z = \frac{3EI}{2l^2}x = -EI\frac{d^2v}{dx^2}$$

Therefore

$$EI\frac{dv}{dx} = -\frac{3EI}{4l^2}x^2 + C_1$$

$$EIv = -\frac{EI}{4l^2}x^3 + C_1x + C_2$$

$$= 0 \quad \text{for} \quad x = 0; \qquad C_2 = 0$$

$$= 0 \quad \text{for} \quad x = l; \qquad C_1 = EI/4$$

Hence

$$EIv = -\frac{EI}{4l^2}x^3 + \frac{EI}{4}x$$

$$v = \frac{x}{4l^2}(l^2 - x^2) \qquad 0 \leqslant x \leqslant l$$

which is the equation to the influence line.

9.4 Maximum effects and envelope diagrams

In section 9.2 we developed a rule to help in positioning loads on a structure to produce the maximum effect. This rule can be used in all cases where the influence line is a simple triangle whatever it is representing. In other cases we can always approach the maximum condition by successive trials. In structural design we frequently need to know the maximum values of say bending moment and shear force at all sections of a beam, for example. In these circumstances, it is convenient to use an 'envelope' diagram. A maximum bending moment envelope is a diagram which shows the maximum value of bending moment which can occur at *every* section of the beam. The envelope is not an influence line, since it relates to actual loads and not a single unit load. Neither is it a bending moment diagram in the usual sense, since the load positions may change from point to point in determining maximum bending moments. The ordinate to an envelope at a point is the maximum value of the effect which can be produced at that point. The load locations will have to be known to compute the ordinate but the ordinate can be used in design without knowing the load positions.

As an illustration of the construction of a maximum bending moment envelope, we shall examine a simply-supported beam AB as in Fig. 9.9(a), carrying a pair of concentrated loads W_1 and

Figure 9.9 Envelope of maximum bending moment

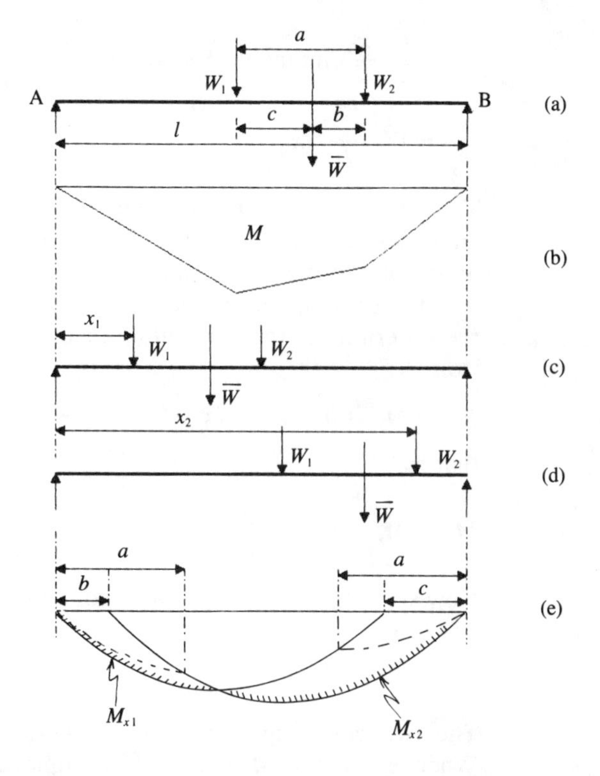

W_2 at a fixed distance apart *a*. Figure 9.9(b) is a typical bending moment diagram and we can deduce that the maximum bending moment at any point in the beam will occur when one of the loads is acting at the point. We simply have to determine *which* load. Suppose load W_1 is at a section distant x_1 from A; the bending moment at x_1 will be

$$M_{x1} = W_1 \frac{l - x_1}{l} x_1 + W_2 \frac{l - x_1 - a}{l} x_1$$

$$= \frac{x_1}{l} \{W_1(l - x_1) + W_2(l - x_1 - a)\}$$

$$= \frac{x_1}{l}\{\text{moment of loads about B}\}$$

$$= \bar{W}\frac{x_1}{l}(l - x_1 - c) \qquad [9.10]$$

where $\bar{W} = W_1 + W_2$ is the resultant load acting at c from W_1 and b from W_2.

Now suppose W_2 to be at a section distant x_2 from A. The bending moment at x_2 is

$$M_{x2} = W_1\frac{x_2 - a}{l}(l - x_2) + W_2\frac{x_2}{l}(l - x_2)$$

$$= \frac{l - x_2}{l}\{\text{moment of loads about A}\}$$

$$= \bar{W}\frac{l - x_2}{l}(x_2 - b) \qquad [9.11]$$

The curves represented by eqns [9.10] and [9.11] are parabolic and are shown in Fig. 9.9(e). The envelope of maximum bending moment is formed from the appropriate part of each parabola so that the envelope is the edge-shaded diagram. We can establish the criterion for the controlling load as follows, at any section x from A:

$$M_{x2} > M_{x1} \quad \text{if} \quad (l - x)(x - b) > x(l - x - c)$$

that is,

$$M_{x2} > M_{x1} \quad \text{if} \quad \frac{x}{l} > \frac{b}{a} \qquad [9.12]$$

Now since $b/a = W_1/(W_1 + W_2)$, condition [9.12] becomes

$$M_{x2} > M_{x1} \quad \text{if} \quad \frac{x}{l} > \frac{W_1}{W_1 + W_2} \qquad [9.13]$$

The two curves are valid only if both loads are on the span. When $x < a$, the M_{x2} parabola is replaced by the dashed curve from $x = 0$ to $x = a$. Similarly, if $x > l - a$ the M_{x1} parabola is replaced by the dashed line in the range $x = l - a$ to $x = l$. This consideration does not however affect the maximum bending moment envelope since it is clear that, for the maximum moment at any point, both loads are on the span, reversed in position if necessary.

Similar considerations can lead to envelope diagrams for maximum shear force in a beam, as in Fig. 9.10(a). If x is a typical point in the beam, then the influence line for shear force at x is given at (b). Now if we consider two concentrated loads W_1 and W_2 at a fixed distance apart a, then for maximum positive shear force at x we can place W_1 or W_2 just to the right of x. If W_1

Figure 9.10 Envelopes of maximum shear force

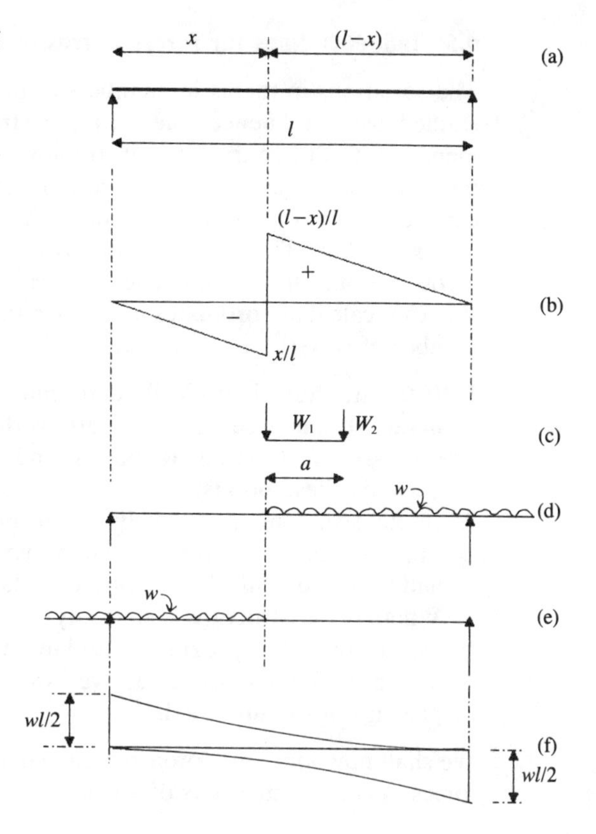

is just to the right of x then both loads are on the positive part of the influence line (unless $x > l - a$). We can establish the criterion for W_2 just to the right of x giving greater shear than W_1 just to the right of x as follows. If the loads are positioned as at diagram (c) then a movement to the left will increase the shear at x by $(W_1 + W_2)a/l$ and decrease the shear by W_1. Hence W_2 at x gives greater shear than W_1 if

$$(W_1 + W_2)\frac{a}{l} > W_1$$

This rule can be extended to any number of loads but becomes rather clumsy. With concentrated loads, load locations for maximum shear can usually be determined by inspection with a few trials.

If the beam carries a uniformly distributed load such that the whole or part of the span can be loaded, then loading as in Fig. 9.10(d) will produce maximum positive shear at x and loading as in (e) will produce maximum negative shear. The envelope of maximum shear is then diagram (f).

9.5 Influence lines for forces in trusses

The variations in forces in members of pin-jointed trusses can be studied using influence lines. In the structural design of such members it is of course crucial to know whether a member is in tension or compression. Under certain circumstances, a member can be in tension under one set of loads and in compression under another. This 'reversal' of stress can occur as loads move across a span, and an influence line can show this very clearly. We can calculate ordinates and construct influence lines in a number of ways:

1. If the structure is statically determinate, it is usually a simple matter to analyse the structure with the unit load placed successively at strategic points and to draw straight lines between these points.
2. If the structure is statically indeterminate, we can use the Muller-Breslau theorem to obtain one or more influence lines and then use statical principles to obtain others.
3. Whether the structure is statically determinate or not, we can use a computer program based on the stiffness method to obtain any influence lines we want by sufficient separate placings of the unit load.

We shall now look at approach 1 and construct influence lines for forces in certain members of a truss.

Example 9.5

The pin-jointed truss in Fig. 9.11 is a 'Warren' girder. It has to carry a uniformly distributed dead load of 10 kN/m and a uniformly distributed live load of 40 kN/m. The live load is longer than the span but either end may be located within the span. We wish to investigate the forces in members 1, 2 and 3. The quickest way to obtain these forces is to consider the truss to be cut by section X–X in Fig. 9.11(a). To find the influence line for P_1, take moments about D to the left or the right. We find that P_1 is compressive for all placings of the unit load and has a maximum value when the unit load is at D:

$$P_{1\,\max} = R_A \frac{8}{2\sqrt{3}} = R_B \times \frac{12}{2\sqrt{3}}$$

$$= \frac{3}{5} \times \frac{8}{2\sqrt{3}} = \frac{12}{5\sqrt{3}} \qquad \text{(compression)}$$

The influence line for P_1 is shown at (b). For P_2 we consider vertical equilibrium of the part of the truss either to the right or to the left of section X–X. If the unit load acts at C then, considering vertical equilibrium of the right-hand part of the

Figure 9.11 Example 9.5: influence lines for forces in a truss

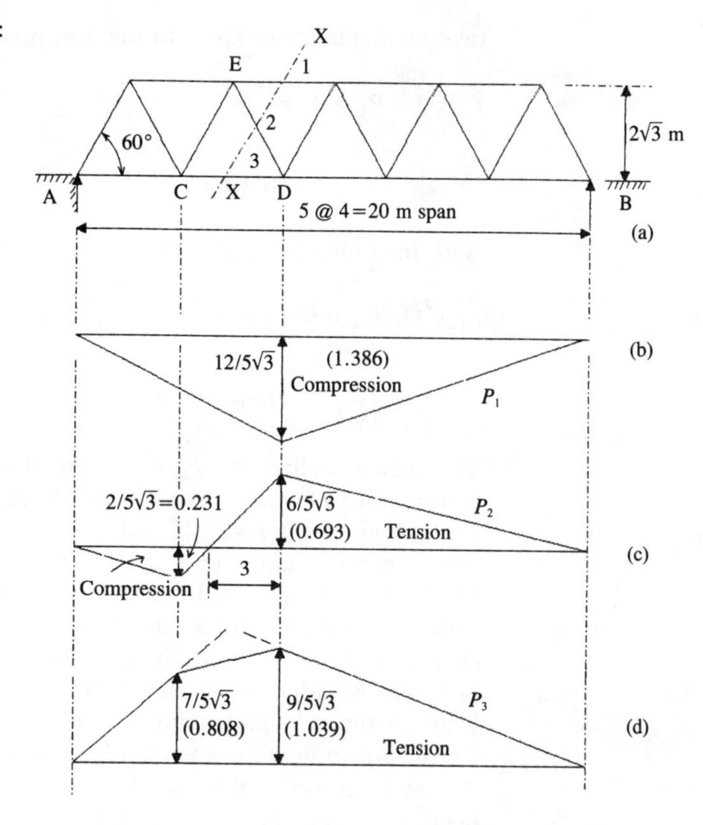

truss,

$$P_2\frac{\sqrt{3}}{2} = R_B = \frac{1}{5}$$

$$P_2 = \frac{2}{5\sqrt{3}} \qquad \text{(compression)}$$

Consideration of the direction of the P_2 force on the right-hand part of the truss requires a downwards arrow at section X–X; thus the member is in compression. Now place the unit load at D and consider vertical equilibrium of the part to the left of X–X:

$$P_2\frac{\sqrt{3}}{2} = R_A = \frac{3}{5}$$

$$P_2 = \frac{6}{5\sqrt{3}} \qquad \text{(tension)}$$

Here the P_2 force must be downwards on the part to the left of X–X, and P_2 is therefore tensile. Thus we see a change of sign in this force as the unit load crosses the span. The influence line for P_2 is shown at (c). For P_3 we again consider section X–X and take moments about E. With the unit load at C,

$$P_3 2\sqrt{3} = R_B \times 14 = \frac{1}{5} \times 14$$

$$P_3 = \frac{7}{5\sqrt{3}} \qquad \text{(tension)}$$

With the unit load at D,

$$P_3 2\sqrt{3} = R_A \times 6 = \frac{3}{5} \times 6$$

$$P_3 = \frac{9}{5\sqrt{3}} \qquad \text{(tension)}$$

The influence line for P_3 is shown at (d) and reveals a new feature not so far met. In using section X–X to find the force P_3, we take moments about E and would normally get a maximum value when the unit load is at E, just as we got a maximum value for P_1 when the unit load was at D. However, since this is what is called a 'through' girder, the load cannot be placed at E and clearly a linear relationship must exist for the influence line between the adjacent ordinates at C and D. This consideration leads to the rule that influence lines for forces in members of trusses are straight between loaded panel points.

Now to consider the actual loads. With distributed loads we need to know the areas under the influence lines. P_1 is in compression for all loads, so we apply dead load plus live load to the whole span and get

$$P_{1\,\max} = (10 + 40) \times \frac{1}{2} \times 20 \times \frac{12}{5\sqrt{3}}$$

$$= 692.8 \text{ kN} \qquad \text{(compression)}$$

For P_2 (maximum tension) we load the whole span with the dead load and only the tension part of the influence line with the live load. The tension area is $3\sqrt{3}$ and the compression area is $-1/\sqrt{3}$. Thus

$$P_{2\,\max}(\text{tension}) = 10(-1/\sqrt{3}) + (10 + 40)(3\sqrt{3})$$

$$= 254 \text{ kN} \qquad \text{(tension)}$$

For $P_{2\,\max}$(compression) we put the live load on the compression part of the influence line only and get

$$P_{2\,\max}(\text{compression}) = 50(-1/\sqrt{3}) + 10(3\sqrt{3})$$

$$= 23.1 \text{ kN} \qquad \text{(tension)}$$

We see that with this applied loading, member 2 does not suffer a reversal of stress; the actual range of force is from 23.1 kN (tension) to 254 kN (tension).

For $P_{3\,max}$ we obtain

$$P_{3\,max} = (10 + 40) \times \text{area under influence line}$$
$$= 50 \times 11.54 = 577\text{ kN} \quad \text{(tension)}$$

Example 9.6

In order to get a comparison with the statically determinate Warren girder of example 9.5, we retain the geometry of the truss but double the length and support it at three points as in Fig. 9.12(a). This truss is now statically indeterminate. The simplest way to obtain influence lines is to adopt a repeated analysis procedure with unit loads placed successively at panel points in the lower member. Using the program PLATRUSS from Part III we can obtain influence lines for P_1, P_2 and P_3 as shown at Fig. 9.12(b), (c) and (d). Now we can make some interesting observations on these results:

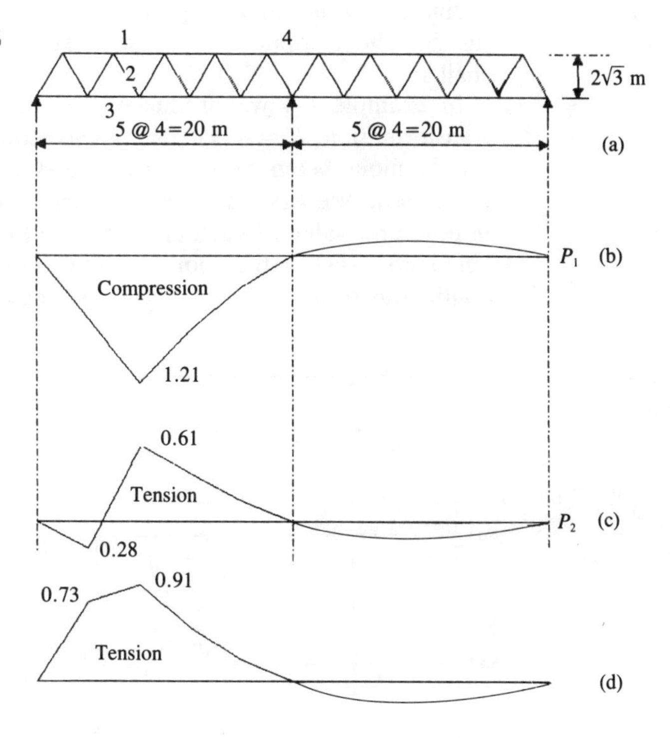

Figure 9.12 Example 9.6

1. The influence lines in the left-hand span have a similar shape to those for the single-span statically determinate truss; compare (b), (c) and (d) in Figs 9.11 and 9.12.
2. The ordinates to the influence lines in Fig. 9.12 are reduced, but not greatly, from those in Fig. 9.11.

3. In each case the influence line in the right-hand span shows a change of sign, which would have the effect of further reducing the forces in the members when a distributed load covers both spans.
4. The removal of member 4 from Fig. 9.12 would make the structure statically determinate and composed of two independent spans.

9.6 Influence lines for statically indeterminate beams

In example 4.2 we illustrated the matrix flexibility method on a five-span continuous beam and applied unit load at sufficient locations on the beam to allow us to draw influence lines for the statically indeterminate moments over the interior supports. In example 4.4 we solved the same problem using the stiffness method. Here we give some further thought to the use of the stiffness method in these circumstances and take a more direct and general approach. The determination of influence line ordinates by multiple application of unit load almost always justifies the computation of an inverse stiffness matrix, as we shall see.

In example 4.4 we produced the stiffness matrix by matrix transformation, $\boldsymbol{K} = \boldsymbol{A}^{\mathrm{T}}\boldsymbol{kA}$. With a structure as uncomplicated as a continuous beam we can form the stiffness matrix directly by inspection. We first identify the degrees of freedom, and then impose unit value of each and evaluate the corresponding forces (the stiffnesses) at the coordinates. Consider the general n-span continuous beam shown in Fig. 9.13. The degrees of freedom are

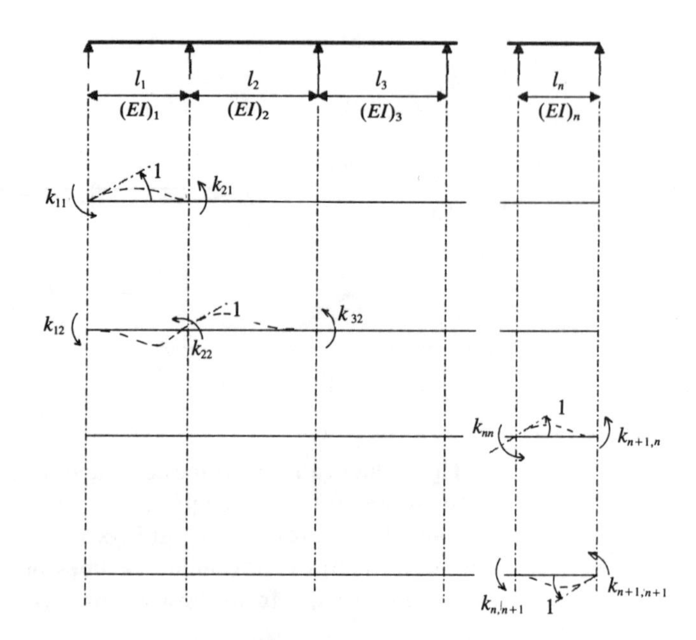

Figure 9.13 n-span continuous beam

the rotations at the supports and we include all rotations, although we could exclude the rotations at the end supports as we showed in example 4.4. By imposing unit rotation in turn at each support as in Fig. 9.13 and defining the individual stiffnesses consistent with these rotations, we can write down the stiffness matrix as

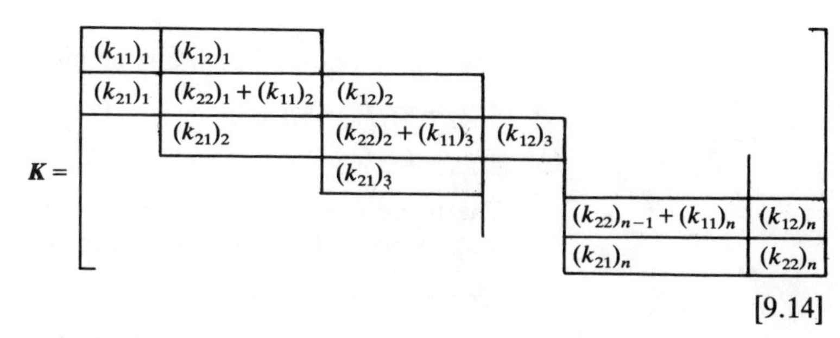

[9.14]

This is a stiffness matrix with a particularly simple structure called 'tridiagonal', and special methods exist for rapid solution or inversion. Now for prismatic spans the stiffnesses are standard:

$$(k_{11})_i = 4(EI/l)_i = (k_{22})_i$$
$$(k_{12})_i = (k_{21})_i = 2(EI/l)_i$$

For non-prismatic spans, the simplest way to compute the stiffnesses is to evaluate flexibilities using numerical integration as in eqns [8.50] and [6.145], repeated here for convenience:

$$k_{11} = \frac{f_{22}}{f_{11}f_{22} - f_{12}^2}$$
$$k_{22} = \frac{f_{11}}{f_{11}f_{22} - f_{12}^2}$$
$$k_{12} = k_{21} = \frac{-f_{12}}{f_{11}f_{22} - f_{12}^2}$$

where

$$f_{11} = \int_0^l \frac{m_1^2}{EI}\,dx$$
$$f_{22} = \int_0^l \frac{m_2^2}{EI}\,dx$$
$$f_{12} = f_{21} = \int_0^l \frac{m_1 m_2}{EI}\,dx$$

The applied loads are formed by calculating equivalent nodal forces from the 'fixed-end' effects. For prismatic beams these can be obtained from Appendix 3. For non-prismatic beams we can again use numerical integration to evaluate end rotations:

$$f_{01} = \int_0^l \frac{m_1 m_0}{EI} \mathrm{d}x$$

$$f_{02} = \int_0^l \frac{m_2 m_0}{EI} \mathrm{d}x$$

These are removed by end fixing moments:

$$M_{F12} = k_{11} f_{01} - k_{12} f_{02}$$

$$M_{F21} = k_{22} f_{02} - k_{21} f_{01}$$

The fixing moments, reversed in sign, constitute the equivalent nodal loading. The solution of the equilibrium equations

$$\boldsymbol{K}\boldsymbol{r} = \boldsymbol{R} \qquad \boldsymbol{r} = \boldsymbol{K}^{-1}\boldsymbol{R}$$

yields the rotations $\boldsymbol{r}$ from which the moments in the beam can be obtained. The fixed-end moments are added at the last stage. Since for influence line computations $\boldsymbol{R}$ will have many columns, corresponding to the number of unit load positions, it will be computationally economical to form $\boldsymbol{K}^{-1}$.

9.7 Influence lines for displacements

Influence lines can be drawn for any 'effect' caused by loading on a structure. So far we have considered influence lines for bending moment, shearing force and axial force; we now consider influence lines for displacement. The beam AB of Fig. 9.14(a) is

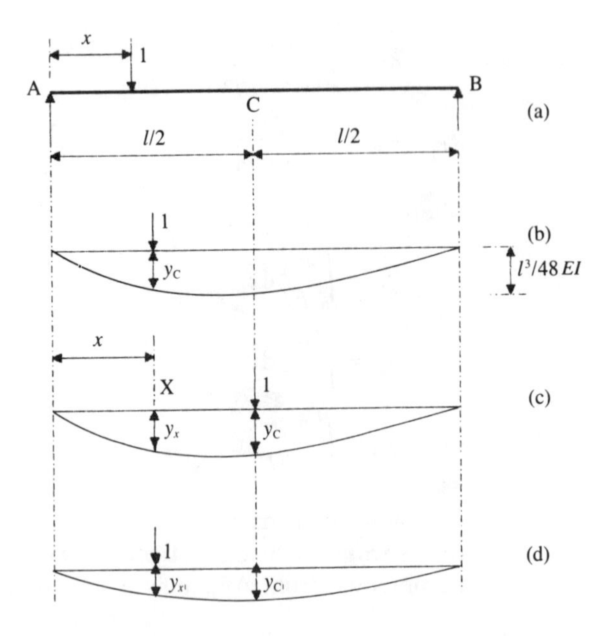

Figure 9.14 Influence line for deflection at centre of simply-supported beam

traversed by a single unit load. Now the 'direct' way to obtain the influence line for y_C, say, is to place a unit load on the beam distant x from A and derive an expression for y_C as a function of x. This can easily be shown to be

$$y_C = \frac{x}{48EI}(3l^2 - 4x^2) \qquad [9.15]$$

This represents the influence line for y_C as shown in diagram (b). An alternative approach is to use the Maxwell-Betti theorem, eqn [9.7]. If we place a unit load at C as in diagram (c), the deflection at C is y_C and that at x is y_x. Now, if we move the unit load to X as in diagram (d), then applying the Maxwell-Betti theorem we obtain

$$(1\text{: at C})(y'_C) = (1\text{: at X})(y_x)$$

Thus the deflection at C for unit load in any position x is given by y_x, which is the deflected shape of the axis of the beam due to unit load at C. Thus we see again that (b) is the influence line for y_C, and we note that the ordinate at C ($x = l/2$) is $l^3/48EI$, which agrees with the standard deflection from Appendix 2.

9.8 An introduction to influence surfaces

We have seen how influence lines can represent effects in structures where the position of the unit load is represented by a single independent variable. In the case of a structure like a plate or shell we would need two independent variables to represent the unit load position. In these circumstances we obtain an influence 'surface' to represent the effect in question.

We now give a brief introduction to this topic by constructing an influence surface for central deflection in a square plate of isotropic material, simply-supported on all four sides, as shown in Fig. 9.15(a). Although we could adopt a 'closed form' solution (Timoshenko 1959) for the central deflection in the plate, we shall nevertheless use a numerical computation which analyses the plate deflections for successive positions of the unit load. A useful mathematical tool for many plate problems is the finite difference method (Ghali, Neville 1978). Plate bending is governed by the biharmonic equation

$$\frac{\partial^4 v}{\partial x^4} + \frac{2\,\partial^4 v}{\partial x^2\,\partial z^2} + \frac{\partial^4 v}{\partial z^4} = \frac{q}{D}$$

or

$$\nabla^4 v = \frac{q}{D} \qquad [9.16]$$

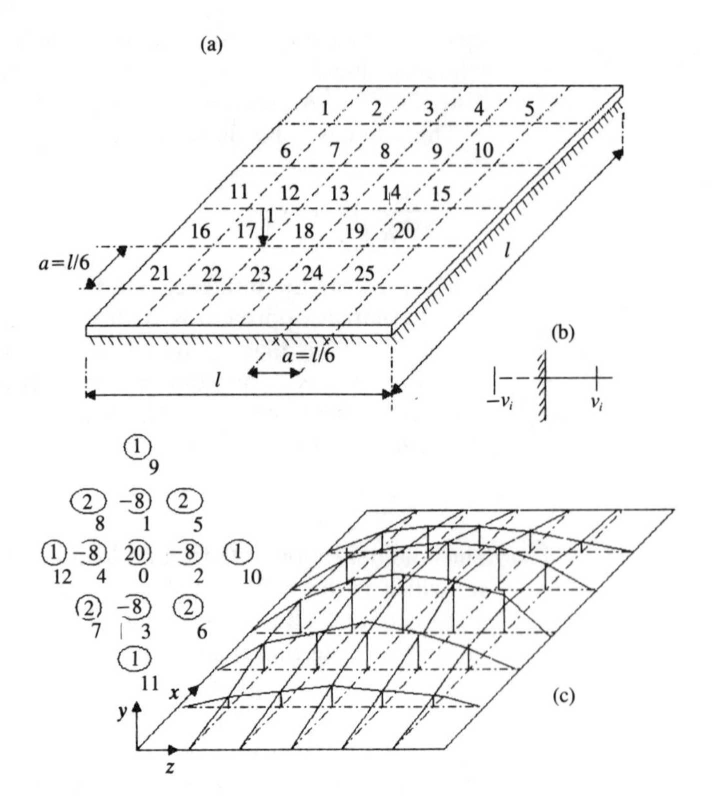

Figure 9.15 Influence surface for central deflection in simply-supported square plate

where
q is the intensity of loading
D is the flexural rigidity of unit width of the plate: $D = Et^3/12(1-v^2)$
t is the thickness of the plate
v is the Poisson's ratio

The finite difference approximation to eqn [9.16] is given in terms of coefficients of displacements v of surrounding nodes by the pattern shown in Fig. 9.15. In this case the boundary condition is the simple one shown in (b). We shall divide the $l \times l$ plate into a 6×6 square mesh having therefore a 5×5 mesh of internal nodes with non-zero deflections. The nodes are numbered 1 to 25 as shown in Fig. 9.15(a).

Now we can write down an equation for each of the nodes 1 to 25 using the finite difference approximation and obtain a set of simultaneous equations. A typical equation is

$$(\nabla^4 v)_0 = \frac{1}{a^4}\{20v_0 - 8(v_1 + v_2 + v_3 + v_4) + 2(v_5 + v_6 + v_7 + v_8) + (v_9 + v_{10} + v_{11} + v_{12})\} = q/D \qquad [9.17]$$

Now for a single unit point load,

$$q = 1/a^2 \quad [9.18]$$

Hence the equations take the form

$$[\boldsymbol{A}][\boldsymbol{v}] = \frac{a^2}{D}[\boldsymbol{I}] \quad [9.19]$$

where $a = l/6$ and $[\boldsymbol{I}] = \boldsymbol{I}_{(25)}$. The coefficient matrix $[\boldsymbol{A}]$ is

$$[\boldsymbol{A}] = \begin{bmatrix} \boldsymbol{A} & \boldsymbol{C} & \boldsymbol{I} & \boldsymbol{O} & \boldsymbol{O} \\ \boldsymbol{C} & \boldsymbol{B} & \boldsymbol{C} & \boldsymbol{I} & \boldsymbol{O} \\ \boldsymbol{I} & \boldsymbol{C} & \boldsymbol{B} & \boldsymbol{C} & \boldsymbol{I} \\ \boldsymbol{O} & \boldsymbol{I} & \boldsymbol{C} & \boldsymbol{B} & \boldsymbol{C} \\ \boldsymbol{O} & \boldsymbol{O} & \boldsymbol{I} & \boldsymbol{C} & \boldsymbol{A} \end{bmatrix}$$

where

$$\boldsymbol{A} = \begin{bmatrix} 18 & -8 & 1 & 0 & 0 \\ -8 & 19 & -8 & 1 & 0 \\ 1 & -8 & 19 & -8 & 1 \\ 0 & 1 & -8 & 19 & -8 \\ 0 & 0 & 1 & -8 & 18 \end{bmatrix}$$

$$\boldsymbol{B} = \begin{bmatrix} 19 & -8 & 1 & 0 & 0 \\ -8 & 20 & -8 & 1 & 0 \\ 1 & -8 & 20 & -8 & 1 \\ 0 & 1 & -8 & 20 & -8 \\ 0 & 0 & 1 & -8 & 19 \end{bmatrix}$$

$$\boldsymbol{C} = \begin{bmatrix} -8 & 2 & 0 & 0 & 0 \\ 2 & -8 & 2 & 0 & 0 \\ 0 & 2 & -8 & 2 & 0 \\ 0 & 0 & 2 & -8 & 2 \\ 0 & 0 & 0 & 2 & -8 \end{bmatrix}$$

$$\boldsymbol{I} = \begin{bmatrix} 1 & 0 & 0 & 0 & 0 \\ 0 & 1 & 0 & 0 & 0 \\ 0 & 0 & 1 & 0 & 0 \\ 0 & 0 & 0 & 1 & 0 \\ 0 & 0 & 0 & 0 & 1 \end{bmatrix}$$

The solution of eqn [9.19] is

$$\boldsymbol{v}_{(25\times25)} = \left(\frac{l}{6}\right)^2 \frac{1}{D}[\boldsymbol{A}]^{-1}$$

and the values of v obtained constitute the ordinates ($\times l^2/36D$) to the influence surface for central deflection, as shown in Fig. 9.15(c).

Exercises

9.1 A uniform cantilever beam of length l is fixed at A and propped at B. If the propping force is R, show that the influence line for R is

$$R = \frac{1}{2}(2 - 3k + k^3)$$

where $k = x/l$ and x is the distance from end B.

9.2 A two-span continuous beam ABC of total length l is provided with vertical support at the ends and at the centre B. Show that the influence line for vertical reaction at B is

$$R_B = 3k - 4k^3$$

where $k = x/l$ and x is the distance from end A.

9.3 A beam of length l is simply-supported at its ends and carries a uniformly distributed load of intensity w and length d ($d < l$). Show that the envelope of maximum bending moment is given by

$$M_{max} = w\,dy\frac{l - y}{l}\left(1 - \frac{d}{2l}\right)$$

where y is measured from either end.

9.4 For the cantilever and suspended span arrangement of Fig. 5.19, construct influence lines for R_A, R_B, M_B and S_C.

CHAPTER

10

Plastic collapse in structures

10.1 Introduction

In Chapter 6 we derived and applied the governing equations for bending of beams. Equation [6.27] gave us the bending stress σ_x at any point (z, y) in a beam cross-section where the rectangular Z and Y axes were principal axes. If we confine our attention to situations where the applied bending moment is only about the Z axis, i.e. $M_y = 0$, then eqn [6.27] becomes

$$\sigma_x = \frac{M_z y}{I_z} \qquad [10.1]$$

If eqn [10.1] is being used in the design context, it is likely that a maximum, design, value of stress σ_x will be specified. In these circumstances it is convenient to use the term 'moment of resistance' M_R, where

$$M_R = \sigma_x I_z / y_{max} \qquad [10.2]$$

$$= \sigma_x Z \qquad [10.3]$$

where Z is the 'section modulus' and σ_x is understood to be the maximum stress for design purposes.

Structural design is now generally carried out according to the 'limit state' concept, in which limits are defined at which the structure is deemed to be unfit for its intended use. These limits usually come under the categories of 'ultimate' and 'serviceability', and the ultimate limit state is now of immediate concern to us in this chapter.

The principal ultimate limit state is that of 'strength', which includes plastic yielding of steel. In Fig. 10.1 we have the idealized stress–strain relationship for a structural steel. The behaviour of the steel is considered to be elastic up to a stress σ_y, the 'yield' stress at which the corresponding strain is ε_y. The slope of the line in this region is Young's modulus E. Values of σ_y for structural steels range from about $250\,N/mm^2$ to $450\,N/mm^2$ depending on the grade of steel used. The strain at

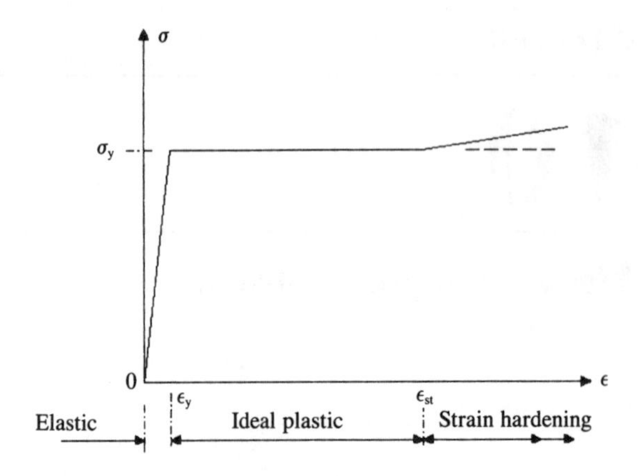

Figure 10.1 Idealized stress–strain relationship for structural steel

onset of yield ε_y ranges from 0.001 to 0.002. The steel can now continue to strain at no increase in stress until a level of strain ε_{st} is reached; thereafter stress increases with increasing strain up to fracture. This last stage is called 'strain hardening' and is usually neglected when applying the 'simple' plastic theory.

We now apply this idealized stress–strain relationship to the bending of the I-section beam shown in Fig. 10.2(a). Under

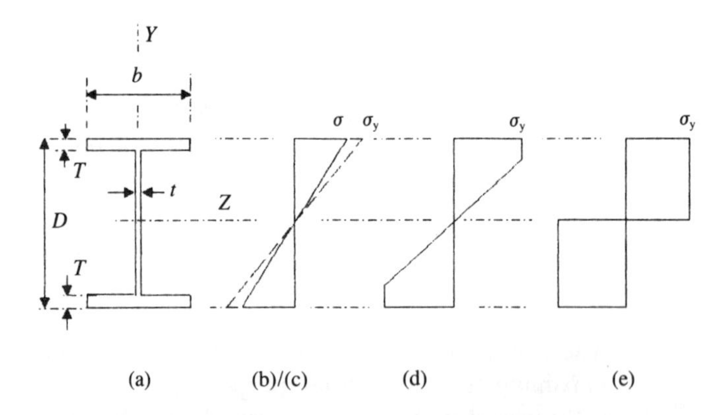

Figure 10.2 Progressive bending of I-section beam into full plasticity

linear elastic conditions ($\sigma < \sigma_y$) the stress distribution over the cross-section is given by eqn [10.1] as a linear function of y. This distribution holds until σ reaches σ_y as in diagram (c). An increase in applied bending moment will not result in an increase in stress at the top and bottom of the cross-section because of the stress–strain relationship adopted. However, yield will penetrate into the flanges until the full depth of each flange is plastic; this state corresponds to diagram (d). Continuing the process, a further increase in applied bending moment will cause plastic yielding to penetrate to the neutral axis. At this stage the beam cross-section is fully plastic and is at the ultimate limit state as far as strength is concerned. We can obtain the moment of resistance in this condition by adding the contributions of flanges and web as follows. For each flange,

$$M_z = bT\sigma_y\left(\frac{D}{2} - \frac{T}{2}\right) = \frac{bT}{2}(D - T)\sigma_y$$

and for the web,

$$M_z = 2t\frac{D - 2T}{2}\sigma_y\frac{D - 2T}{4}$$

$$= \frac{t}{4}(D - 2T)^2\sigma_y$$

Thus the total moment of resistance is

$$M_P = \left[bT(D - T) + \frac{t}{4}(D - 2T)^2\right]\sigma_y \quad [10.4]$$

$$= S\sigma_y \quad [10.5]$$

where S is the 'plastic' modulus of the cross-section. The bending moment M_P is the 'plastic moment of resistance'. Now, if the bracketed term in eqn [10.4] constituting S is examined, we find that

S = first moment of area about the Z axis [10.6]

Actually eqn [10.6] should read, in the general case where we may not have symmetry about the Z axis,

S = first moment of area about the equal area axis [10.7]

since the criterion for locating the appropriate axis is that the resultant thrust on the cross-section must be zero; thus

σ_y (area in tension) = σ_y (area in compression)

The particular I-section we have examined is symmetrical about the Z axis, so the Z axis is the relevant 'equal area' axis.

Example 10.1

Determine the plastic moduli S of the cross-sections shown at (a) and (b) in Fig. 10.3.

For the T-section in diagram (a), the equal area axis is clearly at the underside of the flange. Hence

S = first moment of area about equal area axis

$= 100 \times 20 \times 10 + 100 \times 20 \times 50$

$= 120 \text{ cm}^3$

For the I-section shown at (b), ignoring any fillets and treating

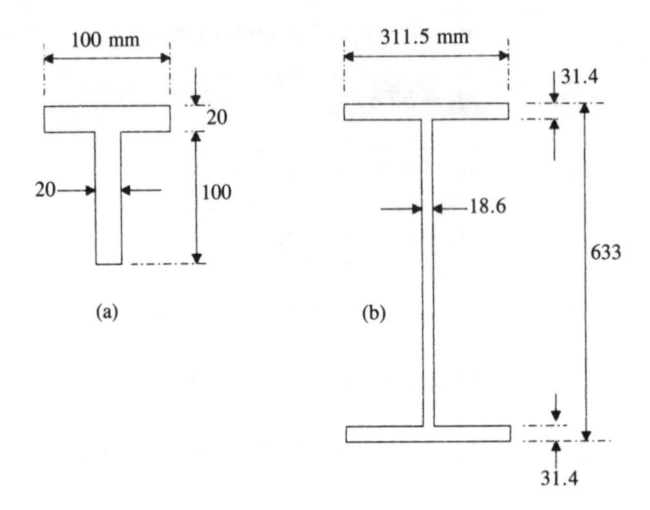

Figure 10.3 Example 10.1

the cross-section as composed of three rectangles,

$$S = 2 \times 311.5 \times 31.4 \times 300.8 + 2 \times \frac{570.2}{2} \times 18.6 \times \frac{570.2}{4}$$

$$= 7396 \text{ cm}^3$$

The ratio of plastic modulus to elastic (section) modulus S/Z is called the 'shape factor'. If we now consider the design of a simply-supported beam, purely from the strength point of view, and assume adequate lateral support to the compression flange, then if the beam has the cross-section shown in Fig. 10.3(b) and $\sigma_y = 265 \text{ N/mm}^2$ the ultimate moment of resistance is

$$M_P = \sigma_y S = 265 \times 7396 \times 10^{-3}$$

$$= 1960 \text{ kN m}$$

If the beam carries a uniformly distributed load of intensity w then the maximum bending moment at the mid-span is $wl^2/8$, and so the maximum (ultimate) load will be

$$w_{ult} = (8/l^2)1960$$

If this load is applied to the beam a 'plastic hinge' will be developed at mid-span and no additional load will be accepted by the beam. The beam is said to have 'collapsed' at the ultimate load w_{ult}.

This simple approach, subject to certain safeguards, forms the basis of the design of steel members in bending in BS 5950. Concrete members in bending can behave in a similar way and develop 'plastic' hinges; however, the behaviour is more complex than that of steel members, and particular attention needs to be paid to the capacity for plastic rotation at the location of the hinge.

10.2 Plastic moment redistribution

In continuous construction the attainment of the ultimate moment M_P at a section does not mean that additional load cannot be carried. This phenomenon is called 'moment re-distribution', and we shall now examine how it occurs.

Consider the propped cantilever beam ACB shown in Fig. 10.4(a). Under purely elastic conditions, the bending moment

Figure 10.4 Moment redistribution in propped cantilever

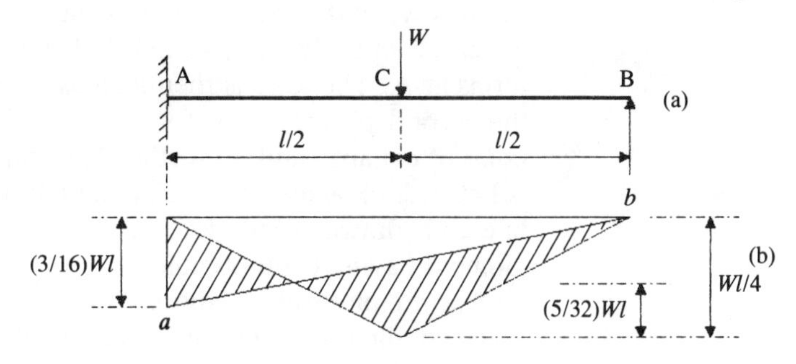

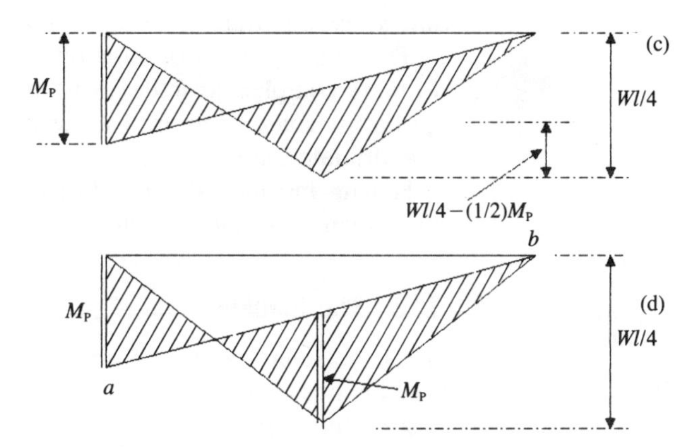

diagram is shown at (b), the maximum bending moment being at the fixed end A:

$$M_A = (3/16)Wl$$

This section of the beam will be the first to reach the fully plastic moment of resistance M_P. It will do so when load W is

$$W_1 = (16/3)(M_p/l) = 5.33M_P/l$$

The corresponding value of the bending moment at C is, from diagram (c),

$$M_C = W_1 l/4 - (1/2)M_P = (5/6)M_p$$

Since $M_C < M_P$, the beam is capable of supporting further load, the additional load being carried as if the beam were simply-supported at A since the moment at A cannot be increased. The maximum amount of additional load is such as to increase M_C until it also reaches M_P. We then have the circumstances shown at (d), where M_P has been reached at A and C. If the corresponding load is W_2 then

$$W_2 l/4 = M_P + (1/2)M_P$$
$$W_2 = 6M_P/l$$

Let us review the sequence of events. The load W is applied and increased until $W = W_1 = 5.33M_P/l$, when the first plastic hinge forms at A. The load is then increased to $W = W_2 = 6M_P/l$, when the second plastic hinge forms at C. The beam is then on the point of collapse and no further load can be carried.

Let us look again at the bending moment diagrams. At (b) the line *ab* is drawn from a knowledge of the end moment at A, this moment having been obtained from elastic theory which is based on known physical conditions of continuity in the beam. The corresponding line *ab* in diagram (d) is drawn on different (non-elastic) principles, using the geometry of the diagram determined by the equality of the moments M_P at A and C. The line *ab* is sometimes called the 'reactant line'.

This is the process of moment redistribution in plastic analysis. In more complex structures with more members the process may go on for some time until a mechanism has been produced and the ultimate load is reached.

Having introduced the subject, it is now time to look at the fundamentals of plastic behaviour and state some basic theorems.

10.3 The fundamentals of plastic behaviour

We have seen that the plastic analysis of a continuous structure can be posed as a problem of arriving at a bending moment distribution which contains sufficient plastic hinges ($M = M_P$) to constitute a mechanism. The mechanism may involve the whole structure or, quite often, only part of it. The question is, how do we arrive at this bending moment distribution? Or, having postulated one, how do we check it? Three fundamental conditions need to be satisfied:

1. A plastic collapse mechanism must have formed ('mechanism' condition).
2. The bending moment diagram must represent an equilibrium state of the structure ('equilibrium' condition).
3. The bending moment diagram must not contain any moment greater than the plastic moment of resistance M_P (the 'yield' condition).

These three conditions are embodied in the 'uniqueness theorem', which states that if the three conditions are met then the collapse load factor is uniquely determined. In practice it is rarely possible to reach a solution directly, and two fundamental theorems are available to help us to reach solutions by indirect means. These are the maximum and minimum principles, which we will now examine.

If we were to postulate a collapse mechanism and, satisfying all equilibrium conditions, to arrive at a load factor λ, then if the yield condition is also satisfied we have found the collapse load factor λ_c. If the yield condition is violated then there is somewhere in the structure a moment greater than M_P, and thus we must conclude that there is another mechanism at a lower load factor. The 'minimum' principle states that the collapse load factor is the minimum one which exists considering all possible mechanisms. Methods based on the use of this principle are comparatively easy to set up. However, it follows that any incorrect mechanism will give a load factor which is too high, indicating that the structure is stronger than it really is. Care is needed in studying the likely mechanisms in any given situation to arrive at, or sufficiently near to, the collapse load factor. Under certain circumstances, the bending moment diagram for a particular mechanism can be drawn and the yield condition checked directly. This is possible when the introduction of plastic hinges renders the structure statically determinate. Load factors produced by postulating a collapse mechanism are *upper* bounds on the collapse load factor, i.e.

$$\lambda \geqslant \lambda_c$$

by the minimum principle.

The converse principle is the 'maximum' principle. If we steadily increase the load on a structure so that the equilibrium and yield conditions are constantly satisfied, then at any stage the load factor λ will be less than λ_c, except that when we have increased the load level to a point where a mechanism forms then

$$\lambda = \lambda_c$$

by the maximum principle. Load factors obtained in this way are *lower* bounds on the collapse load factor. Methods based on the lower bound approach tend to be more complex in application than those based on the minimum principle. However, the two principles can be combined very effectively to establish both an upper and a lower bound. The gap between these two can then be narrowed until the difference is acceptable in the design context.

10.4 Plastic analysis of continuous beams

We shall study the plastic analysis of continuous beams using the reactant bending moment diagram approach already introduced in section 10.2, and also a method based on the principle of virtual work.

We look first of all at a simple two-span beam ABC carrying a uniformly distributed load w, as in Fig. 10.5. Owing to the

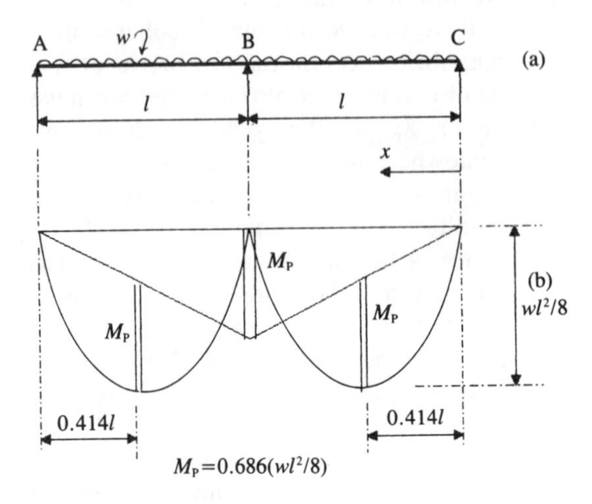

Figure 10.5 Plastic hinges in two-span beam

symmetry of the structure and the applied load we must get a symmetrical collapse mode with a plastic hinge at B and one in each span. We shall show that the plastic hinges are not located at the centre of each span, but close to it. Suppose the load w is such as to cause plastic collapse; then the reactant line is drawn on the free bending moment diagram in such a way that three plastic hinges exist where the bending moment is M_P. The yield condition requires that the span hinges will exist where the distance from the parabola to the reactant line is maximum whilst retaining equality between the M_P moments. This gives us a method of approaching the geometry of the diagram, and indeed we could proceed by trial and error on a scale drawing of the free bending moment diagram. It will, however, be more satisfactory to produce a solution analytically and at the same time to give a standard result which we can use for other problems.

Now at x from C, the free bending moment is $(w/2)x(l-x)$. Hence

$$M_x = \frac{w}{2}x(l-x) - \frac{M_P}{l}x \qquad [10.8]$$

To find $M_{x\,\max}$ between C and B, we take

$$\frac{dM_x}{dx} = -\frac{w}{2}x + \frac{w}{2}(l-x) - \frac{M_P}{l} \qquad [10.9]$$

This is zero for

$$x = \frac{l}{2} - \frac{M_P}{wl} \qquad [10.10]$$

or

$$M_P = \frac{wl^2}{2} - wlx \qquad [10.11]$$

Putting $M_x = M_P$ in eqn [10.8],

$$M_P\left(1 + \frac{x}{l}\right) = \frac{w}{2}x(l - x) \qquad [10.12]$$

Substituting for M_P from eqn [10.11],

$$x^2 + 2xl - l^2 = 0 \qquad [10.13]$$

from which

$$x = l[\sqrt{(2)} - 1]$$
$$\simeq 0.414l \qquad [10.14]$$

Substituting this value of x in eqn [10.11],

$$M_P = 4(3 - 2\sqrt{2})(wl^2/8)$$
$$\simeq 0.686(wl^2/8) \qquad [10.15]$$

The reader will note of course that we have incidentally also solved the problem of the propped cantilever, and might like to check that if, as an approximation, the span hinges are assumed to be at mid-span, then eqn [10.15]

$$M_P = 0.667(wl^2/8) \qquad [10.16]$$

which is in error by less than 3 per cent.

Example 10.2

We now examine another two-span beam, but this time we do not have the advantage of symmetry of the spans. In Fig. 10.6 beam ABC carries a uniformly distributed load on span AB and a centrally placed concentrated load on span BC. The loads have already been factored, and we wish to find suitable steel sections for the two spans. The maximum free bending moments are
In AB:

$52.5 \times 8^2/8 = 420$ kN m

In BC:

$202 \times 5/4 = 253$ kN m

Consider the collapse of each span separately. For span AB with a plastic hinge at B and another at 0.414×8 m from A (see eqn [10.14]) we obtain from eqn [10.16]

$M_{P1} = 0.686 \times 420 = 288$ kN m

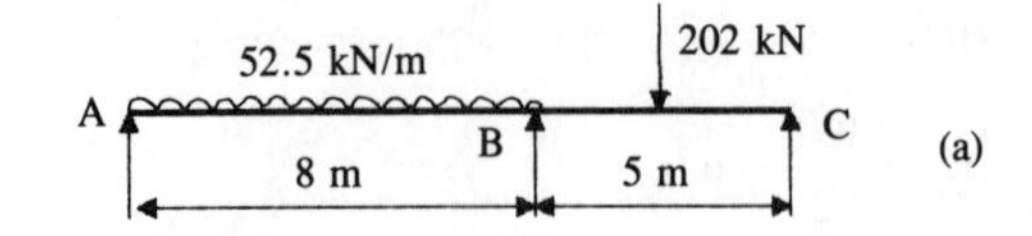

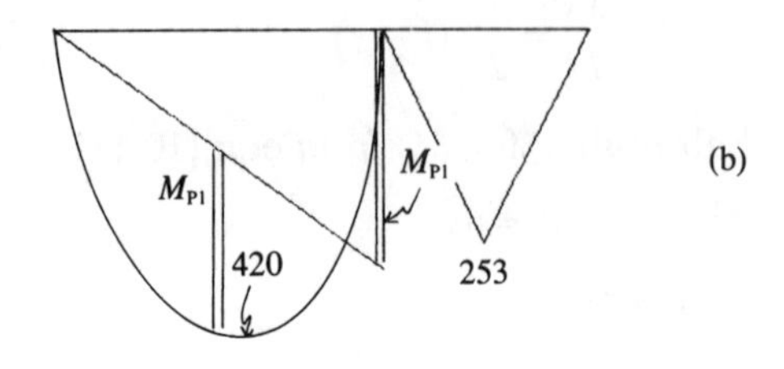

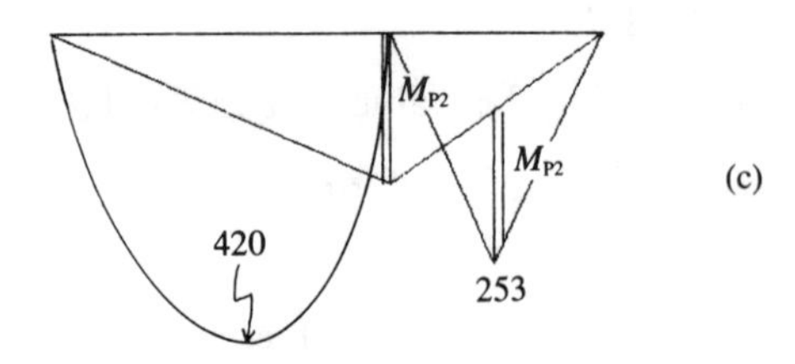

Figure 10.6 Example 10.2

For collapse in BC, the geometry of the free bending moment diagram requires the span plastic hinge to develop under the load. Hence

$$M_{P2} + (1/2)M_{P2} = 253$$
$$M_{P2} = 169 \text{ kN m}$$

The bending moment diagrams corresponding to collapse in AB and collapse in BC are shown in Fig. 10.6(b) and (c) respectively. Now, if we are to use the same section beam for both spans then clearly span AB is critical since M_{P1} is greater than M_{P2}. From steel design tables we could choose a grade 43 steel universal beam $406 \times 178 \times 54$, for which the moment capacity is 289 kN m. If we now adopt a weaker section for span BC, say $406 \times 140 \times 39$, for which the moment capacity is 198 kN m, then we shall need to redesign span AB since the plastic hinge at B will now develop in the weaker beam. The maximum moment in AB, assumed to be at mid-span, is

$$420 - (1/2)198 = 321 \text{ kN m}$$

We could use $406 \times 178 \times 60$, for which the moment capacity is 327 kN m. We see that the saving in weight of steel gained by using different sections is very marginal in these circumstances (about 4 per cent). The least-cost practical solution is probably to adopt a uniform section throughout.

10.5 Plastic analysis of frames

When we come to framed structures, although we can still use a reactant bending moment diagram approach, a method based on the principle of virtual work is usually more efficient. In Fig. 10.7

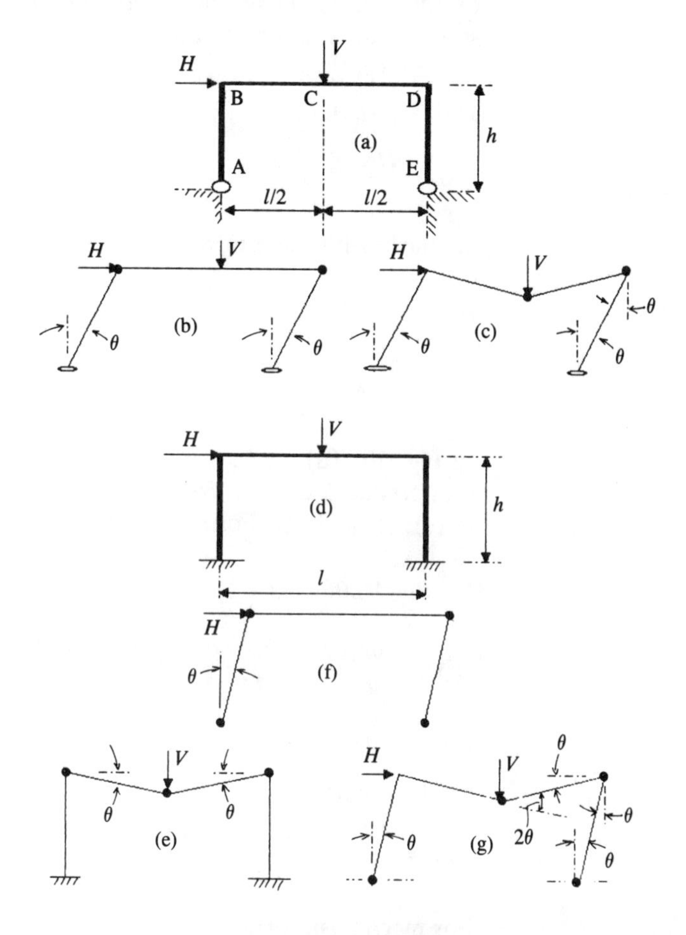

Figure 10.7 Plastic collapse of rectangular portal frames

we have a rectangular rigidly jointed frame with hinged supports in (a) and alternatively with fixed supports in (d). We shall now develop relationships between collapse loads and plastic moments of resistance of the members.

In applying the principle of virtual work we imagine that a plastic collapse mechanism has developed, forming a mechanism in the structure. Although the existence of a mechanism precludes the application of any further load, the structure can theoretically support the load which has produced the mechanism. It is said that the structure is 'on the point of collapse'. If we now imagine that a small movement takes place then we can equate the external work done by the applied loads to the internal work done in the rotations of the plastic hinges. Thus, if we consider the pure sidesway mechanism shown in Fig. 10.7(b), each column rotates through angle θ and the movement of the beam to the right is $h\theta$. The vertical load does not move in its own direction and so does no work. The external work is therefore $Hh\theta$, and the internal work is $2M_P\theta$ due to the rotation of the plastic hinges at B and D. This structure has hinged supports at A and E, so no work is done there. Equating external and internal work,

$$Hh\theta = 2M_P\theta$$

$$M_P = \frac{Hh}{2} \qquad [10.17]$$

Similarly with the alternative mechanism of Fig. 10.7(c),

$$Hh\theta + V\frac{l}{2}\theta = 2M_P 2\theta$$

$$M_P = \frac{Hh}{4} + \frac{Vl}{8} \qquad [10.18]$$

In Fig. 10.7(d) we have the same frame but with direction-fixed supports at A and E. We investigate three mechanisms as shown in (e), (f) and (g). For mechanism (e),

$$V\frac{l}{2}\theta = M_P\theta(1 + 2 + 1)$$

$$M_P = \frac{Vl}{8} \qquad [10.19]$$

For mechanism (f),

$$Hh\theta = 4M_P\theta$$

$$M_P = \frac{Hh}{4} \qquad [10.20]$$

For mechanism (g),

$$Hh\theta + V\frac{l}{2}\theta = 6M_P\theta$$

$$M_P = \frac{Hh}{6} + \frac{Vl}{12} \qquad [10.21]$$

Mechanism (g) is a 'combined' mechanism and it can be obtained by adding mechanisms (e) and (f) as in Table 10.1. Mechanism (e) could be applied to structure (a), but inspection will show that this is really covered by (c) and eqn [10.18].

In what we have just done we have tacitly assumed identical M_P values for beam and columns. If these differ, then all we have to do is to place plastic hinges at joints between beam and column in the *weaker* element.

Table 10.1 Mechanism (g) in Fig. 10.7

Mechanism	External work	Internal work				
		A	B	C	D	E
(e)	$V(l/2)\theta$	0	$M_p\theta$	$2M_p\theta$	$M_p\theta$	0
(f)	$Hh\theta$	$M_p\theta$	$-M_p\theta$	0	$M_p\theta$	$M_p\theta$
(g) = (e) + (f)	$V(l/2)\theta + Hh\theta$			$6M_p\theta$		

Example 10.3

The members of the frame shown in Fig. 10.8(a) have a plastic moment of resistance $M_P = 200$ kN m. We wish to determine the load factor against plastic collapse. We start by finding the load factor for the pure sway mechanism (this is unlikely to be critical but we should check it). Let the load factor be λ; then

$$\lambda \times 15 \times 5\theta = 2 \times 200 \times \theta$$

$$\lambda = 400/75 = 5.33$$

Figure 10.8 Example 10.3

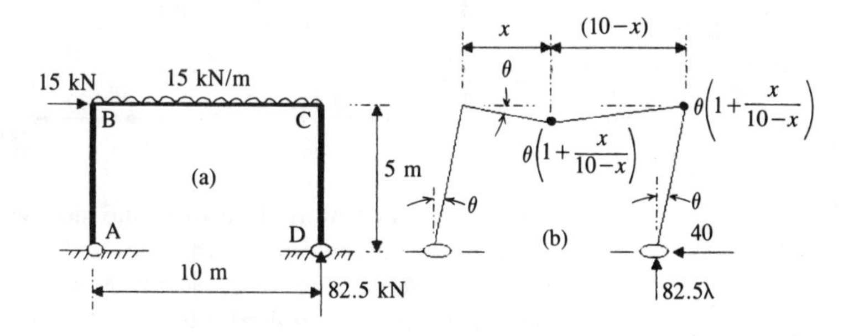

Now we carry out a quick check on the load factor for a combined mechanism as in Fig. 10.8(b) using $x = l/2$ as an approximation:

$$\lambda \times 15 \times 5\theta + \lambda \times 15 \times 10 \times (5/2)\theta = 4 \times 200 \times \theta$$

$$\lambda = 800/450 = 1.78$$

There is no doubt that this is the correct mechanism since $1.78 \ll 5.33$. However, the position of the plastic hinge in BC is not at mid-span, so we shall go through the analysis to find its actual location. The vertical reaction at D is easily found to be $82.5\,\lambda$kN and, since a plastic hinge exists at C, the horizontal reaction at D is $M_P/5 = 40$ kN. From the geometry of the collapse mechanism in diagram (b) we can determine the rotations of the plastic hinges as shown. The work equation is then

$$\lambda \times 15 \times 5\theta + \lambda \times 15 \times 10 \times \frac{x}{2}\theta = 2 \times 200 \times \theta\left(1 + \frac{x}{10 - x}\right)$$

$$\lambda = \frac{53.33}{(1 + x)(10 - x)}$$

λ is a minimum for $x = 4.5$, for which $\lambda = 1.76$. This result is very close to the estimate of 1.78 we made by assuming the plastic hinge to be at mid-span.

Example 10.4

The pitched roof frame of the type shown in Fig. 10.9 is a very suitable structure for plastic design. The most usual mechanism

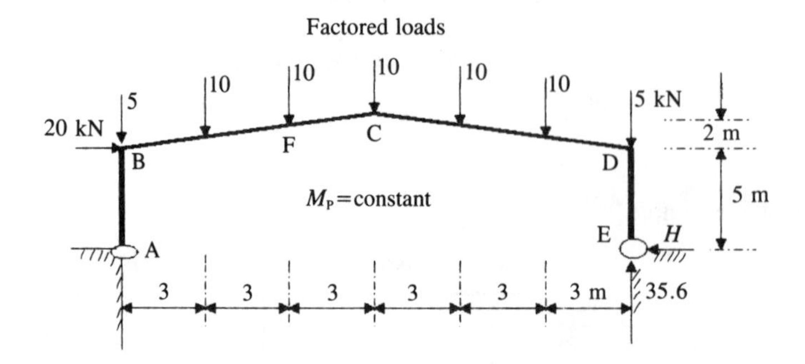

Figure 10.9 Example 10.4

at collapse is a combined vertical and sidesway mechanism with a plastic hinge at the top of the leeward column and just to the windward side of the ridge joint. Now, in reality, the vertical loads will not be applied directly to the rafters but through purlins at fixed distances apart. The effect of this is to cause the plastic hinge in the rafter to be located at a purlin position, usually the one immediately to the windward side of the ridge.

With the plastic hinge positions determined we can proceed straight to a solution. Assuming a plastic hinge at D,

$$H = M_P/5$$

Then, for the bending moment at F ($= M_P$),

$$M_P = 35.6 \times 12 - (M_P/5) \times 6.33 - 10(3 + 6 + 9) - 5 \times 12$$
$$= 82.6 \text{ kN m}$$

The reader might like to check that the corresponding solution assuming a plastic hinge to be located at C is marginally incorrect, and unsafe, at $M_P = 77.3$ kN m.

Example 10.5

We wish to find the collapse load factor for the two-storey frame shown in Fig. 10.10(a). The loads shown are unfactored and the

Figure 10.10 Example 10.5

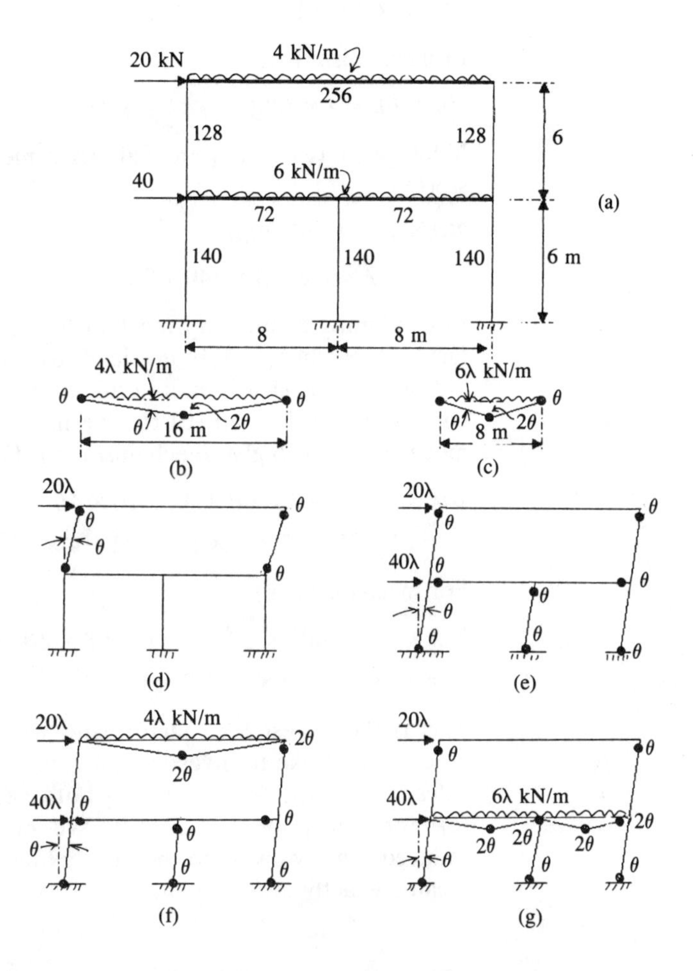

M_P values of the individual members are as shown (kN m units). It is convenient to organize the work by evaluating load factors for 'elementary' mechanisms representing beam and sideway collapses in diagrams (b), (c), (d) and (e), and then consider combined mechanisms such as those shown at (f) and (g). Care is needed to position the plastic hinges in the weaker member where two members of different M_P values meet at a joint. The geometry of each mechanism is defined by representative angle θ and the plastic hinge rotations are then all given in terms of θ. We seek the lowest value of λ, the load factor. Span hinges under the uniformly distributed loads are assumed to be located at mid-span.

In mechanism (b),

$$4\lambda \times 16 \times 4\theta = (2 \times 128 + 2 \times 256)\theta; \qquad \lambda = 3$$

In mechanism (c),

$$6\lambda \times 8 \times 2\theta = 4 \times 72\theta; \qquad \lambda = 3$$

In mechanism (d),

$$20\lambda \times 6\theta = 4 \times 128\theta; \qquad \lambda = 4.27$$

Mechanism (e) is a pure sidesway mechanism involving both storeys:

$$20\lambda \times 12\theta + 40\lambda \times 6\theta$$
$$= (2 \times 128 + 4 \times 140 + 2 \times 72)\theta; \qquad \lambda = 2$$

In combining elementary mechanisms, those with the lower load factors are more likely to lead to the collapse load factor. Sidesway (e) is clearly more critical than (d), so we combine this with beam mechanism (b) to give mechanism (f) and with beam mechanism (c) to give mechanism (g). For mechanism (f),

$$20\lambda \times 12\theta + 40\lambda \times 6\theta + 4\lambda \times 16 \times 4\theta$$
$$= (2 \times 256 + 2 \times 128 + 2 \times 72 + 4 \times 140)\theta; \qquad \lambda = 2$$

For mechanism (g),

$$20\lambda \times 12\theta + 40\lambda \times 6\theta + 2 \times 6\lambda \times 8 \times 2\theta$$
$$= (2 \times 128 + 8 \times 72 + 3 \times 140)\theta; \qquad \lambda = 1.86$$

This is the lowest load factor, and will be accepted as λ_c. The reader might like to investigate a combined mechanism involving sidesway of type (e) along with both beam mechanisms (b) and (c), for which we get $\lambda = 1.9$. We note that $\lambda = 1.86$ will be reduced slightly owing to the plastic hinges in the spans not being located exactly at mid-span.

10.6 Effect of axial forces in members

Axial forces in members carrying bending moments have the effect of changing the stress distribution over the cross-section and, if the force is compressive, producing instability tendencies in the member. Instability in the presence of plasticity is a difficult subject and we will not pursue it here. However, the presence of an axial force, either compressive or tensile, will effectively reduce the plastic moment of resistance of the member. Suppose we have an I-section member as shown in Fig. 10.11 which carries an axial thrust (tension or compression) of P, and suppose that the magnitude of P is such that

$$P = (2y_0 t)\sigma_y \qquad [10.22]$$

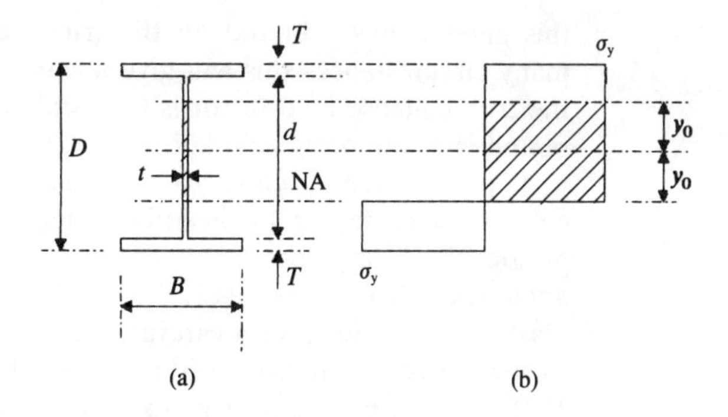

Figure 10.11 Reduction in plastic moment of resistance due to axial force

where $2y_0 \leqslant d$, and where d is the clear depth of web and σ_y is the yield stress. We therefore consider that the neutral axis lies within the depth of the web and not in one of the flanges. Hence the shaded area in Fig. 10.11(b) represents the portion of the (fully plastic) cross-section needed to support the axial load. This area is therefore not available to carry the bending moment, so that M_P of the section is reduced by that of the shaded part of the web, as in diagram (a). Thus the reduced moment M_{PR} is given by

$$M_{PR} = M_P - t(2y_0)^2\sigma_y/4$$

$$M_{PR} = M_P - \frac{A^2}{4t}\sigma_y\left(\frac{P}{P_P}\right)^2 \qquad [10.23]$$

where $P_P = \sigma_y A$ and A is the total cross-sectional area. Equation [10.23] can be rewritten in terms of the reduced plastic section modulus S_R simply by dividing both sides by the yield stress σ_y:

$$S_R = S - \frac{A^2}{4t}\left(\frac{P}{P_P}\right)^2 \qquad [10.24]$$

In BS 5950 the term P/P_P is called n, and values of reduced plastic modulus are quoted in design handbooks for each standard section.

The situation is more complicated if the neutral axis lies in one of the flanges. However, an expression similar to eqn [10.24] can be produced (Moy 1981), and reduced values of plastic modulus are again listed in structural steel handbooks.

10.7 Material and other requirements for plastic design

Although modern structural design is almost wholly based on limit state concepts, including the ultimate strength limit state, true plastic design involves an acceptance of the principle of moment redistribution. Whilst it is a comparatively simple matter to carry out an elastic analysis, identify critical cross-sections in the structure and design these under ultimate loading conditions, this gives a lower bound on the true 'collapse' load factor. In many circumstances this will give a satisfactory design; however, the true collapse load factor is then unknown although the lower bound is a safe design situation.

A progressive method can be used to approach the real collapse load factor by inserting frictionless hinges to replace plastic hinges as these form, and by adopting incremental load applications to the successively modified structure. Methods of plastic analysis deserve a careful study and the reader is referred to several comprehensive texts on the subject (Moy 1981; Horne 1971). It will be useful here to review the conditions for plastic design for steel structures as contained in BS 5950 (clause 5.3):

1. Applied loading should be predominantly static.
2. The stress–strain relationship (see Fig. 10.1) should exhibit a plateau at yield stress with $\varepsilon_{st} \nless 6\varepsilon_y$.
3. The ratio of ultimate tensile strength to yield strength $\nless 1.2$.
4. The material should exhibit adequate elongation capacity.
5. The cross-sectional geometry should be such as to enable the member to develop a plastic hinge with adequate rotation capacity to allow redistribution of moments. Such a cross-section is termed 'plastic'.
6. Torsional restraint should be provided at plastic hinge positions.
7. Web stiffeners are provided at load positions near to plastic hinge locations.
8. Certain restrictions are applied which relate to fabrication in the vicinity of a plastic hinge position.

Structures which are suitable for plastic design include beams (single-span and continuous) and rigidly jointed (particularly pitched roof) frames.

Exercises

10.1 A three-span continuous beam has the outer spans each 8 m long and the centre span 10 m long. The working load on the beam is 10 kN/m and the load factor against plastic collapse is to be 2.0. Determine the M_P value for each span based on simultaneous collapse in all three spans.

10.2 For the general structure of example 10.3, Fig. 10.8, with horizontal load H, vertical load w, span l and height h, show that for collapse

$$M_P = \frac{wl^2}{16}\left(1 + \frac{2Hh}{wl^2}\right)^2$$

10.3 A beam ABC spans 5 m, with support A built in and the beam simply-supported at C. A 50 kN load is carried at B. AB = 2 m. If the collapse load factor is to be 1.75, determine the required M_P for the beam. Use the reactant bending moment diagram method and check the result using the virtual work method.

10.4 The frame shown in Fig. 10.12 has rigid joints and pinned supports. The frame is constructed of a uniform section member which has a fully plastic moment of resistance $M_p = 303$ kN m. Determine the load factor for plastic collapse based on the 'working' loads shown.

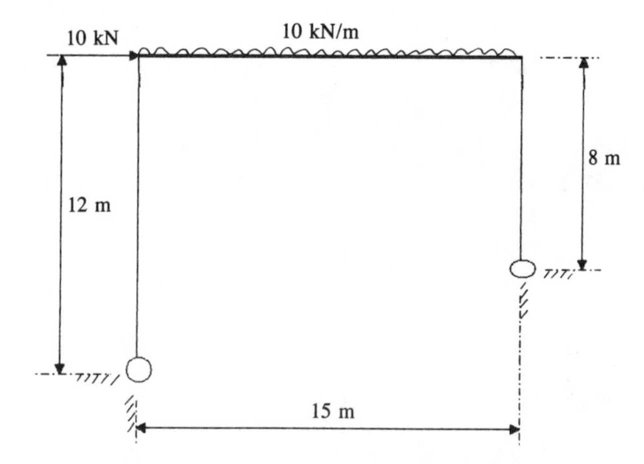

Figure 10.12 Exercise 10.4

10.5 A two-hinged, pitched roof portal frame has a span of 16 m, a height to eaves of 4 m and a rafter rise of 2 m. A horizontal load of 20 kN is applied at the eaves on one side of the frame and a uniformly distributed vertical load of 2 kN/m is applied to the rafters through purlins located at the quarter points of each rafter. If the loads are considered to be factored, determine the minimum value of plastic moment of resistance required.

10.6 In the two-storey rigidly jointed frame shown in Fig. 10.13, the upper frame is provided with diagonal bracing. The members of

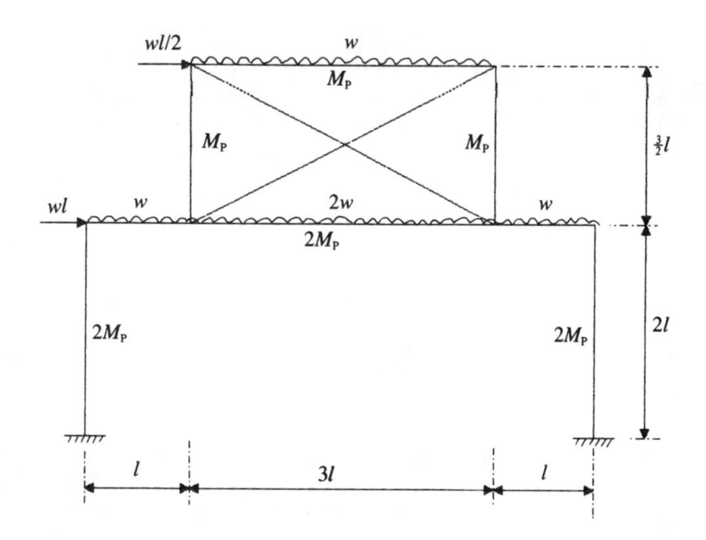

Figure 10.13 Exercise 10.6

the upper frame have a plastic moment of resistance M_P whilst those in the lower frame have $2M_P$. It may be assumed that plastic hinges in members are not located other than at the ends or centres. Given that $M_P = 4wl^2$, and neglecting instability tendencies and the reduction in plastic moment of resistance due to axial forces, determine the collapse load factor for the frame.

CHAPTER

11

Structural vibrations

11.1 Introduction

Vibrations in structures result from the application of 'dynamic' forces, i.e. forces which vary with time. Structural loads are often time-dependent, although in most cases the rate of change of load is slow enough to be neglected and such loads can be regarded as static. Here we are concerned with dynamic loading which causes a structure to vibrate, generating inertia forces. Whether these forces are significant or not depends on the magnitude and frequency of the applied forces and the dynamic response of the structure.

An important characteristic of any structure is the behaviour under 'free' vibrations. These are vibrations which take place after a disturbing force is removed and the structure is allowed to vibrate freely. In practice such vibrations are 'damped' by friction, air resistance and other factors, but the undamped free vibrations of a structure represent an important theoretical concept.

Certain types of structure may be susceptible to vibration, particularly any structure carrying moving loads (such as bridges and crane girders) or structures supporting rotating or reciprocating machinery. Non-steady wind forces and earthquakes are well-known examples of natural dynamic forces on structures. In circumstances where dynamic behaviour is significant, structural design needs to minimize or make acceptable the effects of vibration.

The fundamental mathematical basis of dynamic analysis is Newton's second law of motion:

$$\text{force} = \text{mass} \times \text{acceleration} \qquad [11.1]$$

We shall spend some time developing applications of this principle and the solution of the resulting differential equations. First we state some important definitions:

Amplitude is the maximum displacement from the mean position.
Period is the time for one complete cycle of vibration.
Frequency is the number of vibrations in unit time (cycles per second or hertz).
Forced vibration results from external, time-dependent, forces.
Free vibration occurs on removal of any disturbing forces.
Damping is the progressive reduction in amplitude of vibration due to (a) internal molecular friction, (b) loss of energy due to friction in slip in joints, and (c) resistance to motion in air or other fluid (drag). The type of damping usually predominating in structural vibrations is called 'viscous' damping, in which the force resisting motion is proportional to the velocity. This type of damping adequately represents (a) and (c).
Degrees of freedom are the independent displacements or 'coordinates' necessary to completely define the deformed state of the structure at any instant in time.

When a single coordinate is sufficient to define the displaced position of any point in the structure, then the structure is said to have a single degree of freedom. A continuous structure with a distributed mass such as a beam has an infinite number of degrees of freedom; however, we can approximate the real situation by concentrating the mass at fixed positions where we define coordinates. This is called the 'lumped mass' approximation. Accuracy can always be improved by adopting more lumped mass positions, and corresponding coordinates. Some typical lumped mass approximations are shown in Fig. 11.1. The

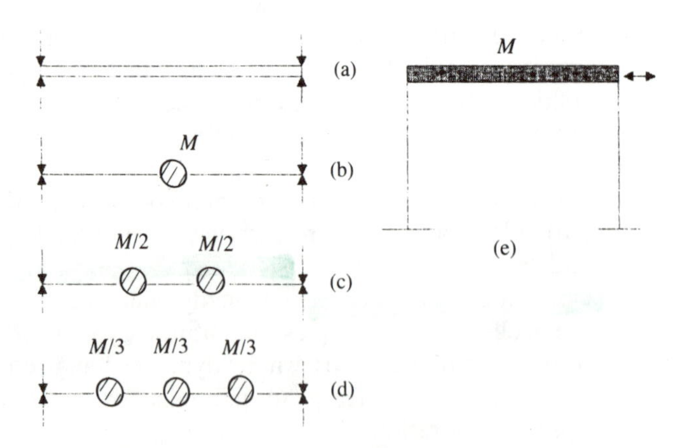

Figure 11.1 Lumped mass systems

structural properties, cross-sectional area, flexural rigidity and so on are unaltered; it is only the distribution of mass which is approximated. A beam with distributed mass, as in diagram (a), can be approximated by lumped mass systems as at (b), (c) and (d). Clearly the single-mass system (b) will not be expected to give as accurate a result as (c) or (d). In buildings the major part of the total mass will generally be concentrated in the floors, so a lumped mass as shown in diagram (e) can be adopted where the vibrations are predominantly in horizontal coordinates. The accuracy of analysis will depend not only on the number of masses adopted but also on the actual distribution of the masses. Best results are obtained if a 'consistent mass matrix' is adopted similar to that of section 8.4.

11.2 Free vibrations of single degree of freedom systems

Consider the simply-supported beam shown in Fig. 11.2(a) and assume that the entire mass of the beam is lumped at mid-span.

Figure 11.2 Free vibration of single lumped mass beam

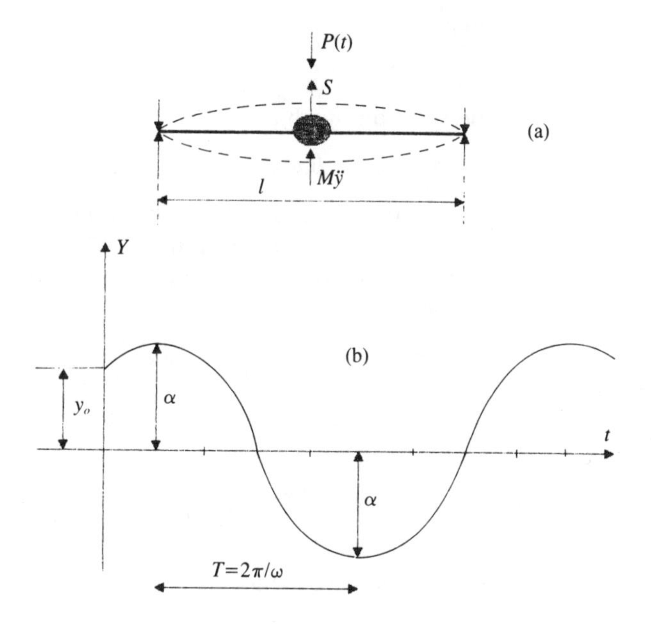

What we are now to do will give an approximate solution for a beam with distributed mass but an exact solution for a massless beam with a mass M applied at the centre.

Suppose a disturbing force $P(t)$ is applied as shown; the inertia force is $M\ddot{y}$, where $\ddot{y} = \mathrm{d}^2y/\mathrm{d}t^2$ is the acceleration. The force resisting motion, arising from the stiffness of the beam, is S. Thus, applying eqn [11.1],

net force = mass × acceleration

$$P(t) - S = M\ddot{y}$$

or

$$M\ddot{y} + S = P(t) \qquad [11.2]$$

When the disturbing force $P(t)$ is removed, free vibrations result and eqn [11.2] becomes

$$M\ddot{y} + S = 0 \qquad [11.3]$$

The force S is a function of the stiffness of the beam k and the displacement y:

$$S = ky \qquad [11.4]$$

Hence

$$M\ddot{y} + ky = 0 \qquad [11.5]$$

Putting

$$\omega^2 = k/M \qquad [11.6]$$

then

$$\ddot{y} + \omega^2 y = 0 \qquad [11.7]$$

the solution of which is

$$y = A \sin \omega t + B \cos \omega t \qquad [11.8]$$

The solution is 'harmonic': the value of y at an arbitrary time t is the same as the value at time $t + T$, where T is the period of the vibration. ω is the 'natural circular frequency' such that the period is

$$T = 2\pi/\omega \qquad [11.9]$$

Thus

$$\begin{aligned} y &= A \sin \omega t + B \cos \omega t \\ &= A \sin \omega\left(t + \frac{2\pi}{\omega}\right) + B \cos \omega\left(t + \frac{2\pi}{\omega}\right) \end{aligned} \qquad [11.10]$$

The 'natural frequency' is

$$f = 1/T = \omega/2\pi \qquad [11.11]$$

The velocity of the vibrating mass is

$$\begin{aligned} \dot{y} &= \mathrm{d}y/\mathrm{d}t \\ &= A\omega \cos \omega t - B\omega \sin \omega t \end{aligned} \qquad [11.12]$$

The integration constants A and B can be determined by applying the initial conditions of the problem. For example, if $y = y_0$ at $t = 0$ then from eqn [11.8] $B = y_0$. If the initial velocity is $\dot{y}(0)$ then from eqn [11.12] $A = \dot{y}(0)/\omega$. Hence

$$y = \frac{\dot{y}(0)}{\omega} \sin \omega t + y_0 \cos \omega t \qquad [11.13]$$

$$\dot{y} = \dot{y}(0) \cos \omega t - \omega y_0 \sin \omega t \qquad [11.14]$$

The amplitude of vibration α is given by the maximum value of y

from eqn [11.13]:

$$\alpha = \sqrt{\left[y_0^2 + \frac{\dot{y}(0)^2}{\omega^2}\right]} \qquad [11.15]$$

An alternative form of eqn [11.13] is

$$y = \alpha \sin(\omega t + \beta) \qquad [11.16]$$

where

$$\beta = \tan^{-1}[y_0\omega/\dot{y}(0)] \qquad [11.17]$$

and α is given by eqn [11.15].

The periodic relationship between displacement y and time t is shown in Fig. 11.2(b).

We can now obtain the solution for the simply-supported, single lumped mass beam by observing that the relevant stiffness is $k = 48EI/l^3$. Hence, in eqn [11.6],

$$\omega^2 = k/M = 48EI/Ml^3$$

$$\omega = \sqrt{(48EI/Ml^3)}$$

and in eqn [11.9]

$$T = 2\pi/\omega = 2\pi\sqrt{(Ml^3/48EI)}$$

This value is about 40 per cent higher than the exact value based on the distributed mass of the beam.

Example 11.1

The rectangular frame ABCD shown in Fig. 11.3 has columns with flexural rigidity *EI*. The beam is assumed to be very stiff

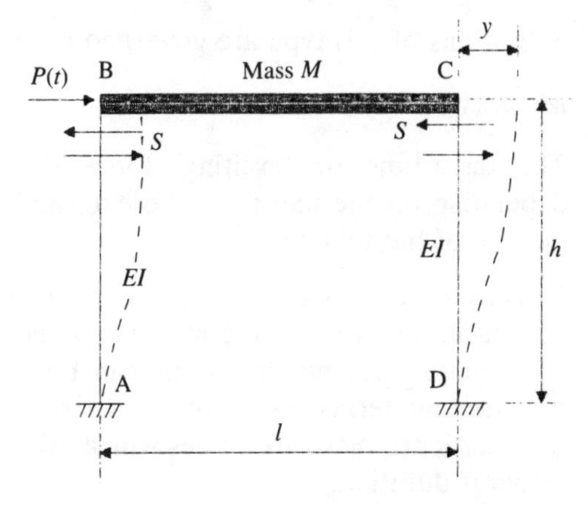

Figure 11.3 Example 11.1

compared with the columns, so we can assume negligible rotations of the joints B and C. The entire mass M is concentrated in the beam, and we can make the usual assumption that the change of axial length of the columns is negligible. Thus there is a single degree of freedom, which is the horizontal displacement of the beam y. The restoring forces are the shear forces S at the tops of the columns:

$$S = 12EIy/h^3$$

On removal of the disturbing force $P(t)$ the equation of motion is

$$M\ddot{y} + 2S = 0$$

$$M\ddot{y} + 24EIy/h^3 = 0$$

Hence from eqn [11.6] the natural circular frequency is

$$\omega = \sqrt{(24EI/h^3M)}$$

and from eqn [11.9] the period is

$$T = 2\pi/\omega = 2\pi\sqrt{(Mh^3/24EI)}$$

The displacement y at time t can be expressed using eqn [11.13]:

$$y = \frac{\dot{y}(0)}{\omega}\sin \omega t + y_0 \cos \omega t$$

where in this case $\dot{y}(0) = 0$. Hence

$$y = y_0 \cos \omega t$$

and the amplitude is, from eqn [11.15], $\alpha = y_0$.

11.3 Forced vibrations of single degree of freedom systems without damping

Vibrations of this type are governed by eqn [11.2]:

$$M\ddot{y} + ky = P(t)$$

The disturbing or 'exciting' force $P(t)$ may take any form depending on the nature of the externally applied forces, which may be of the following types:

1. Harmonic: this is the type of disturbing force produced by out-of-balance rotating machinery. Forces which are periodic in nature but not harmonic can be represented as a sum of harmonic terms using Fourier series.
2. Transient: these are forces which are applied suddenly, or for short duration.
3. Random: for example, forces generated by gusts of wind.

Forces in category 3 must be handled statistically and require special treatment. We shall look briefly at the dynamics of forces in categories 1 and 2.

11.3.1 *Harmonic force*

The harmonic disturbing force is

$$P(t) = P_0 \sin \Omega t \qquad [11.18]$$

where P_0 is the maximum value of $P(t)$, and Ω is the circular frequency of the force or the 'impressed' frequency.

The governing equation for undamped vibrations is

$$M\ddot{y} + ky = P_0 \sin \Omega t \qquad [11.19]$$

The solution is

$$y = A \sin \omega t + B \cos \omega t + \frac{P_0}{k} \frac{1}{1 - \Omega^2/\omega^2} \sin \Omega t \qquad [11.20]$$

The first two terms of eqn [11.20] represent the free vibrations and the third term the 'forced' vibrations. Generally the free vibrations can be ignored in comparison, so we can take

$$y = \frac{P_0}{k} \frac{1}{1 - \Omega^2/\omega^2} \sin \Omega t \qquad [11.21]$$

If a force $P_0 \sin \Omega t$ were to be applied statically, then the displacement would be $y = (P_0/k) \sin \Omega t$. Hence the term $1/(1 - \Omega^2/\omega^2)$ is a 'magnification' factor which is applied to the static displacement to obtain the corresponding dynamic displacement. The absolute value of this factor is shown plotted in Fig. 11.4.

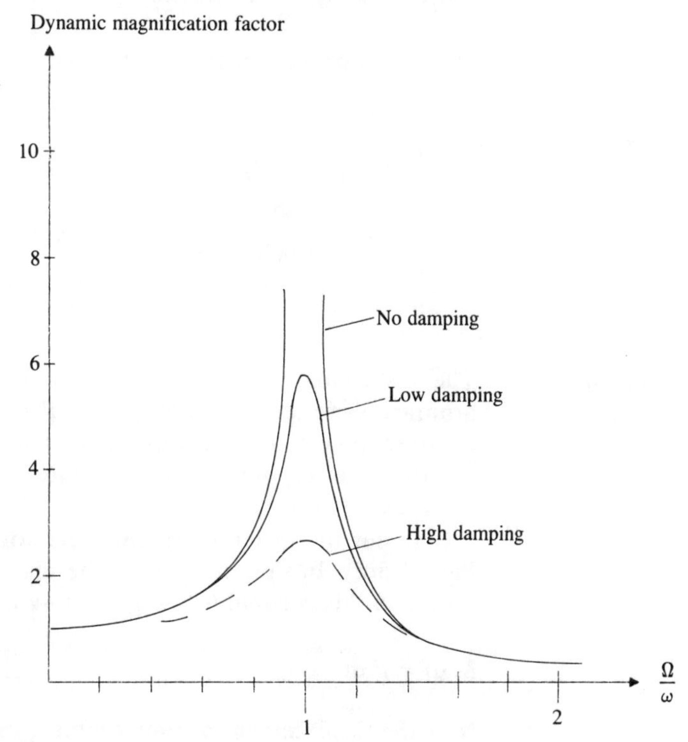

Figure 11.4 Dynamic magnification factor for harmonic disturbing force

The factor increases from unity at $\Omega/\omega = 0$ to infinity at $\Omega/\omega = 1$; the latter condition is called 'resonance', and occurs when the disturbing frequency is equal to the natural frequency of the structure. Theoretically displacements become infinite at resonance but in practice damping will reduce the amplitude, although this may still be beyond acceptable limits. At higher values of the ratio Ω/ω the magnification factor decreases and tends to zero.

Example 11.2

A fixed-end beam carries a central weight of 1000 kN. The beam spans 1.6 m and has a flexural rigidity $EI = 1600\ \text{kN m}^2$ and negligible self-weight.

(a) Calculate the natural frequency.
(b) Neglecting damping and assuming a disturbing force $P(t) = 500 \sin 40t$ kN, determine the maximum steady-state displacement of the beam. $g = 9.81\ \text{m/s}^2$.

(a) The stiffness k is the force for unit displacement:

$$k = 2 \times 12EI/(0.8)^3 = 75\,000\ \text{kN/m}$$

$$\omega = \sqrt{\left(\frac{k}{M}\right)} = \sqrt{\left(\frac{75\,000}{1000/9.81}\right)} = 27.125\ \text{rad/s}$$

(b) The maximum displacement occurs when $\sin 40t = \pm 1$. Thus

$$y_{\max} = \frac{P_0}{k}\frac{1}{1 - \Omega^2/\omega^2}(\pm 1)$$

$$= \pm\frac{500}{75\,000}\frac{1}{1 - 40^2/27.125^2} \times 10^3 = \pm 5.68\ \text{mm}$$

11.3.2 Transient force, undamped system

The disturbing force causing forced vibrations may have an arbitrary variation with time, as in Fig. 11.5(a). For example a constant force F may be applied suddenly to a mass M as in Fig. 11.5(b). The response of the structure in this situation is termed 'transient'.

Consider the general non-harmonic force–time relationship of Fig. 11.5(a). In time interval $\mathrm{d}t$ the mass M receives an impulse $P\,\mathrm{d}t$, and this is equal to the gain in momentum $M\,\mathrm{d}\dot{y}$:

$$M\,\mathrm{d}\dot{y} = P\,\mathrm{d}t \qquad [11.22]$$

Now the displacement at time t after a velocity $\mathrm{d}\dot{y}$ is imposed is,

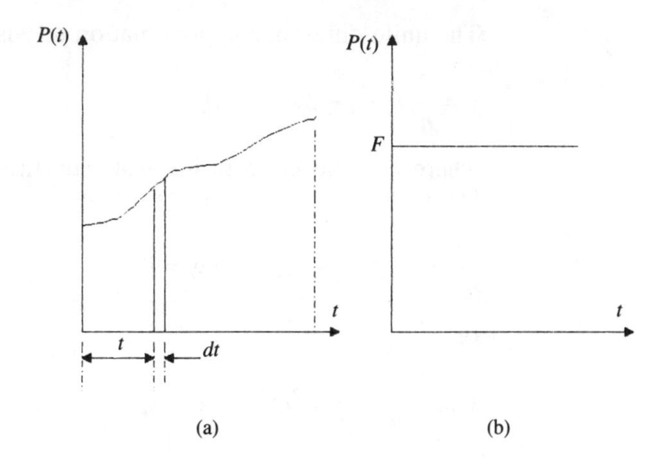

Figure 11.5 Non-harmonic disturbing forces

from eqn [11.13],

$$\mathrm{d}y = \frac{\mathrm{d}\dot{y}}{\omega} \sin \omega t$$

$$= \frac{P}{M} \frac{\mathrm{d}t}{\omega} \sin \omega t$$

The total displacement at time $t = t_1$ is

$$y(t_1) = \int_0^{t_1} \frac{P(t)}{M\omega} \sin \omega(t_1 - t)\, \mathrm{d}t \qquad [11.23]$$

This integral is usually evaluated numerically. However, if $P(t)$ is a simple function, for example a constant force F, then we can evaluate eqn [11.23] directly. Suppose that F is suddenly applied to a mass at rest ($y_0 = 0$), as in Fig. 11.5(b). Then, using eqn [11.23],

$$y(t_1) = \int_0^{t_1} \frac{F}{M\omega} \sin \omega(t_1 - t)\, \mathrm{d}t$$

$$= \frac{F}{M\omega^2}(1 - \cos \omega t_1)$$

Hence the maximum displacement is

$$y_{\text{max}} = \frac{2F}{M\omega^2} = \frac{2F}{k}$$

Since F/k is the static displacement caused by F, we see that y_{max} due to sudden application of F is twice the static displacement.

An alternative procedure on the computer with difficult $P(t)$ functions is to use finite difference approximations to solve the differential eqn [11.2]:

$$M\ddot{y} + ky = P(t)$$

The finite difference approximation to $\ddot{y}$ is

$$\ddot{y}_i = \frac{1}{h^2}(y_{i-1} - 2y_i + y_{i+1}) \qquad [11.24]$$

where h is the uniform interval. Substituting eqn [11.24] in eqn [11.2],

$$\frac{M}{h^2}(y_{i-1} - 2y_i + y_{i+1}) + ky_i = P_i$$

or

$$y_{i+1} = \frac{h^2}{M}P_i - (h^2\omega^2 - 2)y_i - y_{i-1} \qquad [11.25]$$

Equation [11.25] is applied recurrently at equal intervals of time h to produce the displacement–time relationship of the vibration. We shall illustrate the procedure by repeating example 11.2 with the harmonic disturbing force replaced by $P(t) = \text{constant} = 500$ kN.

Example 11.3

The data is the same as for example 11.2 but with $P(t) = \text{constant} = 500$ kN. We have $M = 1000/9.81$ and $k = 75\,000$ kN/m, so

$$\omega^2 = k/M = 75\,000 \times 9.81/1000 = 735.75$$

We start the process with $h = 0.05$ s; so $h^2P/M = 0.0\,122\,625$ and $(h^2\omega^2 - 2) = -0.1606$. Therefore eqn [11.25] becomes

$$y_{i+1} = 0.0\,122\,625 + 0.1606y_i - y_{i-1}$$

We now construct Table 11.1 line by line and plot the resulting

Table 11.1 Deflections for example 11.3

i	t	y_{i-1}	y_i	y_{i+1}
0	0	0	0	0.012 26
1	0.05	0	0.012 26	0.014 23
2	0.1	0.012 26	0.014 23	0.002 288
3	0.15	0.014 23	0.002 288	−0.001 60
4	0.2	0.002 288	−0.001 60	0.009 718
5	0.25	−0.001 60	0.009 718	0.015 42
6	0.3	0.009 718	0.015 42	0.005 021
7	0.35	0.015 42	0.005 021	−0.002 351
8	0.4	0.005 021	−0.002 351	0.006 864
9	0.45	−0.002 351	0.006 864	0.015 72
10	0.5	0.006 864	0.015 72	0.007 922

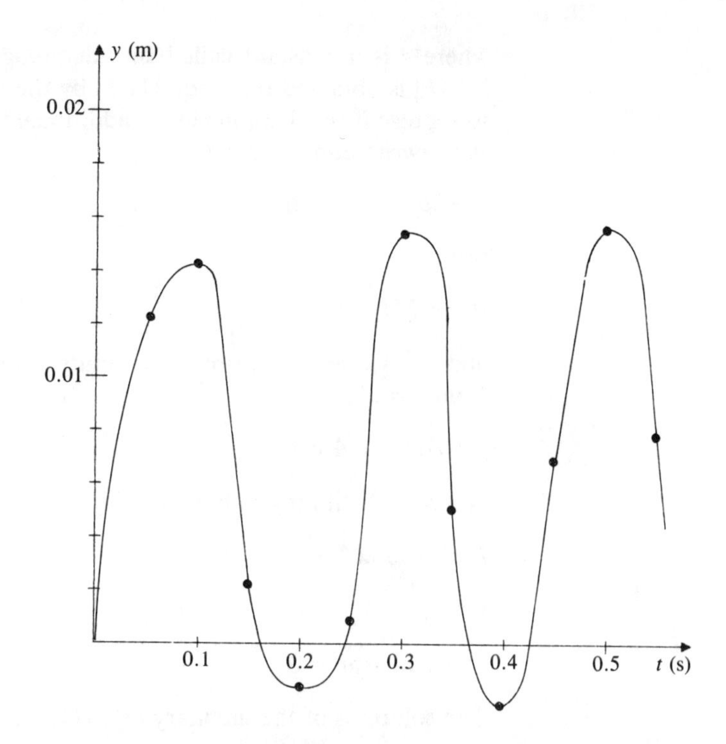

Figure 11.6 Example 11.3

values of y_i as in Fig. 11.6. We can deduce that the maximum deflection is approximately 0.016 m. By comparison, the exact solution in this case is

$$y_{max} = 2 \times \text{static deflection} = 2F/k$$
$$= 2 \times 500/75\,000 = 0.01\,333 \text{ m}$$

With the time interval selected, the method has overestimated the maximum displacement by 20 per cent. However, if we now repeat the computation with $h = 0.02$ s (which is left as an exercise for the reader) we obtain $y_{max} = 0.0139$ m, which is very close to the exact solution.

The finite difference method requires extreme care with the arithmetic since any errors made are carried forward and may become magnified. This is a very good reason for transferring the work to the computer.

11.4 Free vibrations with viscous damping

Damping is the dissipation of energy in a vibrating structure. In engineering structures, damping is generally taken to be proportional to the velocity. This type of damping is called 'viscous' damping, and the governing equation for free vibrations is

$$M\ddot{y} + c\dot{y} + ky = 0 \qquad [11.26]$$

where c is a constant called the 'damping coefficient'. Equation [11.26] is obtained from eqn [11.5] by the addition of the term $c\dot{y}$ to represent the damping as an additional force opposing motion. We rewrite eqn [11.26] as

$$\ddot{y} + 2\beta\dot{y} + \omega^2 y = 0 \qquad [11.27]$$

where

$$\beta = c/2M \qquad [11.28]$$

and $\omega^2 = k/M$ as before. The solution of eqn [11.27] is in the form $y = e^{\lambda t}$, i.e.

$$y = A_1 e^{\lambda_1 t} + A_2 e^{\lambda_2 t} \qquad [11.29]$$

Hence, substituting in eqn [11.27],

$$\lambda^2 e^{\lambda t} + 2\beta\lambda e^{\lambda t} + \omega^2 e^{\lambda t} = 0$$

or

$$\lambda^2 + 2\beta\lambda + \omega^2 = 0 \qquad [11.30]$$

The solutions of the auxiliary eqn [11.30] are

$$\left.\begin{aligned} \lambda_1 &= -\beta + \sqrt{(\beta^2 - \omega^2)} \\ \lambda_2 &= -\beta - \sqrt{(\beta^2 - \omega^2)} \end{aligned}\right\} \qquad [11.31]$$

The constants of integration A_1 and A_2 in eqn [11.29] are determined by the initial conditions.

Three distinct cases arise in the evaluation of eqns [11.31], since the expression $\beta^2 - \omega^2$ can be positive, zero or negative depending on the relative values of β^2 and ω^2.

Case 1: $\omega^2 < \beta^2$

Here $\beta^2 - \omega^2$ is always positive and $<\beta^2$; hence the roots λ_1 and λ_2 are real and negative. The solution eqn [11.29] is then

$$y = e^{-\beta t}[A_1 e^{\sqrt{(\beta^2-\omega^2)}t} + A_2 e^{-\sqrt{(\beta^2-\omega^2)}t}] \qquad [11.32]$$

The relationship between displacement y and time t given by eqn [11.32] has the form shown in Fig. 11.7(a). After a disturbance causing an initial displacement y_0 at $t = 0$, the displacement gradually returns to zero and no vibrations result. This situation is called 'overdamping'. From eqns [11.6] and [11.28], since $\beta^2 > \omega^2$,

$$c^2/4M^2 > k/M$$
$$c > 2\sqrt{(Mk)} = c_c$$

where c_c is described in the following case.

Figure 11.7 Effect of damping on free vibrations: (a) overdamped disturbance (no vibration), $c > 2\sqrt{(Mk)}$ (b) underdamped vibrations, $c < 2\sqrt{(Mk)}$

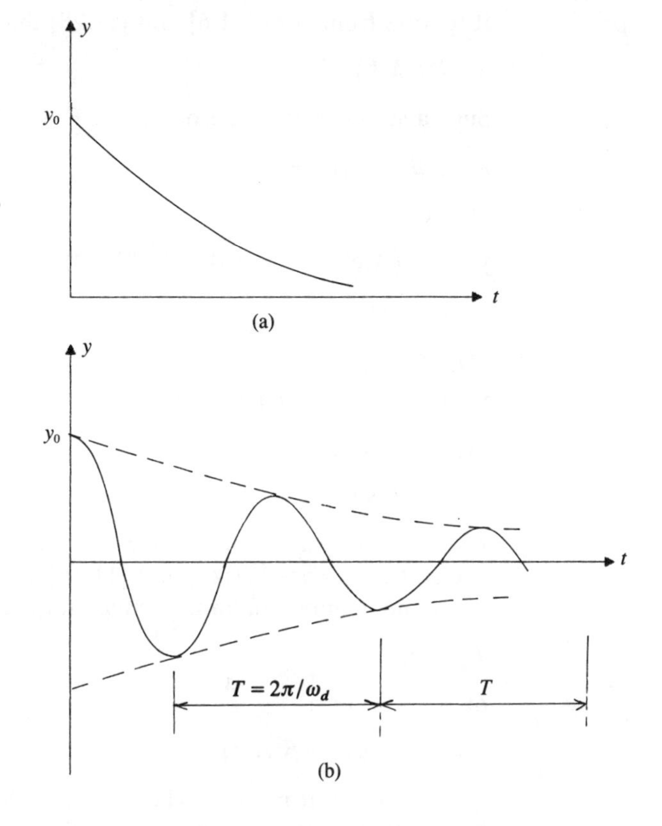

Case 2: $\omega^2 = \beta^2$

Substituting in eqn [11.30] we obtain equal, negative roots of the auxiliary equation, i.e.

$$\lambda_1 = \lambda_2 = -\beta$$

Hence

$$y = e^{-\beta t}(A_1 + A_2 t) \qquad [11.33]$$

Now eqn [11.33] has the same form as that in Fig. 11.7(a) and so there are no vibrations. However, this case has a special significance in the theory of vibrations. From eqns [11.6] and [11.28], since $\omega^2 = \beta^2$,

$$c = 2\sqrt{(Mk)} = c_c \qquad [11.34]$$

This particular value of c is the 'critical damping coefficient' c_c and is the value of the damping coefficient on the boundary between vibratory and non-vibratory motion. It is a useful measure of the damping capacity of a structure, which is usually expressed as a percentage of the critical damping coefficient.

Case 3: $\omega^2 > \beta^2$

It follows from eqns [11.6] and [11.28] that, since $\omega^2 > \beta^2$,

$$c < 2\sqrt{(Mk)} = c_c \quad [11.35]$$

Such a structure is 'underdamped'. From eqn [11.31],

$$\lambda = -\beta \pm \mathrm{i}\sqrt{(\omega^2 - \beta^2)} \quad [11.36]$$

Hence

$$y = \mathrm{e}^{-\beta t}(A_1 \mathrm{e}^{\mathrm{i}\sqrt{(\omega^2-\beta^2)}t} + A_2 \mathrm{e}^{-\mathrm{i}\sqrt{(\omega^2-\beta^2)}t}) \quad [11.37]$$

or, putting

$$\omega_d^2 = \omega^2 - \beta^2, \quad [11.38]$$

$$y = \mathrm{e}^{-\beta t}(B_1 \cos \omega_d t + B_2 \sin \omega_d t) \quad [11.39]$$

or, alternatively,

$$y = C\mathrm{e}^{-\beta t} \sin (\omega_d t + \gamma) \quad [11.40]$$

where C and γ are new arbitrary constants.

Comparing eqns [11.40] and [11.16] we see that $\omega_d = \sqrt{(\omega^2 - \beta^2)}$ is the 'damped natural circular frequency', and

$$T = 2\pi/\omega_d \quad [11.41]$$

or

$$T = 2\pi/\omega\sqrt{(1 - \beta^2/\omega^2)} \quad [11.42]$$

It is evident from eqn [11.42] that the period is constant; however, the amplitude decreases with time due to the damping. This is shown in Fig. 11.7(b). The decrease in amplitude is called 'decay', and is such that the ratios of amplitudes at time intervals equal to the period are constant, i.e.

$$y(t)/y(t+T) = \mathrm{e}^{\beta T} \quad [11.43]$$

Now

$$\log \mathrm{e}^{\beta T} = \beta T = \delta \quad [11.44]$$

where δ is the 'logarithmic decrement', which is a useful measure of damping capacity.

The 'percentage critical damping' is given by

$$100c/c_c = 100\beta/\omega = 100\delta/\omega T \quad [11.45]$$

Percentage critical damping ranges from about 1 per cent to about 10 per cent depending on the material of the structure and the foundation. Modern structures tend to exhibit lower damping than older structures.

Example 11.4

If the fixed-end beam of example 11.2 has 10 per cent critical damping, calculate the damped natural frequency and the damped natural period. Compare these with the corresponding undamped values from example 11.2.

We have

$$c/c_c = 0.1 = \beta/\omega$$

$$\omega = 27.125 \text{ rad/s}$$

Therefore $\beta = 2.7125$. Then

$$\omega_d = \sqrt{(\omega^2 - \beta^2)} = \sqrt{(27.125^2 - 2.7125^2)} = 26.99 \text{ rad/s}$$

and

$$T = 2\pi/\omega\alpha = 0.233 \text{ s}$$

11.5 Free vibrations of multiple degree of freedom structures

The governing equation for free, undamped vibrations of a single degree of freedom structure was derived as eqn [11.5] and the solution was expressed in the form of eqn [11.16]. A similar governing equation will apply to each degree of freedom in a multiple degree of freedom structure, so we simply rewrite the equations in matrix form:

$$\boldsymbol{M}\ddot{\boldsymbol{y}} + \boldsymbol{K}\boldsymbol{y} = \boldsymbol{O} \qquad [11.46]$$

$$\boldsymbol{y} = \sin(\omega t + \beta)\boldsymbol{\alpha} \qquad [11.47]$$

where $\boldsymbol{M}$, the 'mass' matrix, is a diagonal matrix of the lumped masses at the coordinates; $\boldsymbol{y}$ and $\ddot{\boldsymbol{y}}$ are the column matrices of displacements and second derivatives of displacements respectively; $\boldsymbol{K}$ is the stiffness matrix; and $\boldsymbol{\alpha}$ is a matrix of amplitudes. Now if we differentiate eqn [11.47] twice we get

$$\ddot{\boldsymbol{y}} = -\omega^2\boldsymbol{y} \qquad [11.48]$$

and hence eqn [11.46] becomes

$$\omega^2\boldsymbol{M}\boldsymbol{y} = \boldsymbol{K}\boldsymbol{y} \qquad [11.49]$$

Premultiplying by $\boldsymbol{M}^{-1}$,

$$\boldsymbol{M}^{-1}\boldsymbol{K}\boldsymbol{y} = \omega^2\boldsymbol{y} \qquad [11.50]$$

This will be recognized as a standard eigenvalue equation. The solution produces n eigenvalues (ω^2) and n corresponding eigenvectors (mode shapes), where n is the number of degrees of freedom. Now we could proceed to develop a numerical solution of eqn [11.50]; alternatively we can re-form this equation in terms of flexibility rather than stiffness. If we premultiply eqn [11.49] by the flexibility matrix $\boldsymbol{F}$, where $\boldsymbol{F} = \boldsymbol{K}^{-1}$, then

$$\boldsymbol{F}\boldsymbol{M}\boldsymbol{y} = \frac{1}{\omega^2}\boldsymbol{y} \qquad [11.51]$$

The solution of eqn [11.51] can be achieved by any standard technique; however, a numerical method based on matrix iteration is very suitable for engineering structures and is very appropriate to the computer. The iterative solution we shall develop will produce the largest eigenvalue $1/\omega^2$ and hence the lowest, fundamental mode, frequency of free vibration. The process commences with an estimated eigenvector $\boldsymbol{y}_0$. Then

$$\boldsymbol{FMy}_0 \simeq \frac{1}{\omega^2}\boldsymbol{y}_0$$

Putting $\boldsymbol{FMy}_0 = \boldsymbol{y}_1$, then

$$\boldsymbol{y}_1 \simeq \frac{1}{\omega^2}\boldsymbol{y}_0$$

We cannot form $\boldsymbol{y}_0/\boldsymbol{y}_1$ since each $\boldsymbol{y}$ is a column matrix. So we take the ratio of corresponding elements in $\boldsymbol{y}_0$ and $\boldsymbol{y}_1$ and form

$$\omega_1^2 = y_0/y_1$$

It is best to use the numerically greatest y for this purpose. Continuing the process,

$$\boldsymbol{FMy}_1 = \boldsymbol{y}_2 \simeq \frac{1}{\omega^2}\boldsymbol{y}_1$$

giving

$$\omega_2^2 = y_1/y_2$$

Again,

$$\boldsymbol{FMy}_2 = \boldsymbol{y}_3 \simeq \frac{1}{\omega^2}\boldsymbol{y}_2$$

giving

$$\omega_3^2 = y_2/y_3$$

The process converges to the largest eigenvalue $(1/\omega^2)$, and hence to the lowest frequency corresponding to the fundamental mode. The general expression for the iterative procedure is

$$\boldsymbol{y}_{i+1} = \frac{1}{\omega^2}\boldsymbol{y}_i = \boldsymbol{FMy}_i \qquad [11.52]$$

and at any stage,

$$\omega_i^2 = y_{i-1}/y_i \qquad [11.53]$$

Example 11.5

We shall apply the iterative process of eqn [11.52] to a simple cantilever beam of mass m (per unit length) distributed uniformly

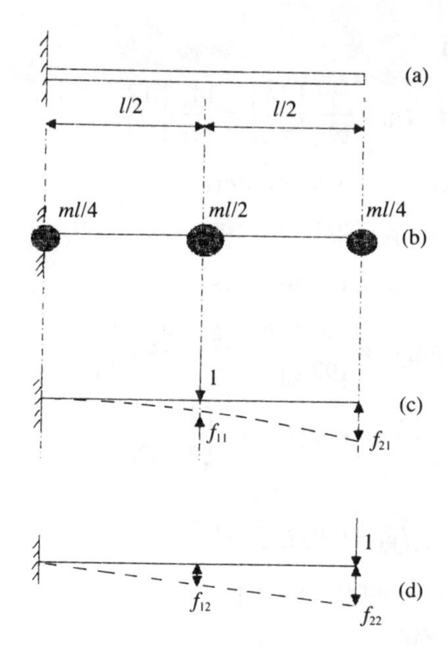

Figure 11.8 Example 11.5

over its length. We lump the mass of the cantilever at the tip and mid-span as shown in Fig. 11.8. This will probably be a rather crude representation, so we shall compare our result with the exact solution. The mass matrix is

$$\boldsymbol{M} = \frac{ml}{4}\begin{bmatrix} 2 & 0 \\ 0 & 1 \end{bmatrix}$$

and the flexibilities are easily shown to be

$$f_{11} = l^3/24EI$$
$$f_{12} = f_{21} = 5l^3/48EI$$
$$f_{22} = l^3/3EI$$

Hence

$$\boldsymbol{F} = \frac{l^3}{48EI}\begin{bmatrix} 2 & 5 \\ 5 & 16 \end{bmatrix}$$

and

$$\boldsymbol{FM} = \frac{ml^4}{192EI}\begin{bmatrix} 4 & 5 \\ 10 & 16 \end{bmatrix} = \frac{\gamma}{192}\begin{bmatrix} 4 & 5 \\ 10 & 16 \end{bmatrix}$$

where $\gamma = ml^4/EI$.

In choosing an initial vector y_0 we are concerned only with *shape*, since only relative values can be computed. Take

$$y_0 = \begin{bmatrix} 1 \\ 2 \end{bmatrix}$$

Then

$$y_1 = FMy_0 = \frac{\gamma}{192}\begin{bmatrix}14\\42\end{bmatrix} = \frac{14\gamma}{192}\begin{bmatrix}1\\3\end{bmatrix}$$

Hence after a single iteration,

$$\omega_1^2 = y_0/y_1 = 2/3(14\gamma/192) = 9.143EI/ml^4$$

A second iteration gives

$$y_2 = FMy_1 = \frac{\gamma}{192}\begin{bmatrix}4 & 5\\10 & 16\end{bmatrix}\frac{14\gamma}{192}\begin{bmatrix}1\\3\end{bmatrix}$$

$$= 0.007\,216\gamma^2\begin{bmatrix}1\\3.0526\end{bmatrix}$$

and

$$\omega_2^2 = y_1/y_2 = 9.93EI/ml^4$$

A third iteration gives

$$y_3 = FMy_2$$

$$= 0.000\,724\gamma^3\begin{bmatrix}1\\3.055\end{bmatrix}$$

and

$$\omega_3^2 = y_2/y_3 = 9.96EI/ml^4$$

We can conclude that the process has converged, so we take

$$\omega^2 = 9.96EI/ml^4$$

The exact value is $12.49EI/ml^4$, so our result is too low. It could be improved by using more degrees of freedom or by constructing an improved (consistent) mass matrix. The reader might like to repeat this example with three degrees of freedom (masses $ml/3$, $ml/3$ and $ml/6$) and obtain the improved result, $\omega^2 = 11.2EI/ml^4$.

Example 11.6

A 20 m high tower is divided into five sections for the purpose of approximate vibration analysis. The mass is lumped at five levels as shown in Fig. 11.9, and the values are given in Table 11.2. *EI* values at the levels have been computed and are also given in the table. The tower is fixed at the base and is unstayed. Determine the fundamental mode frequency and the period of oscillation for undamped free vibrations.

We need the flexibility matrix with flexibilities computed at each of the levels 1 to 5. We could use the numerical integration method of section 6.9 to do this. However, if we have a plane frame program available we could use this to compute both

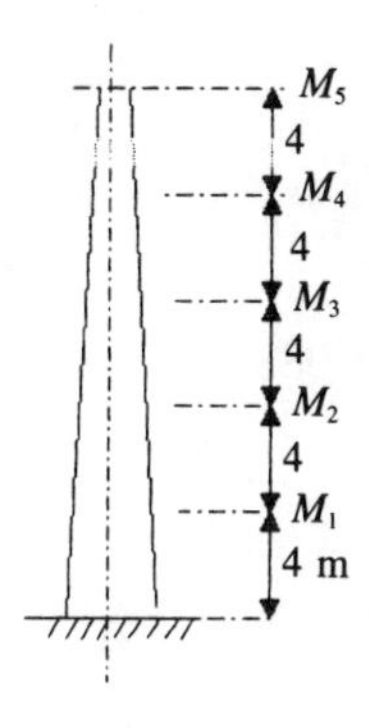

Figure 11.9 Example 11.6

Table 11.2 Data for example 11.6

Level (i)	M (kg)	EI (kN m^2) $\times 10^6$	Mean EI of segment (N m^2) level i to level $i+1$ $\times 10^9$
0	*	18.4	15.24
1	18 322	12.08	9.81
2	14 477	7.54	5.98
3	11 088	4.42	3.403
4	8 141	2.386	1.768
5	2 827	1.1505	

* Not required

direct and cross flexibilities by repeated application of a unit load; the tower will be divided into five segments, each segment constituting a member for the plane frame program. To the purist this may appear a crude way to derive a flexibility matrix, but it is quite effective computationally. Each segment is taken to have a uniform EI value equal to the mean between the adjacent levels. In this way we obtain the flexibility matrix:

$$\boldsymbol{F} = 10^{-9}\begin{bmatrix} 1.400 & 3.500 & 5.598 & 7.698 & 9.797 \\ 3.500 & 11.972 & 21.532 & 31.092 & 40.652 \\ 5.598 & 21.532 & 45.382 & 71.016 & 96.650 \\ 7.698 & 31.092 & 71.016 & 124.350 & 180.810 \\ 9.797 & 40.652 & 96.650 & 180.810 & 289.570 \end{bmatrix}$$

The first column of $\boldsymbol{F}$ is obtained by applying unit load at level 1, the second column by applying unit load at level 2, and so on.

Now

$$\boldsymbol{M} = \begin{bmatrix} 18\,322 & & & & \\ & 14\,477 & & & 0 \\ & & 11\,088 & & \\ & 0 & & 8141 & \\ & & & & 2827 \end{bmatrix}$$

Hence

$$\boldsymbol{FM} = 10^{-6}\begin{bmatrix} 25.65 & 50.67 & 62.07 & 62.67 & 27.70 \\ 64.13 & 173.3 & 238.7 & 253.12 & 114.92 \\ 102.6 & 311.7 & 503.2 & 578.14 & 273.2 \\ 141.0 & 450.1 & 787.4 & 1012.3 & 511.2 \\ 179.5 & 588.5 & 1071.7 & 1472.0 & 818.6 \end{bmatrix}$$

Keeping in mind the likely shape of the deflected tower, we start the process with an eigenvector:

$$y_0 = \begin{bmatrix} 1 \\ 3 \\ 6 \\ 10 \\ 15 \end{bmatrix}$$

Then

$$y_1 = \boldsymbol{FM}y_0$$

$$= 10^{-6}\begin{bmatrix} 1\,592.3 \\ 6\,271.2 \\ 13\,936.3 \\ 24\,006.7 \\ 35\,374.2 \end{bmatrix} = 10^{-3} \times 1.5923 \begin{bmatrix} 1 \\ 3.94 \\ 8.75 \\ 15.08 \\ 22.22 \end{bmatrix}$$

and

$$y_2 = \boldsymbol{FM}y_1$$

$$= 10^{-9} \times 1.5923 \begin{bmatrix} 2\,329 \\ 9\,206 \\ 20\,523 \\ 35\,428 \\ 52\,263 \end{bmatrix} = 10^{-6} \times 3.708 \begin{bmatrix} 1 \\ 3.95 \\ 8.81 \\ 15.2 \\ 22.4 \end{bmatrix}$$

and

$y_3 = \boldsymbol{FMy}_2$

$$= 10^{-12} \times 3.708 \begin{bmatrix} 2\,346 \\ 9\,273 \\ 20\,674 \\ 35\,693 \\ 52\,657 \end{bmatrix} = 10^{-9} \times 8.699 \begin{bmatrix} 1 \\ 3.95 \\ 8.81 \\ 15.2 \\ 22.4 \end{bmatrix}$$

Clearly the process has converged, so we can take $\boldsymbol{y} = \boldsymbol{y}_3$. Hence

$$\omega^2 = y_2/y_3 = 10^3 \times 3.708/8.699 = 426$$

Thus

$$\omega = 20.6 \text{ rad/s}$$

$$f = \omega/2\pi = 3.28 \text{ Hz}$$

$$T = 1/f = 0.304 \text{ s}$$

The numerical procedure just described is ideally suited to the computer. Some further consideration would need to be given to a suitable general convergence criterion.

Exercises

11.1 A cantilever of negligible self-weight, span l and flexural rigidity EI, carries a mass M at the free end. Show that the natural period of free vibrations is given by

$$T = 2\pi\sqrt{(Ml^3/3EI)}$$

11.2 A simply-supported beam of span 4 m carries a central weight of 100 kN. The beam has a flexural rigidity $EI = 15\,000$ kN m^2 and may be considered to have negligible self-weight. Find the natural frequency of free vibration.

11.3 A portal frame consists of a horizontal beam AB supported by columns of height 3 m at A and 4 m at B. The columns are fixed at their feet, have negligible mass and have $EI = 25\,000$ kN m^2. The total mass of 2000 kg is considered to be lumped at the beam level and the beam is considered to be rigid. There is assumed to be no damping. (a) Determine the period of natural vibration. (b) If the beam is given a transverse displacement of 10 mm and then released at time $t = 0$, derive an expression for the resulting sidesway displacement. (c) If the frame is subjected to a horizontal disturbing force $P(t) = 100 \sin 5t$ kN at beam level, determine the maximum steady-state sway displacement of the beam. (d) If the percentage critical damping is 12 per cent, calculate the damped natural circular frequency and the damped natural period.

11.4 A simply-supported uniform cross-section beam has a mass m per unit length. Adopting a lumped mass representation at the three internal quarter span points, use the matrix iterative scheme of section 11.5 to obtain the fundamental mode frequency and compare the result with the 'exact' value $\omega^2 = \pi^4 EI/ml^4$.

CHAPTER

12

Approximate methods and verification techniques

12.1 Introduction

The reliability of the results of a structural analysis depends on a number of factors:

1. Correct data
2. Accurate handling of the data, including the correct interpretation and manipulation of units
3. Suitability of the method employed, including the validity of the assumptions made
4. Correctness of the arithmetic carried out, including an assessment of any loss of accuracy due to rounding
5. Reliability of output, i.e. the numbers and what they purport to be
6. Correct interpretation of the output, including signs and units.

Computerized analysis based on the stiffness method is, fortunately, a very reliable and generally straightforward method of analysis in which arithmetical rounding errors are usually negligible even with very large structures. Furthermore the data preparation is usually easy to carry out; but here lies a danger since even the simplest data requires meticulous checking before transmission to the computer. Human beings can be very careless in these circumstances! It is crucial to develop a careful approach at each stage and to check all the requirements set out above. Experience helps a great deal and, if using a strange computer package, it is important to carry out some validation and familiarization exercises similar to those given in Part III. The purpose is to ensure that the interpretation of the program specification in the preparation and input of the data and the handling of output is correct. The computer cannot do this for us. It will simply do what it is told to do whether this makes practical sense or not.

Even with the most careful preparation and handling of the

data, there may remain some doubt about the validity of the results, and in these circumstances it is important to have available some other tools for assessing reliability of output from the computer. In this chapter we shall examine some approximate methods and verification techniques which we can and should apply to computer output if in any doubt about its validity. Some of the methods will involve what is no more than a 'rough-and-ready' approach simply to provide a very approximate level of a representative result; other methods will be able to provide a higher level of approximation but may require more effort to do so. Sometimes we will use the computer to check itself – the ideal way – but more often our checking will require hand calculation. Checking should be directed at the more critical values of stress resultants and displacements. There is no point in verifying results which are clearly not critical to the design of the structure. Finally, and this cannot be overemphasized, we must absolve the computer from the responsibility and accept this ourselves.

12.2 Moment distribution

Distribution methods of analysis have been very popular and extremely useful since the 1940s. They have tended to be displaced by matrix methods based on stiffness concepts suitable for computers. In the educational context, much less time is now spent on the method of moment distribution, and indeed it is sometimes argued that it no longer justifies a place in the curriculum. One educational advantage of the method was that, since it is based on clearly seen physical concepts, the student could better 'see' what was happening in the process of analysis, and thereby develop a 'feel' for the method. Until adequate educational devices, based on the use of the computer, have been developed, there is still some merit in acquiring a facility with moment distribution.

It should be understood that moment distribution is not strictly an 'approximate' method. It is essentially an iterative solution of the equations of rotational equilibrium at the joints of the structure and as such can be continued to produce any desired degree of accuracy. We include the method in this chapter since it now lies more appropriately with other methods used in parallel with computers.

Moment distribution is a useful method for structures with relatively few nodes and members, for example continuous beams and single-storey frames. We include a brief exposition of the method here as a useful 'hand' method for checking computer output, and also, perhaps more importantly, to provide assistance to the student in developing a 'feel' for structural analysis.

Figure 12.1 Moment distribution in two-span beam

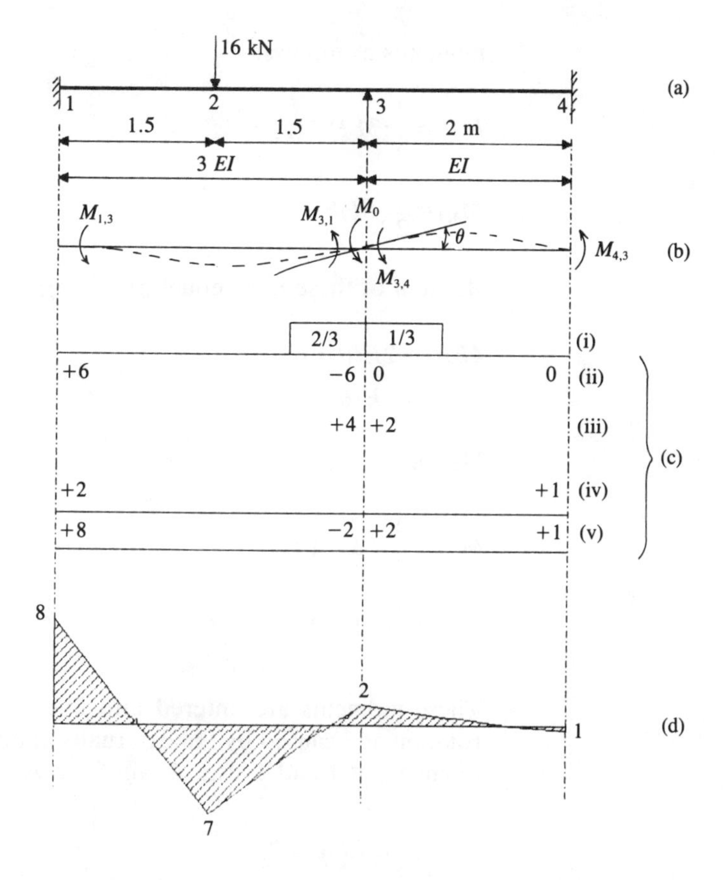

Suppose we have a simple two-span beam as shown in Fig. 12.1. The process commences by assuming that all support nodal freedoms are constrained. In Fig. 12.1(a) this means that we constrain the rotation of the beam at node 3. The other nodal freedoms at 1, 3 and 4 are already constrained by the support conditions. Since span 1–3 is loaded, we shall need a constraining moment at 3 to prevent rotation. This is the 'fixed-end' moment, which in this case is $Wl/8 = 16 \times 3/8 = 6$. A fixed-end moment will also be required at support 1. The signs of these constraining moments are positive at 1 and negative at 3 with the Z axis positively directed towards the viewer. No fixed-end moments are introduced into span 3–4 since this span is not loaded. The fixed-end moments are entered into line (c)(ii) of the figure, and we can see that there is a lack of rotational equilibrium at node 3 where an internal moment −6 exists in the beam 1–3. To restore equilibrium we apply an external moment +6 to the beam at 3, the rotational constraint being removed while we do this. This externally applied moment is shown as M_0 in Fig. 12.1(b). The result of this will be a rotation of the beam, say θ, at node 3 and a consequent sharing of the moment M_0 between the two spans. From the standard results of Appendix 2 we can quantify the moments as follows:

$$M_{3,1} = \frac{4}{3}(3EI)\theta$$

$$M_{3,4} = \frac{4}{2}(EI)\theta$$

the sum of these must equal M_0 ($= 6$); thus

$$4EI\theta + 2EI\theta = 6$$
$$EI\theta = 1$$

Hence

$$M_{3,1} = \frac{4}{3} \times 3 \times 1 = 4$$

$$M_{3,4} = \frac{4}{2} \times 1 = 2$$

These moments are entered into line (c)(iii) of the figure. The rotation at 3 caused by the externally applied moment M_0 induces moments at 1 and 4 also. Again from Appendix 2, these are

$$M_{1,3} = \frac{2}{3}(3EI)\theta = 2$$

$$M_{4,3} = \frac{2}{2}(EI)\theta = 1$$

These 'carry-over' moments are shown in line (c)(iv) of the figure. We have now restored equilibrium, so we can sum to obtain the total moment at each point. The figure is therefore completed by entering totals in line (c)(v).

Now if we examine what we have done, we see that the applied moment M_0 has been 'distributed' between the two spans and that the distributed moments are

Span 1:

$$\left[\frac{(I/l)_1}{(I/l)_1 + (I/l)_2}\right](M_0)$$

Span 2:

$$\left[\frac{(I/l)_2}{(I/l)_1 + (I/l)_2}\right](M_0)$$

The terms in the square brackets are called 'distribution factors', and for the current problem these are

Span 1:

$$\frac{3/3}{3/3 + 1/2} = 2/3$$

Span 2:

$$\frac{1/2}{3/3 + 1/2} = 1/3$$

These factors were entered in line (c)(i) of the figure and are used in the arithmetical process of moment distribution at node 3. The out-of-balance moment at 3 is −6, so a moment of +6 is applied, and the consequent distributed moment going into 3–1 is

$$2/3 \times 6 = 4$$

and that going into 3–4 is

$$1/3 \times 6 = 2$$

Note that the distribution factors sum to unity since the whole moment M_0 is distributed.

The bending moment diagram is shown in Fig. 12.1(d) and is drawn on the tension side of the beam.

In this simple example the process has been concluded with only one distribution. In general the 'release' of one joint will result in an out-of-balance situation at other joints and the moment distribution will be continued until the process converges to an acceptable degree. The process can start at any unconstrained joint but convergence will be quicker if one starts at the joint with the greatest initial out-of-balance. As each joint is released and a rotation takes place, the externally applied moment is distributed into the members meeting at the joint, which is then in a state of rotational equilibrium. The distribution factors at a joint determine how this moment is shared between the members. For example in Fig. 12.2(a) the external moment

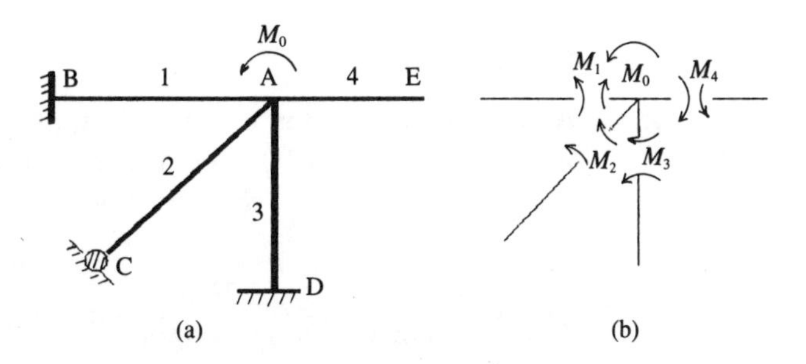

Figure 12.2 Distribution factors in moment distribution

M_0 will be shared between the four members as follows:

$$M_1 = 4\left(\frac{EI}{l}\right)_1 \theta$$

$$M_2 = 3\left(\frac{EI}{l}\right)_2 \theta \qquad [12.1]$$

$$M_3 = 4\left(\frac{EI}{l}\right)_3 \theta$$

$$M_4 = 0$$

The coefficient 3 is used for M_2 because member 2 is hinged at the end remote from joint A. No moment is absorbed by member 4 since this member offers no resistance to a rotation of joint A. Now

$$M_0 = M_1 + M_2 + M_3 + M_4$$

$$= 4\theta\left[\left(\frac{EI}{l}\right)_1 + \frac{3}{4}\left(\frac{EI}{l}\right)_2 + \left(\frac{EI}{l}\right)_3\right] \qquad [12.2]$$

Eliminating θ from eqns [12.1] and [12.2]:

$$M_1 = M_0\left(\frac{EI}{l}\right)_1 \Big/ \left[\left(\frac{EI}{l}\right)_1 + \frac{3}{4}\left(\frac{EI}{l}\right)_2 + \left(\frac{EI}{l}\right)_3\right] \qquad [12.3]$$

$$M_2 = M_0\frac{3}{4}\left(\frac{EI}{l}\right)_2 \Big/ \left[\left(\frac{EI}{l}\right)_1 + \frac{3}{4}\left(\frac{EI}{l}\right)_2 + \left(\frac{EI}{l}\right)_3\right] \qquad [12.4]$$

$$M_3 = M_0\left(\frac{EI}{l}\right)_3 \Big/ \left[\left(\frac{EI}{l}\right)_1 + \frac{3}{4}\left(\frac{EI}{l}\right)_2 + \left(\frac{EI}{l}\right)_3\right] \qquad [12.5]$$

Thus each distributed moment is obtained, as we have already seen, by multiplying the applied moment M_0 by the appropriate distribution factor. We note the use of the factor 3/4 for members pinned at the remote end, and remind ourselves that the carry-over factor is 1/2 to any remote end (permanently or temporarily) fixed in direction.

Example 12.1

A three-span continuous beam is shown in Fig. 12.3(a) (this is structure 1F in Part II). The distribution factors are

At B:

$$\text{BA}:\text{BC} = \frac{I/4}{I/4 + 2I/5} : \frac{2I/5}{I/4 + 2I/5} = 0.385 : 0.615$$

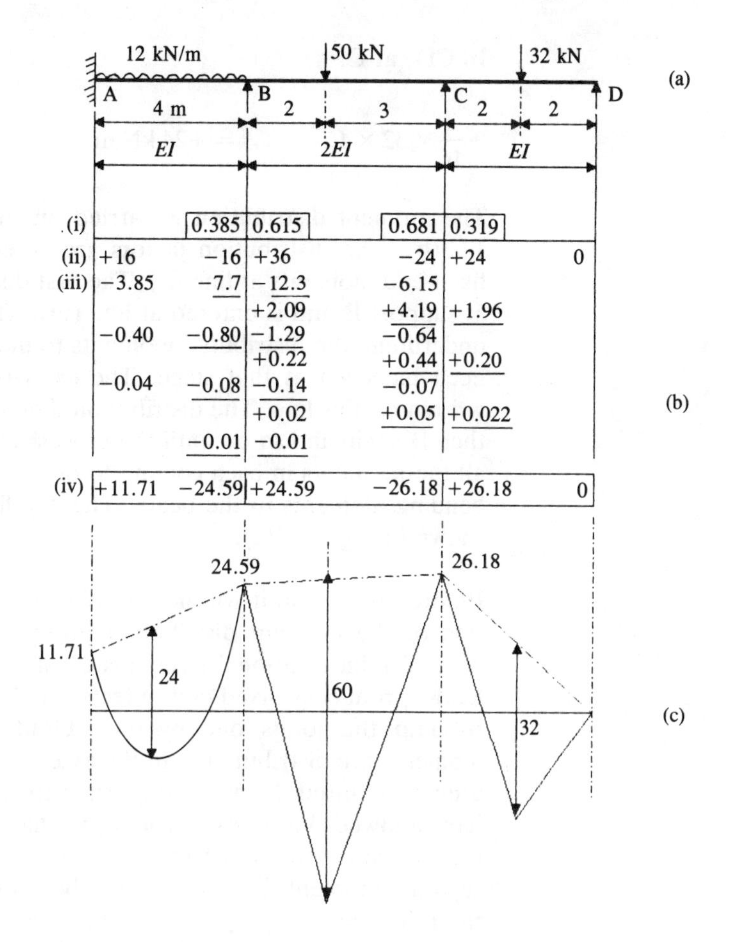

Figure 12.3 Example 12.1

At C:

$$CB:CD = \frac{2I/5}{2I/5 + (3/4)(I/4)} : \frac{(3/4)(I/4)}{2I/5 + (3/4)(I/4)}$$

$$= 0.681 : 0.319$$

The fixed-end moments are, from Appendix 3,

In AB:

$$\pm 12 \times \frac{4^2}{12} \qquad = \pm 16 \text{ kN m}$$

In BC, at B:

$$+\frac{50 \times 2 \times 3}{5} \times \frac{3}{5} \qquad = +36 \text{ kN m}$$

at C;

$$-\frac{50 \times 2 \times 3}{5} \times \frac{2}{5} \qquad = -24 \text{ kN m}$$

In CD, at C:

$$+\frac{3}{16}\times 32\times 4 \qquad = +24\ \text{kN m}$$

The moment distribution is carried out in tabular form in Fig. 12.3(b). The distribution factors are entered at line (i) and the fixed-end moments at line (ii). The first distribution is carried out at support B and is entered at line (iii). There is a convention of underlining the distributed moments to indicate that the joint has been balanced at that stage. The carry-over moments are also written on this line. The distribution then continues at support C, then B again and so on until the process converges. We then sum all the moments in each column to obtain the final values for the bending moments in the beam. The bending moment diagram is shown in Fig. 12.3(c).

Before moving on it will help to observe the physical interpretation of the moment distribution process in that we commence with all joint rotational freedoms constrained; then we apply the loads producing fixed-end effects; and then we release and reclamp the joints one by one. Gradually the applied joint moments are distributed until the external, rotational, constraints are not required. Now it is important to realize that, although we have allowed the joints to rotate, we have not allowed any joint translation. If this is necessary, and it often is, we carry out a separate moment distribution and then combine the results using the principle of superposition. The procedure of handling 'sway' in moment distribution will be best illustrated by an example.

Example 12.2

We shall use the method of moment distribution to analyse the frame shown in Fig. 12.4(a). The distribution factors are

BA : BC = 1/3 : 2/3

CB : CD = 2/3 : 1/3 noting $(3/4)(I/l)$ for CD

The fixed-end moments in the beam are

$$wl^2/12 = \pm 30\times 4^2/12 = \pm 40\ \text{kN m}$$

These fixed-end moments are entered in Fig. 12.4(b) and a moment distribution is carried out, producing the column moments given in line (f)(iii). The reader should check this by carrying out the distribution. Now we examine the equilibrium of the beam in the horizontal direction. We have prevented sidesway movement of the beam, so we shall need a force F(b)

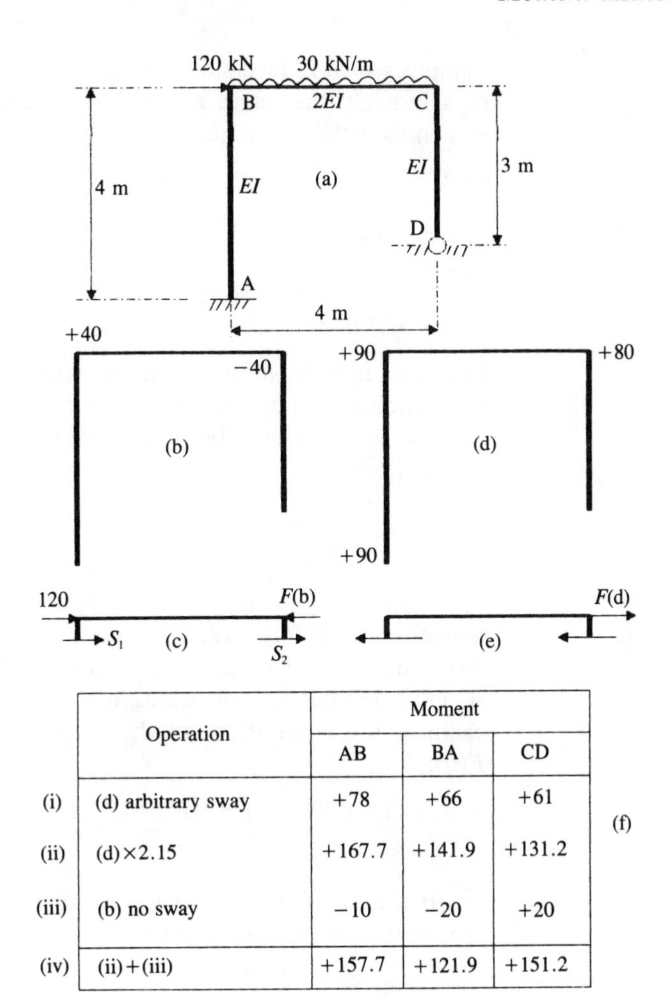

	Operation	Moment		
		AB	BA	CD
(i)	(d) arbitrary sway	+78	+66	+61
(ii)	(d)×2.15	+167.7	+141.9	+131.2
(iii)	(b) no sway	−10	−20	+20
(iv)	(ii)+(iii)	+157.7	+121.9	+151.2

Figure 12.4 Example 12.2

which we can now calculate:

$$F(\mathrm{b}) = 120 + S_1 + S_2$$
$$= 120 + (1/4)(10 + 20) + (1/3)(-20)$$
$$= 120.8 \text{ kN}$$

The horizontal shear forces S_1 and S_2 in the columns are obtained by taking moments about the bottoms of the columns. Care is needed with the signs; with the positive Z axis directed towards the viewer, positive bending moments in the plane of the frame appear anticlockwise. Now the force F(b) does not exist and so must be removed. Clearly the beam will move to the right and we can achieve this by applying a force such as F(d) in Fig. 12.4(e). It is clear that F(d) and F(b) should be equal; however, we move indirectly to this condition by applying an arbitrary value of F(d). We do this by applying internal moments to the columns which induce sidesway. Since the sidesway displacement Δ is sensibly the same at the top of each column, then from Appendix 3 the moments will be

In AB:

$$M_{AB} = M_{BA} = 6EI/\Delta/4^2$$

In CD:

$$M_{CD} = 3EI\,\Delta/3^2$$

Now although the moments are arbitrary, they must be correctly proportioned relatively. If we were to take $M_{AB} = M_{BA} =$ 90 kN m, which would be a convenient figure to work with, then we must take

$$M_{CD} = 90 \times \frac{3EI\Delta/3^2}{6EI\Delta/4^2} = 80 \text{ kN m}$$

Introducing these moments into the columns as fixed-end moments as in Fig. 12.4(d), we carry out a second moment distribution and get the results in line (f)(i). Again the reader should carry out this distribution to verify the results. We again examine the equilibrium of the beam and compute a value for F(d):

$$F(\text{d}) = (1/4)(78 + 66) + (1/3)(61)$$
$$= 56.3$$

So to achieve $F(\text{b}) = F(\text{d})$ we need to scale the results of the second distribution by a factor

$$\lambda = 120.8/56.3 = 2.15$$

The moments in line (f)(i) are therefore multiplied by λ at line (ii) and added to those in line (iii) from the first distribution. The final moments are in line (iv).

Extensions and refinements of the method of moment distribution are possible for multiple degrees of freedom in sidesway and for special treatment of symmetrical structures.

12.3 Equilibrium checks

An application of the stiffness method of analysis involves the solution of the equilibrium equations at all the specified degrees of freedom. It follows that if these equations are correctly set up and solved correctly, then the results – the nodal displacements – are reliable. Conversely if we are given the results of an analysis but not the loads applied, we should be able to apply equilibrium principles to determine what loads *were* applied. This can be a useful approach to take if there is some doubt as to the loads that the computer has used in the analysis. We take as an example structure 1F in Part II. In span 1–2, using the values given in the bending moment diagram

$$wl^2/8 - (1/2)(11.71 + 24.18) = 6.06$$

$$w = 12 \text{ kN/m}$$

In span 2-4,

$$W \times 2 \times 3/5 = 34.76 + (24.58 \times 3 + 26.18 \times 2)/5$$

$$W = 49.98 \text{ kN} \qquad (\text{cf. } 50)$$

In span 4–6,

$$W \times 4/4 = 18.9 + 1/2 \times 26.18$$

$$W = 32 \text{ kN}$$

Here we have simply used the values of the stress resultants from the analysis to set up equilibrium equations in the directions of the applied loads to determine those loads. The type of load and load position were evident from the analysis in this case; if this is not so then some further investigation may be needed.

It is frequently useful to check the overall equilibrium of a structure by summing the horizontal, vertical and moment reactions and comparing these with the applied loads. We illustrate this by checking the equilibrium of the applied loads and reactions of structure 9F in Part II.

Check horizontal equilibrium:

$$40 + 40 = H_1 + H_4 + H_7$$

$$= 20 + 40 + 20 = 80 \text{ kN} \quad (\text{OK})$$

Check vertical equilibrium:

$$0 = V_1 + V_4 + V_7$$

$$= 36 + 0 - 36 = 0 \quad (\text{OK})$$

Check rotational equilibrium: take moments about node 1:

$$40 \times 8 + 40 \times 16 = M_{1,2} + M_{4,5} + M_{7,8} + V_4 \times 8 + V_7 \times 16$$

$$= 96 + 192 + 96 + 0 \times 8 + 36 \times 16$$

$$960 = 960 \quad (\text{OK})$$

Example 12.3

Calculate the intensity of the distributed load used in the analysis of structure 11T in Part II. Consider the vertical equilibrium of one-half girder:

$$T \times 10/\surd(10^2 + 20^2) + P_{1,2} \times 3/\surd(3^2 + 2^2) = w \times 20$$

Now the sample results do not give us the appropriate value of

$P_{1,2}$ in task 1, but we can proceed using the results from task 3 with $T = 0$:

$$20w = 2163 \times 3/\sqrt{(3^2 + 2^2)} + 2w$$
$$w = 99.98 \text{ kN/m} \quad (\text{cf. } 100)$$

Example 12.4

For the structure shown in Fig. 12.5(a) a partially completed bending moment diagram is shown at (b). Calculate the applied

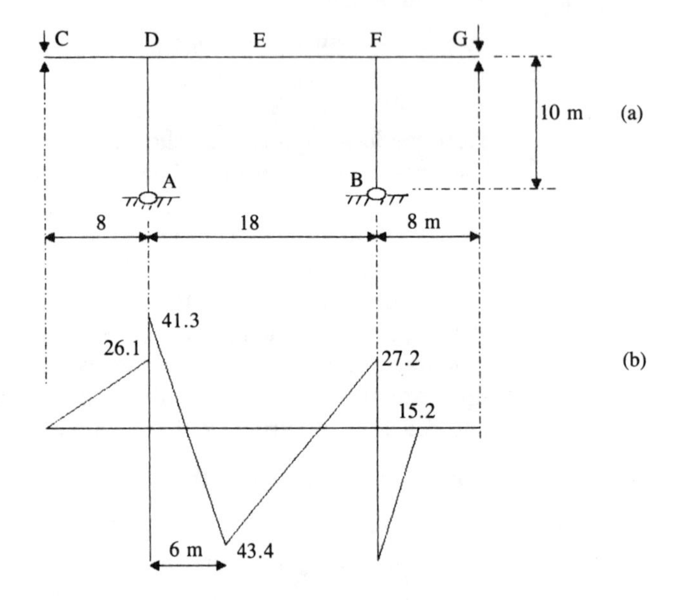

Figure 12.5 Example 12.4

loading on the structure; calculate M_{FG} and M_{DA}; and complete the bending moment diagram.

There is clearly a vertical load W at a point 6 m from D in DEF. From the bending moment diagram for DEF,

$$W \times 6 \times 12/18 = 1/18(27.2 \times 6 + 41.3 \times 12) + 43.4$$
$$W = 20 \text{ kN}$$

Hence the total vertical reactions must be 20 kN. Now

$$R_{VC} = 26.1/8 = 3.263 \quad (\text{down})$$
$$R_{VA} = 3.263 + \text{shear at D from DF}$$
$$= 3.263 + (20 \times 12/18 + 41.3/18 - 27.2/18)$$
$$= 17.38 \text{ kN} \quad (\text{up})$$

Considering rotational equilibrium of F,

$$M_{GF} - 27.2 + 15.2 = 0$$
$$M_{FG} = 12.0 \text{ kN m}$$

and hence

$$R_{VG} = 12.0/8 = 1.5\text{ kN} \quad \text{(down)}$$

Considering vertical equilibrium

$$R_{VB} = 20 + 3.263 + 1.5 - 17.38$$
$$= 7.38\text{ kN} \quad \text{(up)}$$

Lastly we examine joint D and observe that, for rotational equilibrium

$$M_{DA} + 41.3 - 26.1 = 0$$
$$M_{DA} = -15.2\text{ kN m}$$

Since $M_{DA} = -M_{FB}$ we can conclude that R_{HA} and R_{HB} are equal and opposite. Since C and G are roller mounted, we can conclude that no horizontal loads have been applied to the structure.

Example 12.5

A continuous beam ABCD has three identical spans of 10 m supported with purely vertical restraint. The beam has been analysed by computer for a uniformly distributed load of 6 kN/m over all three spans. The following results have been produced by the computer:

$$M_B = 70\text{ kN m} \quad \text{(hogging)}$$
$$M_C = 20\text{ kN m} \quad \text{(hogging)}$$
$$R_{VA} = 23\text{ kN}$$
$$R_{VB} = 72\text{ kN}$$
$$R_{VC} = 27\text{ kN}$$
$$R_{VD} = -2\text{ kN}$$

The results are not symmetrical and are therefore suspect. Determine what error has been made.

Clearly something is wrong in span CD, since

$$M_C = 20 = R_{VD} \times 10$$

It appears that the load has been omitted from this span. If we examine the other spans, for example B,

$$10R_{VA} = 6 \times 10 \times 5 - 70 = 230$$
$$R_{VA} = 23$$

and, for ABC,

$$10R_{VB} = 6 \times 20 \times 10 - 20 - 23 \times 20$$
$$R_{VB} = 72$$

We conclude that the computer output is consistent with the load having been omitted from span CD.

12.4 Upper and lower bounds

It is usually possible to apply crude checks to computer output to determine whether the results lie in a practical range. In some circumstances upper and lower bounds can be placed on specific critical values of, say, a bending moment. For example in a beam within a structural frame, the end restraint provided by the joints will reduce the simply-supported bending moments near the centre of the beam. It follows that the maximum free bending moment will be an upper bound on the maximum 'span' moment near mid-span. For the end moments, an upper bound will usually be consistent with the fixed-end moment; however, in practice this moment may be increased if the joint rotation, affected by an adjacent span, is counter to the direction of free rotation. A lower bound on span moment can be taken with full end restraint from the joints, and a lower bound on the end moments will be zero corresponding to the simply-supported state. The principle of bounds on values of stress resultants and displacements related to extremes of support conditions is a useful one in assessing the practicality of computer output.

Example 12.6

A 6 m span beam AB carries a uniformly distributed load of 4 kN/m. The beam is part of an extensive structural frame with some stiffness in the joints. A computer analysis gives $M_A = -0.2$ kN m and $M_B = -16.5$ kN m. Compare the maximum span and end moments with estimated upper and lower bounds.

An upper bound on the mid-span moment is

$$wl^2/8 = 4 \times 6^2/8 = 18 \text{ kN m}$$

and a lower bound is

$$wl^2/8 - wl^2/12 = wl^2/24 = 6 \text{ kN m}$$

For the end moment, we obtain an upper bound of

$$wl^2/12 = 12 \text{ kN m}$$

and a lower bound of zero.

We can show the results in Table 12.1. The upper bound we have adopted is not a good predictor in this case, since a joint rotation in the frame has increased one of the end moments above the level of the fixed-end moment. Nevertheless the approach gives some guidance on what we might expect from a computer analysis and should be sufficient to prompt further investigation if anomalies appear.

Table 12.1 Bounds on predicted moments in the beam of example 12.6

	Upper bound (kN m)	Actual (kN m)	Lower bound (kN m)
Span moment	18	9.65	6
End moment	12	16.5	0

Example 12.7

Establish bounds on the cable tension in structure 4F in Part II.

The upper bound on T will be when we assume that the beam has no flexural stiffness, in which case

$$T = 1000\sqrt{(20^2 + 15^2)}/15 = 1667 \text{ kN}$$

The lower bound on T, corresponding to a beam with infinite flexural rigidity, is $T = 0$. The actual value of T in the circumstances of this structure is 952.4 kN, which is clearly a practicable value.

Example 12.8

Calculate upper and lower bounds on the vertical deflection of node 3 in the continuous beam of structure 1F in Part II.

To obtain an upper bound, assume the beam to be simply-supported at nodes 2 and 4. Then

$$R_{V2} = (3/5) \times 50 = 30 \text{ kN}$$

and, using eqn [6.41],

$$\begin{aligned} 2EI(\mathrm{d}^2v/\mathrm{d}x^2) &= -M_z \\ &= 30x - 50[x-2] \\ 2EI(\mathrm{d}v/\mathrm{d}x) &= 15x^2 - 25[x-2]^2 + A \\ 2EIv &= 5x^3 - (25/3)[x-2]^3 + Ax + B \\ &= 0 \quad \text{for} \quad x = 0; \qquad B = 0 \\ &= 0 \quad \text{for} \quad x = 5; \qquad A = -80 \end{aligned}$$

At $x = 2$,

$$2EIv = 5(2)^3 - 80 \times 2$$

$$v = -\frac{60}{EI}$$

To obtain a lower bound, assume that the span 2–4 is

direction-fixed at its ends; then

$$M_{F2} = 50 \times 2 \times (3/5) \times (3/5) = 36$$

$$M_{F4} = 24$$

$$S_{V2} = (1/5)(50 \times 3 + 36 - 24) = 32.4 \text{ kN}$$

$$2EI(\mathrm{d}^2v/\mathrm{d}x^2) = -36 + 32.4x - 50[x-2]$$

$$2EI(\mathrm{d}v/\mathrm{d}x) = -36x + 16.2x^2 - 25[x-2]^2 + A$$

$$2EIv = -18x^2 + 5.4x^3 - (25/3)[x-2]^3 + Ax + B$$

$$= 0 \quad \text{for} \quad x = 0; \qquad B = 0$$

$$= 0 \quad \text{for} \quad x = 5; \qquad A = 0$$

At $x = 2$,

$$2EIv = -18 \times 4 + 5.4 \times 8$$

$$v = -14.4/EI$$

The actual deflection from the computer analysis is $-22/EI$. In comparing this with the lower bound we should look at the senses of the computed rotations at 2 and 4 to ensure that they are compatible with this result.

12.5 Points of inflexion

Points of inflexion are positions in structural members where there is a change in sense of the curvature, for example from hogging to sagging in a beam or from concave to the left to concave to the right in a column. At these points the bending moment is zero and we could, theoretically, insert hinges at such points. If we know the location of a point of inflexion we can write an equilibrium equation by taking moments about the point. Sometimes we know the location approximately and can obtain a useful approximation to the analysis. Care is needed since the adjacent stress resultants are quite sensitive to the position of the point of inflexion. We know, for example, that a point of inflexion exists in a propped cantilever beam, and if the beam carries a uniformly distributed load w then the point of inflexion is at $l/4$ from the fixed end. The propping force is $(3/8)wl$ and the moment at the fixed end is $wl^2/8$. Now if we estimate the position of the point of inflexion as say at mid-span, then we find that the propping force is reduced to $wl/4$ and the fixed-end moment increased to $wl^2/4$, so our estimate has not produced very accurate results. A simple sketch of the deflected shape of the beam would have suggested to us that the point of inflexion was nearer to the fixed end, and this would have given us an improved estimate.

In some circumstances we can deduce positions of points of inflexion from conditions of symmetry or skew symmetry in our structure and can therefore proceed to an exact analysis in these cases. Sometimes we know that a point of inflexion lies fairly close to a certain position and can then proceed to an approximate analysis. This situation often exists in multi-storey frames, as we shall see in the next example.

Example 12.9

Here we carry out an approximate analysis of the three-storey frame of example 5.10. Under lateral loading we have a skew-symmetric situation with equal moments, of the same sign, left and right of the centre line. The situation is slightly disturbed if we take axial straining of the members into account. We can therefore conclude that points of inflexion exist at mid-span of each beam. We also know that points of inflexion exist in each column and we shall assume that these occur at mid-height in each case.

The situation is shown in Fig. 12.6(a). Consider first the shears

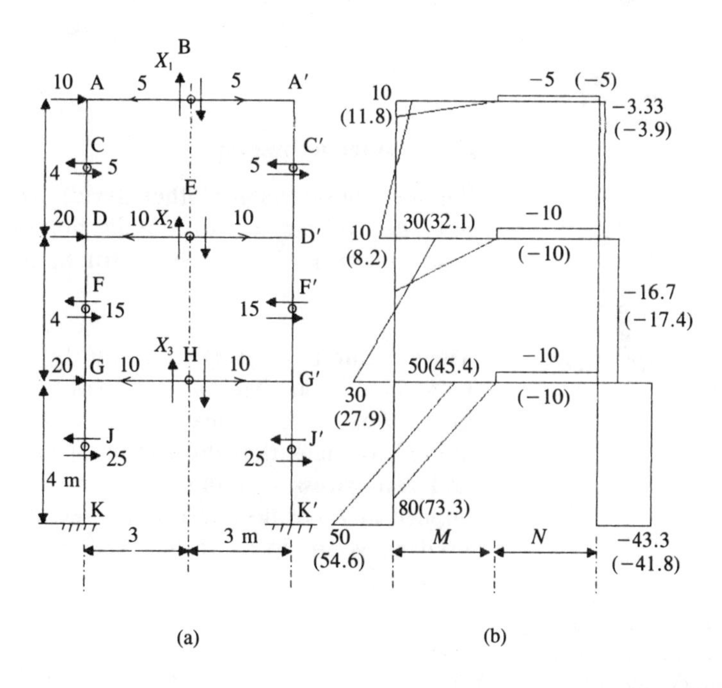

Figure 12.6 Example 12.9: analysis of multi-storey frame based on points of inflexion

at the points of inflexion in the columns and start at the top storey. The external horizontal shear of 10 kN must be shared equally by the two columns, so the shear at the points of inflexion is 5 kN. Similarly in the second storey the shear is $(1/2)(10+20) = 15$ kN and in the lower storey $(1/2)(10+20+20) = 25$. Note that we have not yet used our assumption as to the locations of the points of inflexion in the columns, but we are about to do so. At the points of inflexion in the beams there will be vertical shear forces; let us call these X_1, X_2 and X_3 as shown. Now we can find X_1 by taking moments about the point of inflexion in the topmost column for the part ABC (or A′BC′):

$$X_1 \times 3 = (10 - 5) \times 2$$
$$X_1 = 10/3$$

Similarly for CDEF (or C′D′EF′):

$$X_2 \times 3 = (20 - 10) \times 2 + 5 \times 4$$
$$X_2 = 40/3$$

For FGHJ the reader should show that

$$X_3 = 80/3$$

We can now determine all the stress resultants and draw any distribution diagrams we need. The bending moment and thrust diagrams are shown in Fig. 12.6(b), each being drawn on one-half of the frame. Corresponding values of the stress resultants from an exact analysis are shown in brackets for comparison. Generally the approximate method gives good results.

12.6 Other techniques

We now consider some other devices which we can use to verify computer output or to obtain approximate values of stress resultants or displacements in structures.

12.6.1 *Expectation testing*

The structural analysis carried out by a computer can sometimes be verified by applying a 'test' load to the structure which should produce expected results. An extreme case is to apply no load at all and to check that the computer produces zero displacements and zero stress resultants. Alternatively a single load can be applied at a position and in a direction where the effect is very predictable and local. In Fig. 12.7(a) a vertical load W is applied

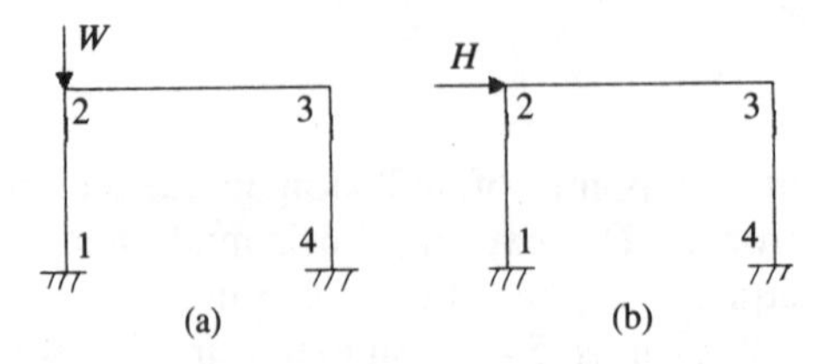

Figure 12.7 Test loading of structures to verify computer output

directly at node 2. With relatively small axial strains, the analysis should show a compression W in 1–2 and zero or very small stress resultants in 2–3 and 3–4. In Fig. 12.7(b) a horizontal load

H applied to a symmetrical frame should produce skew symmetry in the results with a point of inflexion at mid-span of the beam. Symmetry is often a useful guide in assessing the validity of the output of the computer, as is the principle of superposition in that loads *W* and *H* acting together should produce the same results as (a) + (b).

12.6.2 Deflected shapes

Here we take the values of the nodal displacements from the computer and 'sketch' the deformed shape of the structure or part of it. The purpose is to ensure that the computed displacements make physical sense. If it is difficult or impossible to make practical sense of the displacements then we must suspect the results, and the data and the output must be given further examination.

In Fig. 12.8 a two-storey frame is shown. The computer

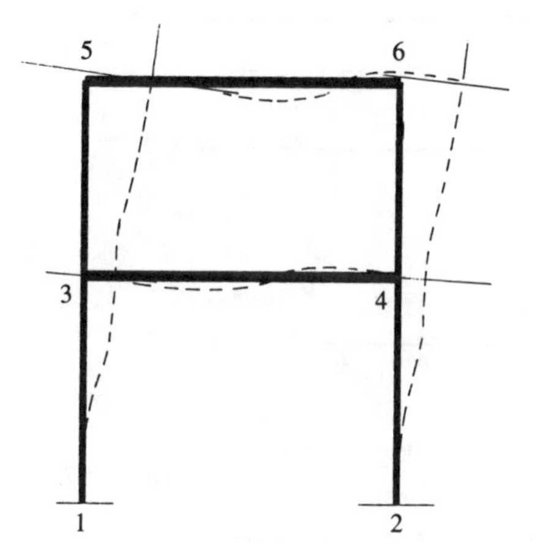

Figure 12.8 Sketch of deflected form

analysis gives clockwise rotations for joints 3, 4, 5 and 6 and sway displacements to the right for beams 5–6 and 3–4. The corresponding deformed shape is shown dashed and the signs of the bending moments should agree with the curvatures shown. Note also the points of inflexion.

12.6.3 Estimating reactions

It might appear that approximate values for stress resultants could be obtained in certain circumstances by estimating the values of statically indeterminate reactions. Although this approach can sometimes give reasonable results it can be seriously misleading and should therefore only be used with caution if at all. Consider structure 5F in Part II and assume that the horizontal reactions at 1 and 4 are each equal to one-half of the applied load. Assume also that the vertical reactions are not affected by the column moments at 1 and 4. Then the vertical reactions are each 50 kN (up at 4 and down at 1) and thus

$$M_{3,4} = -50 \times 5 + 50 \times 10 = 250 \text{ kN m}$$

This compares with the correct value of 167.8 kN m.

12.6.4 *Replacement of distributed loads*

Comprehensive structural analysis programs will usually handle a number of different types of load including uniformly and non-uniformly distributed loads. If we encounter a load which our package will not handle, say a non-uniformly distributed load, then we can approximate this quite acceptably as point loads at additional nodes.

Consider the beam shown in Fig. 12.9(a). The beam is

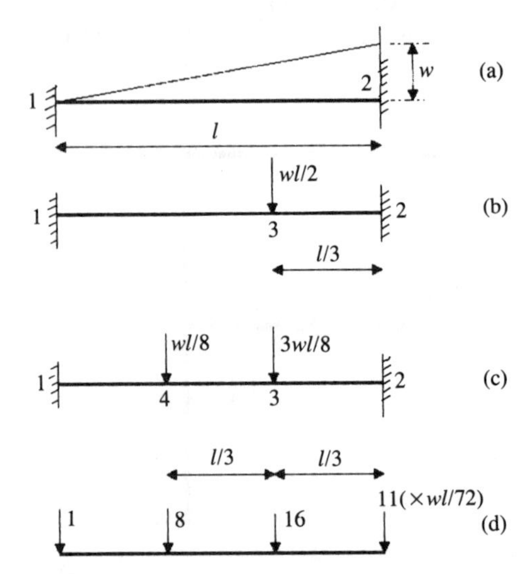

Figure 12.9 Replacement of distributed load by point loads

fixed-ended and carries a distributed load varying in intensity from zero at end 1 to w at end 2. It can be shown that the exact values of the end moments are

$$M_1 = wl^2/30$$

$$M_2 = wl^2/20$$

Let us see what we get for these moments if we replace the distributed load by point loads. Suppose we place the total load $(1/2)wl$ at the centre of gravity of the load as in Fig. 12.9(b). Then we can evaluate M_1 and M_2 as standard cases of fixed-end moments from Appendix 3:

$$M_1 = wl^2/27$$

$$M_2 = (2/27)wl^2$$

The errors are 10 per cent in M_1 and 32 per cent in M_2.

We might improve the representation by using two additional nodes as in Fig. 12.9(c) with point loads $wl/8$ and $(3/8)wl$. We then get

$$M_1 = (5/108)wl^2$$

$$M_2 = (7/108)wl^2$$

The errors are now 28 per cent in M_1 and 23 per cent in M_2. In all cases the end moments are overestimated. Better results will be obtained if the applied load is shared, with some proportions going to the end nodes 1 and 2 as shown in Fig. 12.9(d). We then get

$$M_1 = (8/243)wl^2$$

$$M_2 = (10/243)wl^2$$

The errors are now 1.25 per cent and 21 per cent, but additional nodes will improve this.

12.6.5 Structural simplification

Sometimes an approximate solution can be obtained by some simplification of detail of the structure whilst retaining the fundamental structural form. The ten-storey single-bay frame shown in Fig. 12.10(a) is basically a form of cantilever, and this

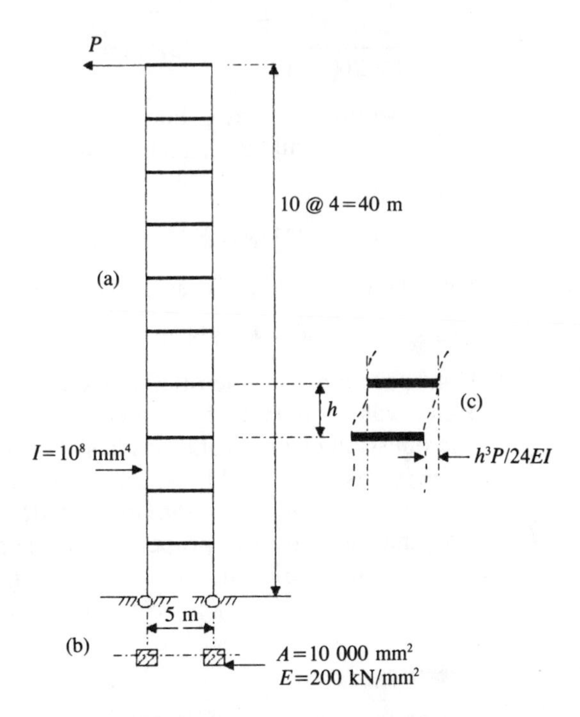

Figure 12.10 Structural simplification of multi-storey frame

suggests an approach to an approximate analysis. The contribution to sidesway displacement under the force P comes from three principal sources:

1. Horizontal shear deformations of the panels
2. Axial deformations of the columns
3. Rotation of the joints.

If the flexural stiffness of the beams is significantly greater than that of the columns, then we could ignore source 3 and develop a solution allowing for sources 1 and 2 only.

The cross-sectional area of the columns is 10 000 mm², so we take the cross-section of the substitute cantilever as shown in Fig. 12.10(b). Hence the second moment of area of this is

$$
\begin{aligned}
I &= 2 \times 10\,000 \times 2500^2 \\
&= 1250 \times 10^8 \text{ mm}^4
\end{aligned}
$$

The deflection at the tip of the cantilever due to axial deformation in the columns is therefore

$$
\begin{aligned}
\Delta_c &= Ph^3/3EI \\
&= \frac{P \times 40\,000^3}{3 \times 200 \times 1250 \times 10^8} = 0.853P \text{ mm} \qquad (P \text{ in kN units})
\end{aligned}
$$

The deflection due to shear deformation (sidesway) of the columns is, from Fig. 12.10(c),

$$
\Delta_s = \frac{10 \times 4000^3 \times P}{24 \times 200 \times 10^8} = 1.333P \text{ mm}
$$

(Here we use I for the columns, which is 10^8 mm⁴.)

The total estimated deflection is

$$
\begin{aligned}
\Delta &= \Delta_c + \Delta_s \\
&= (0.853 + 1.333)P = 2.186P \text{ mm}
\end{aligned}
$$

Thus the estimated stiffness of the frame in the direction of P is

$$
P/\Delta = 10^3/2.186 = 457 \text{ kN/m}
$$

This approximate value of stiffness should be compared with the 'exact' value obtained for structure 10F in Part II (302 kN/m). The omission of joint rotations has clearly increased the apparent stiffness, as we should expect. The relative contributions of sources 1, 2 and 3 to the total deflection can be varied by changing the structural properties of beams and columns. This is an ideal parametric study for the computer.

Exercises

12.1 The computer output for the analysis of the structure shown in Fig. 12.4(a) is given below. The data used was $E = 200$ kN/mm², $A = 10^4$ mm² for all members, $I = 10^8$ mm⁴ for the columns and

$I = 2 \times 10^8$ mm^4 for the beam. The moments are in kN m units.

$M_{AB} = 157.6$ $M_{BA} = 121.6$ $M_{CD} = 150.6$

$\theta_B = -3.598 \times 10^{-3}$ $\theta_C = -1.043 \times 10^{-3}$ $\theta_D = -1.233 \times 10^{-2}$

$\Delta_{BC} = 25.8$ mm

Use these results to compute the applied loads and to sketch the deflected shape of the structure.

12.2 In the frame shown in Fig. 12.11 the beams GH and DE may be considered to have infinite flexural rigidity, whereas the columns

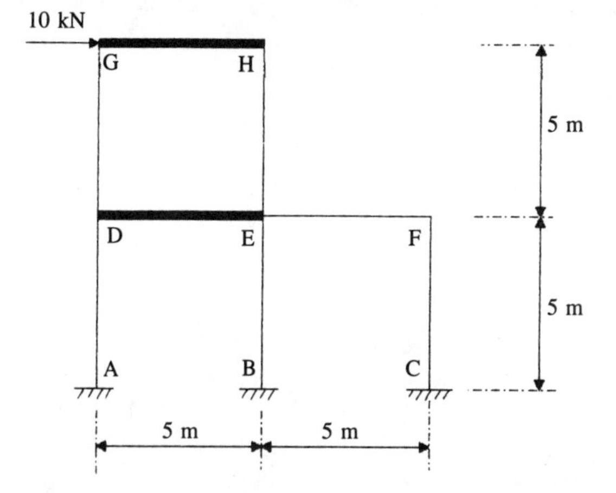

Figure 12.11 Exercise 12.2

and the beam EF have $EI = 1000$ kN m^2. Calculate the horizontal deflection of the beam GH. If for rapid analysis the beam EF is also considered to have infinite EI, what would be the error in the calculated deflection?

12.3 Use moment distribution to calculate the bending moments in the continuous beam of structure 1F in Part II.

PART

II

Workshop

The purpose here is to help the reader to consolidate knowledge of structural theory and to assist in the develoment of skills of application and analysis. The approach will be from a number of directions, the underlying intention throughout being to develop the facilities of both computer analysis and 'hand' analysis in a complementary style. It is important also to develop a critical approach at all times and to question the output of every computer analysis: does it look right, is it reliable, can I understand it, does it make sense? If the answer to any of these questions is 'no' then something must be done to establish confidence. In Chapter 12 we examined several ways of assessing results by approximate and other methods; we must now go on to apply these as appropriate to the circumstances of any particular structure.

In carrying out the tasks set in the workshop, computer analysis can be by any suitable program package. However, programs are provided in Part III for both plane truss (PLATRUSS) and plane frame (PLANFRAM) analyses, and the structures to be studied are designated either T for truss or F for frame. PLANFRAM will allow the introduction of pin-ended members, simply by allocating zero or small EI to such members, so it can be used for mixed structures.

Verification of 'hand' calculations

In this approach the solution is worked by hand and the result is verified by computer. It matters not at all that the computer will be used to solve an extremely simple problem; it is a very willing and economical worker! Generally the methods employed will be 'exact' in character, so the results should agree. Differences will therefore reflect an error, and so must be explored until agreement is reached. Checks should be made on

1. Fundamentals of theory used in 'hand' method
2. Application of theory

3. Consistency of units employed
4. Sign conventions
5. Arithmetic

and, since the fault may be in the computer results,

6. The input data
7. The interpretation of the output.

Assessment of approximate methods

Under this heading, problems will be tackled to obtain solutions by approximate methods. In some cases the approximation will be 'crude' in order to get a 'feel' for the credibility of the assumptions made. The assessment should include some differentiation between 'critical' and 'non-critical' results, since the validity of an approximate method must be measured on the more critical aspects of the analysis. Significant differences might indicate an error in the computer results due to error in input (frequently), misinterpretation of output, or misunderstanding of the basis of the method employed by the program. Both PLATRUSS and PLANFRAM in Part III are provided with validation exercises designed to test the working of the programs and the user's understanding of the data input and intepretation of output. The validation exercises can of course be used with any analysis package if validation or practice is needed. It cannot be overemphasized that with all computer output the question should always be asked: does it look reasonable?

Parametric studies

Once sufficient experience has been gained and the reader has developed confidence in the software employed, parametric studies should be undertaken on structures to study the influences of

1. Geometry
2. Member cross-sections
3. Material properties
4. Lack of fit
5. Environmental effects
6. Addition and removal of structural elements
7. Changes in support conditions.

Studies of structural types

These include

1. Structures with pinned joints, rigid joints, and both
2. Structures with sloping members

3. Structures with cables
4. Structures with non-rigid supports, usually treated as springs; PLANFRAM contains a facility for these.

Step-by-step analysis

A typical sequence in this learning approach would be

1. Determine reactions and check with computer.
2. Determine forces and/or bending moments in members and check with computer one by one.
3. Calculate shear forces and check.
4. Calculate displacements and check.

Both PLATRUSS and PLANFRAM have output routines constructed so that the user can access results in this way.

A note on the use of units

In PLATRUSS and PLANFRAM any units can be used providing they are used consistently. If kN and mm are used at input, then output will be in the same units: bending moments will be kN mm, for example. Whereas in hand calculations in structural analysis much use is made of relative values of EI and EA from member to member, in the computer it is better always to use real practical values of the quantities. Relative values can be used but great care is needed, for example, if we are handling ratios I/A, since this ratio has the dimensions $(\text{length})^2$ and will change if the units are changed.

The tasks given for the various structures are not exhaustive and the reader is encouraged to study beyond what is asked whenever there seems something interesting to be investigated.

Sign conventions

The sign conventions used in the workshop are the same as those adopted elsewhere in the book and defined in Fig. 8.9. Since the positive Z axis is directed towards the viewer, moments and rotations viewed in the plane of the paper (the X–Y plane) appear anticlockwise if positive. A change in sign convention for applied vertical loads only is introduced in the computer programs in order to make vertically downwards loads positive. Prompts are displayed by the programs to remind the user.

Truss structures

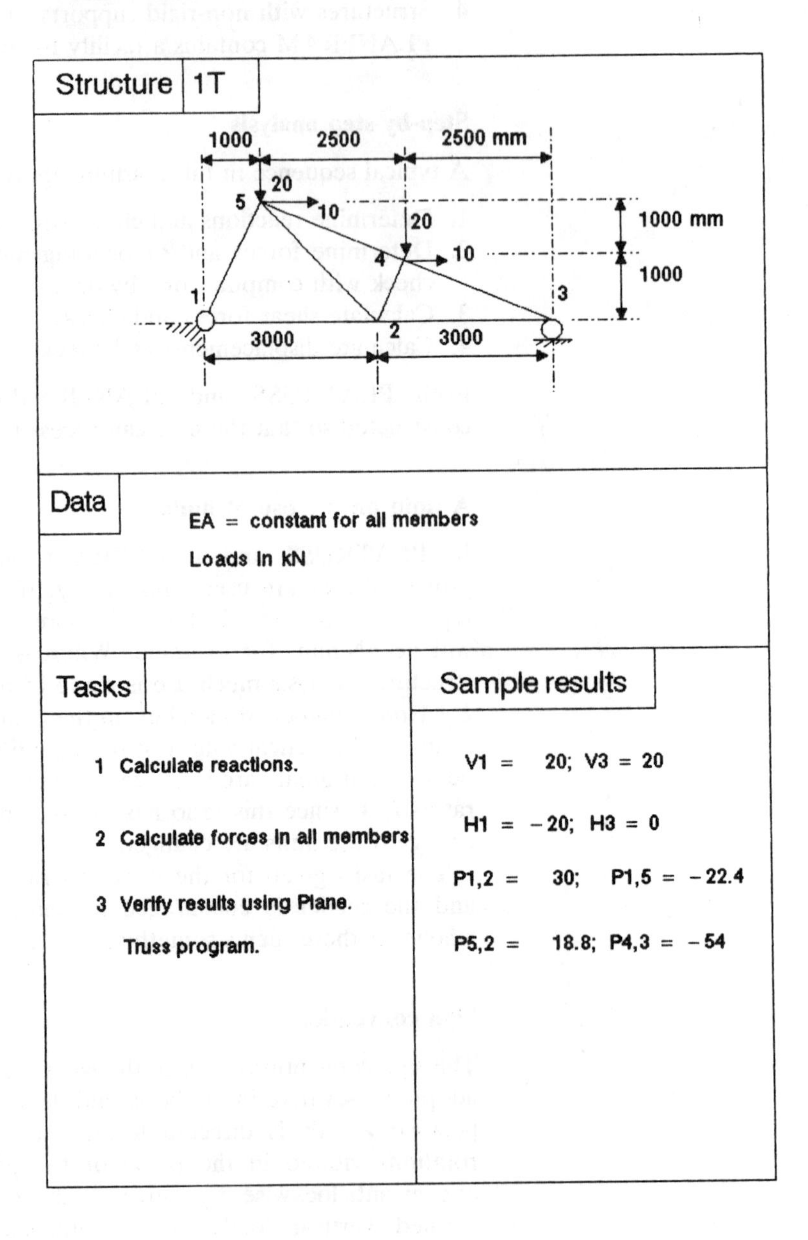

Structure	2T

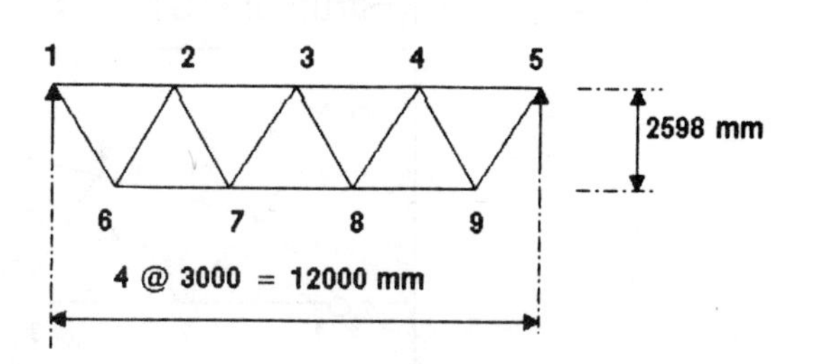

Data

For all members; E = 200 kN/mm2 ; A = 4000 mm2

Tasks	Sample results
1 Find forces in members for unit vertical load at 2.	P2,3 = −0.722; P7,3 = 0.289 P7,8 = 0.577
2 Repeat for unit vertical load at 3.	P2,3 = −0.866; P7,3 = −0.577 P7,8 = 1.155
3 Construct influence lines for P2,3 , P7,3 and P7,8 and compare with computed results.	

Structure 3T

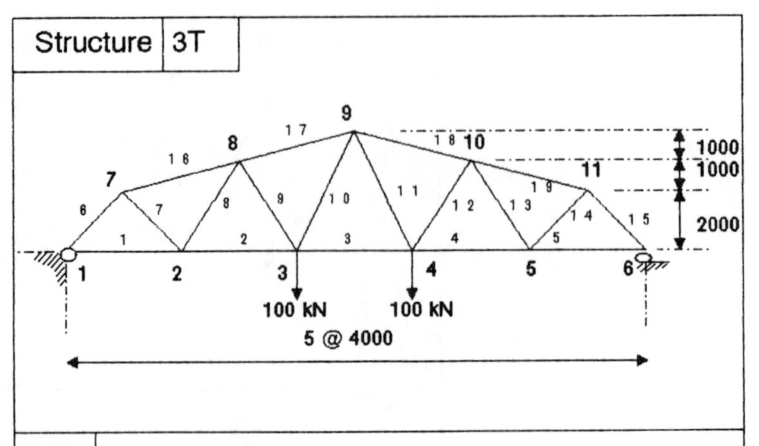

Data

EA = constant for all members

Tasks

1 Calculate forces in all members by hand.

2 Confirm results by computer.

Sample results

P1 = 100; P2 = 200

P6 = −141.42; P7 = 84.85

P10 = 63.89; P17 = −235.6

P3 = 200

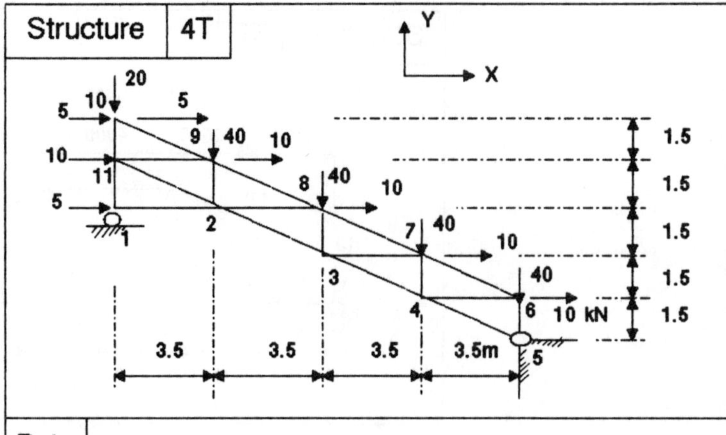

Data

EA = 10000 kN for all members

Tasks

1. Calculate reactions at 1 and 5.
2. Calculate forces in all members by hand.
3. Confirm results using computer.

Sample results

V1 = 60.7; V5 = 119.3; H5 = −60

P1,2 = −5; P11,2 = 108.8

P11,9 = −110; P10,9 = −5.44

P1,11 = −60.7; P11,10 = −17.86

Structure 5T

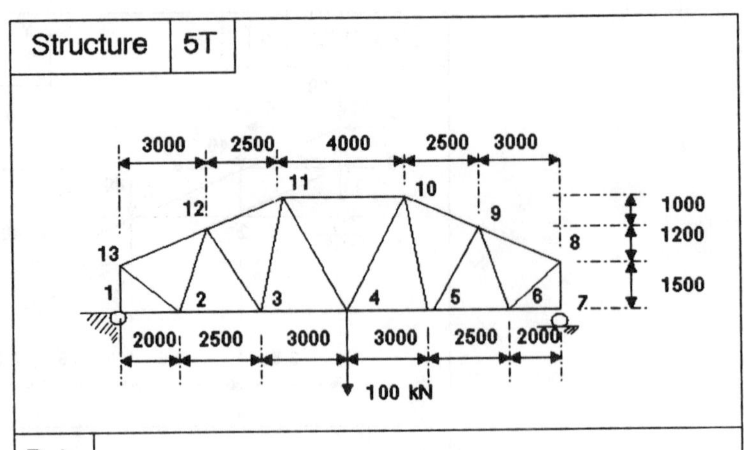

Data

For all members, E = 200 kN/mm2, A = 2500 mm2

Tasks

1 Calculate all reactions and members forces by hand.

2 Verify results using computer.

Sample results

V1 = 50; H = 0

P1,2 = 0; P2,3 = 55.6

P3,4 = 74.3; P13,2 = 54.3

P2,12 = −34.77

Structure 6T

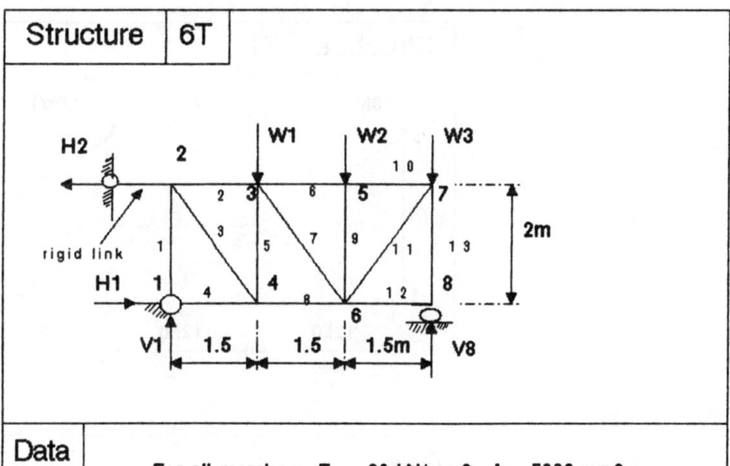

Data

For all members, E = 20 kN/mm2; A = 5000 mm2

Tasks

1 Given forces in members below, calculate by hand, W1,W2,W3 and reactions.
2 Verify by computer.

Member	force(kN)	Member	force
1	−11.2	2	−5.69
3	14.0	4	−2.708
5	−11.2	6	−6.60
7	1.505	8	5.694
9	−10	10	−6.60
11	11	12	0
13	−13.8		

Sample results

W1 = 10 kN

W2 = 10 kN

W3 = 5 kN

V1 = 11.2

V8 = 13.8

H1 = 2.708

H2 = 2.708

Structure	7T

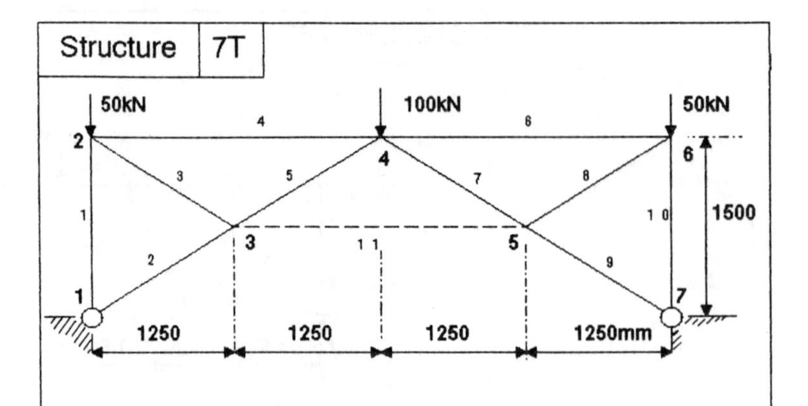

Data

For all members, EA = 10 000 kN

Tasks	Sample results
1 Calculate reactions and forces in all members.	V1 = V7 = 100; H1 = H7 = 83.3 P1 = P10 = − 50; P3 = P4 = P6 = P8 = 0 P2 = P5 = P7 = P9 = − 97.18
2 Remove horizontal restraint at 7 and insert member 3,5. Recalculate reactions and member forces.	P11 = 166.7 P2 = P9 = 0 P5 = P7 = − 97.18
3 Repeat 2 with horizontal restraint at 7 re imposed.	H1 = H7 = 73.32 P1 = − 56;P11 = 20.03

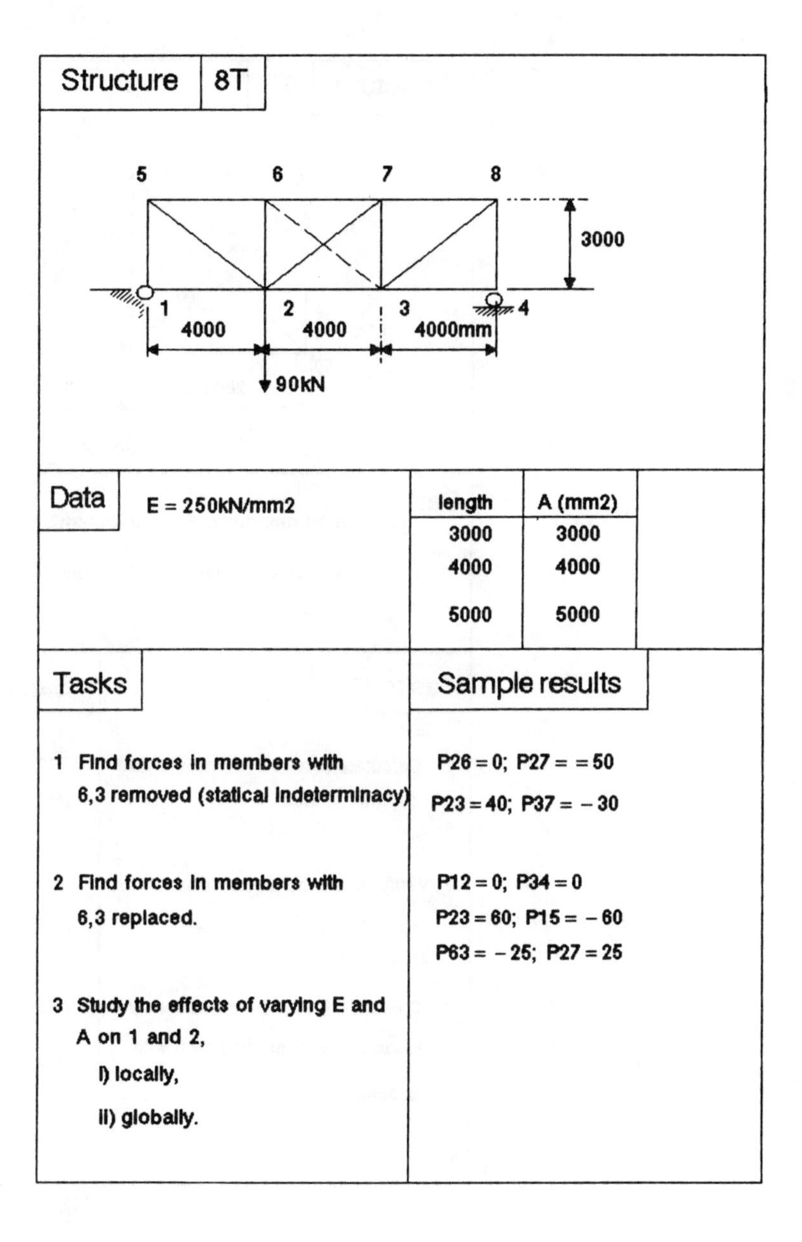

Structure 8T

Data

E = 250kN/mm2

length	A (mm2)
3000	3000
4000	4000
5000	5000

Tasks	Sample results
1 Find forces in members with 6,3 removed (statical indeterminacy)	P26 = 0; P27 = = 50 P23 = 40; P37 = − 30
2 Find forces in members with 6,3 replaced.	P12 = 0; P34 = 0 P23 = 60; P15 = − 60 P63 = − 25; P27 = 25
3 Study the effects of varying E and A on 1 and 2, i) locally, ii) globally.	

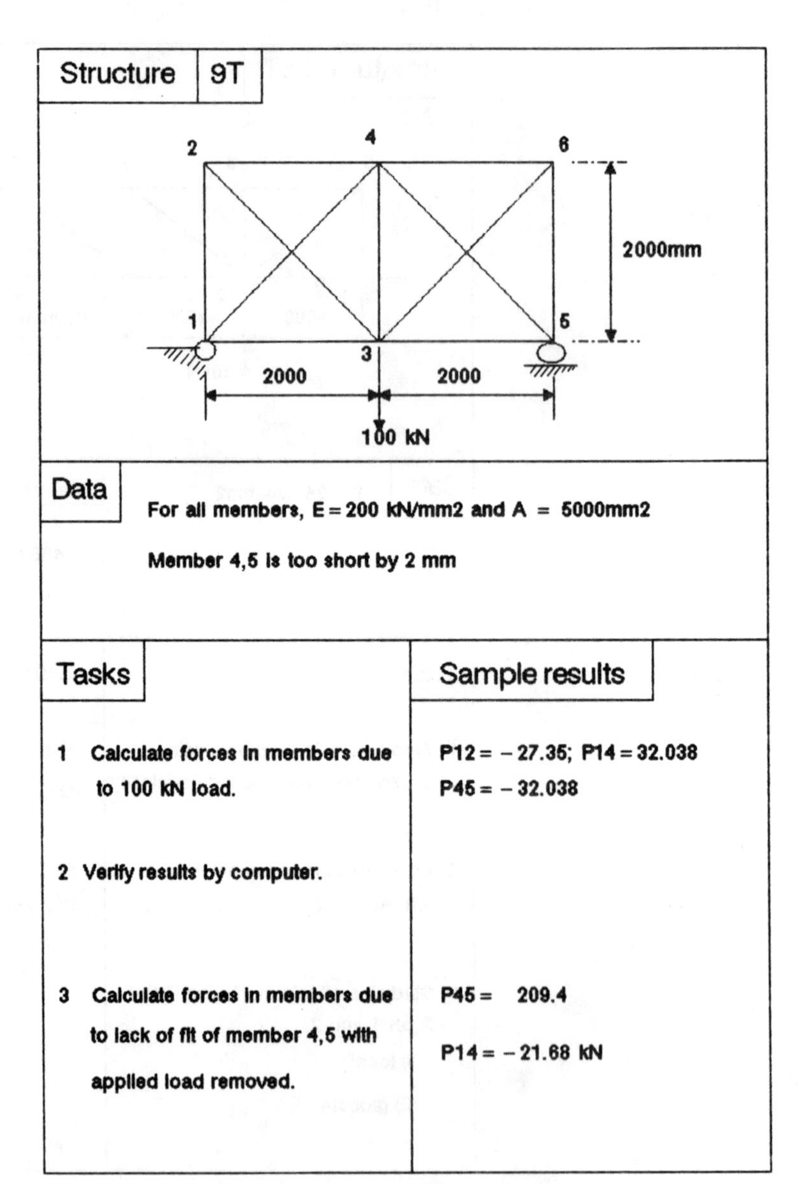
Structure
9T
2
4
6
2000mm
1
5
3
2000
2000
100 kN
Data
For all members, E = 200 kN/mm2 and A = 5000mm2
Member 4,5 is too short by 2 mm
Tasks
Sample results
1 Calculate forces in members due to 100 kN load.
P12 = − 27.35; P14 = 32.038
P45 = − 32.038
2 Verify results by computer.
3 Calculate forces in members due to lack of fit of member 4,5 with applied load removed.
P45 = 209.4
P14 = − 21.68 kN

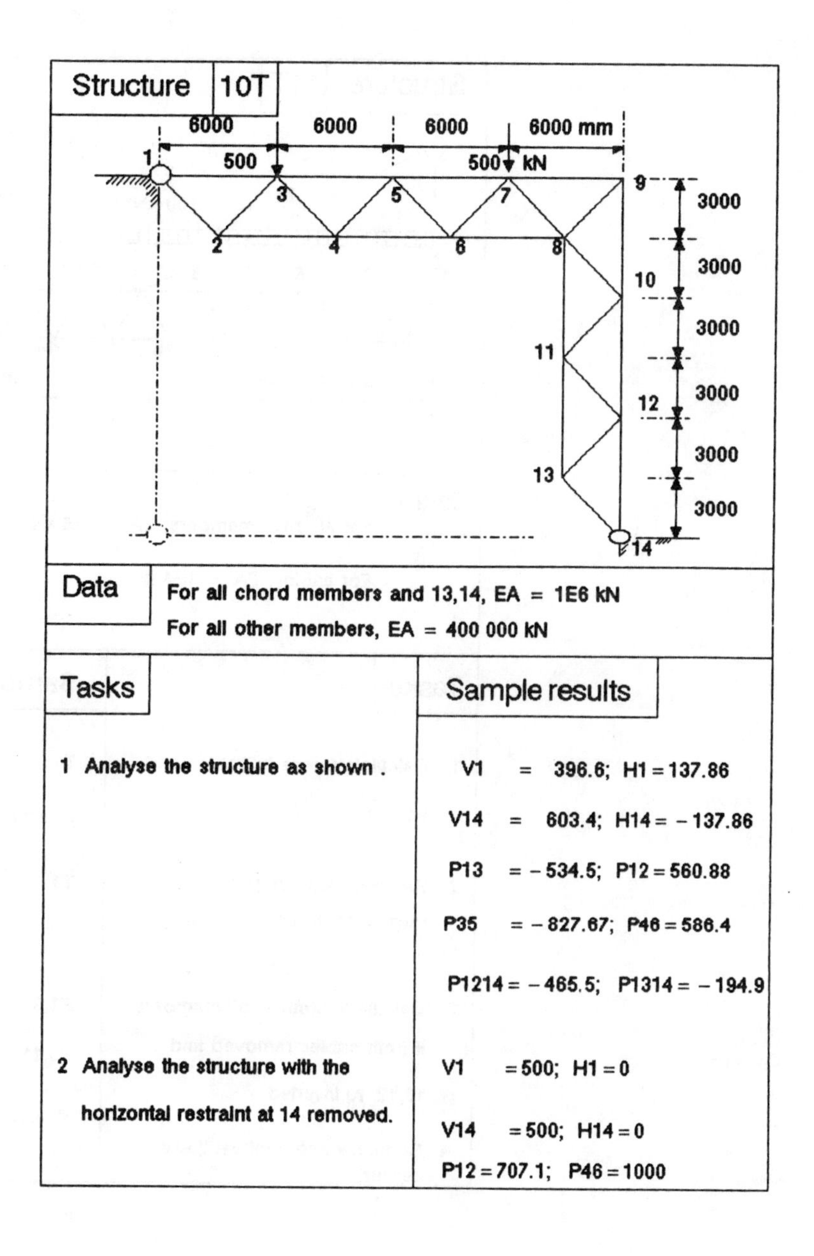
Structure
10T
6000
6000
6000
6000 mm
500
500 kN
1
2
3
4
5
6
7
8
9
10
11
12
13
14
3000
3000
3000
3000
3000
3000
Data
For all chord members and 13,14, EA = 1E6 kN
For all other members, EA = 400 000 kN
Tasks
1 Analyse the structure as shown .
2 Analyse the structure with the horizontal restraint at 14 removed.
Sample results
V1 = 396.6; H1 = 137.86
V14 = 603.4; H14 = − 137.86
P13 = − 534.5; P12 = 560.88
P35 = − 827.67; P46 = 586.4
P1214 = − 465.5; P1314 = − 194.9
V1 = 500; H1 = 0
V14 = 500; H14 = 0
P12 = 707.1; P46 = 1000

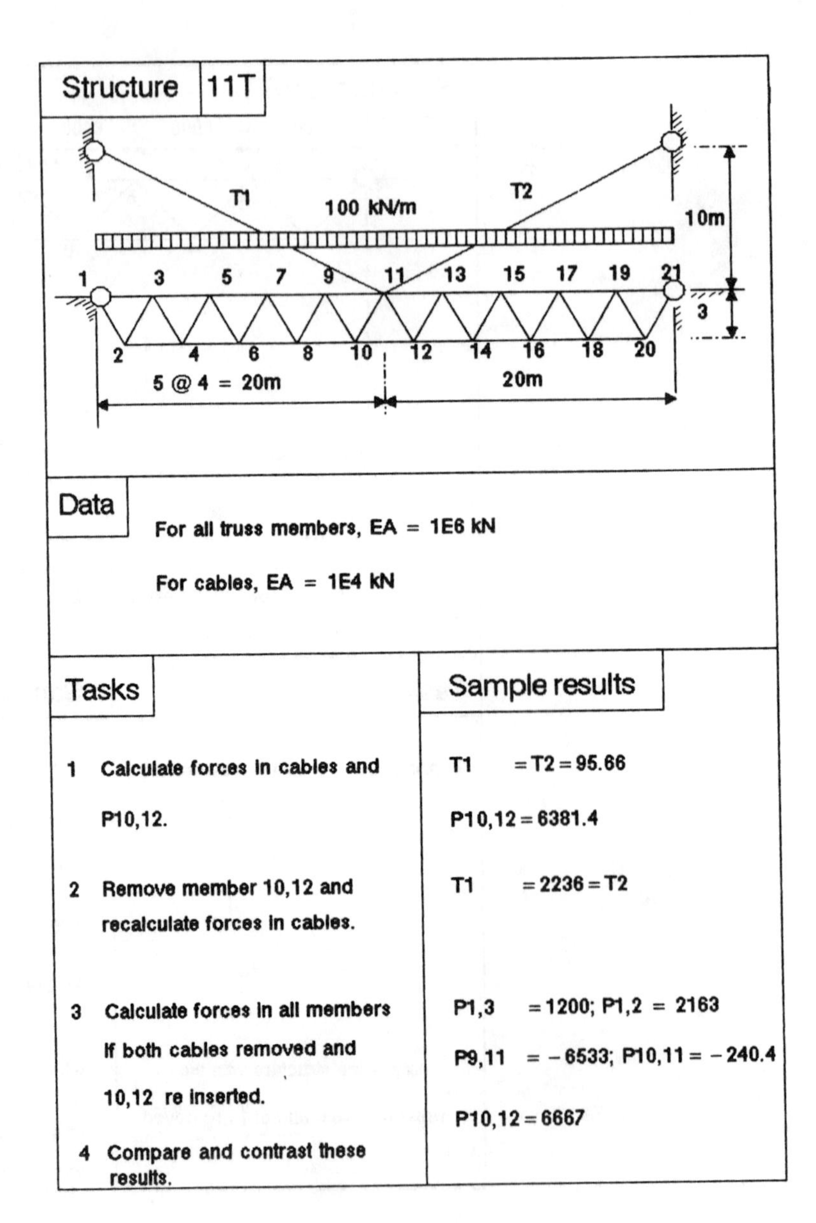

Data

For all truss members, EA = 1E6 kN

For cables, EA = 1E4 kN

Tasks	Sample results
1 Calculate forces in cables and P10,12.	T1 = T2 = 95.66 P10,12 = 6381.4
2 Remove member 10,12 and recalculate forces in cables.	T1 = 2236 = T2
3 Calculate forces in all members if both cables removed and 10,12 re inserted.	P1,3 = 1200; P1,2 = 2163 P9,11 = −6533; P10,11 = −240.4 P10,12 = 6667
4 Compare and contrast these results.	

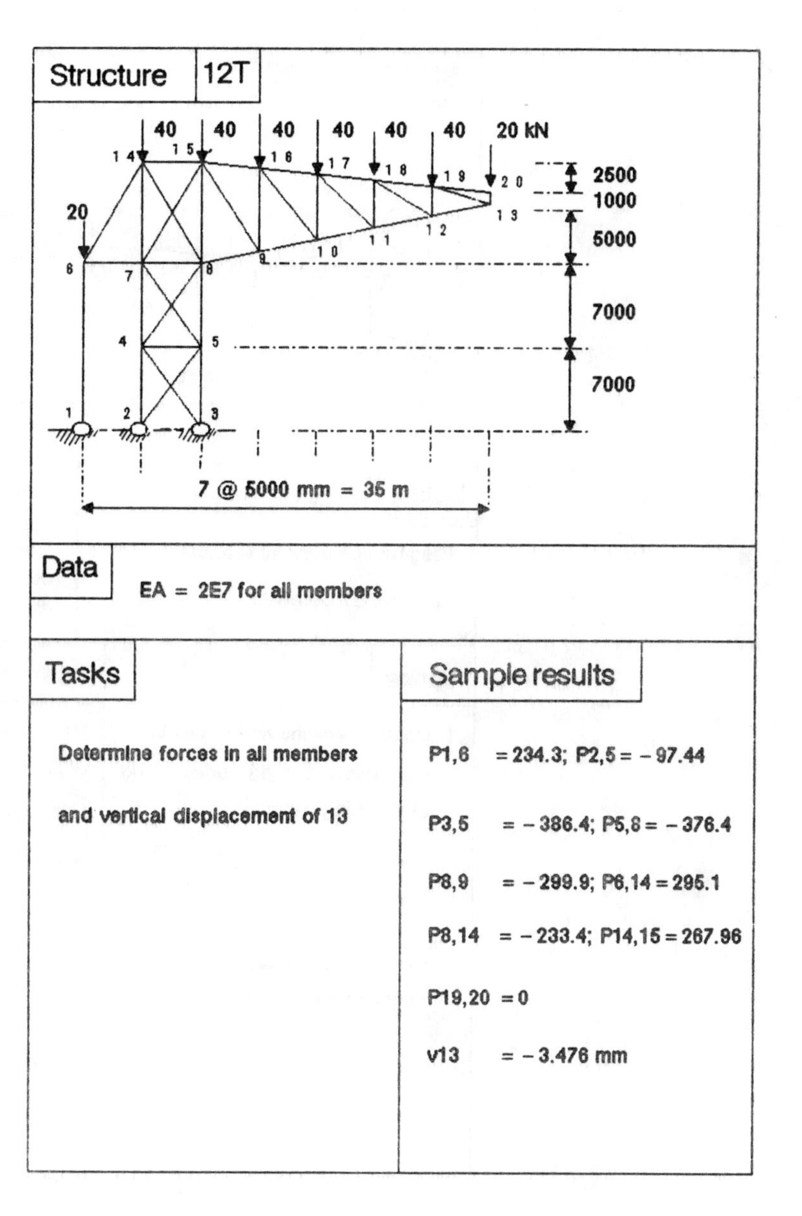
Structure
12T
40
40
40
40
40
40
20 kN
20
2500
1000
5000
7000
7000
7 @ 5000 mm = 35 m
Data
EA = 2E7 for all members
Tasks
Determine forces in all members
and vertical displacement of 13
Sample results
P1,6 = 234.3; P2,5 = − 97.44
P3,5 = − 386.4; P5,8 = − 376.4
P8,9 = − 299.9; P6,14 = 295.1
P8,14 = − 233.4; P14,15 = 267.96
P19,20 = 0
v13 = − 3.476 mm

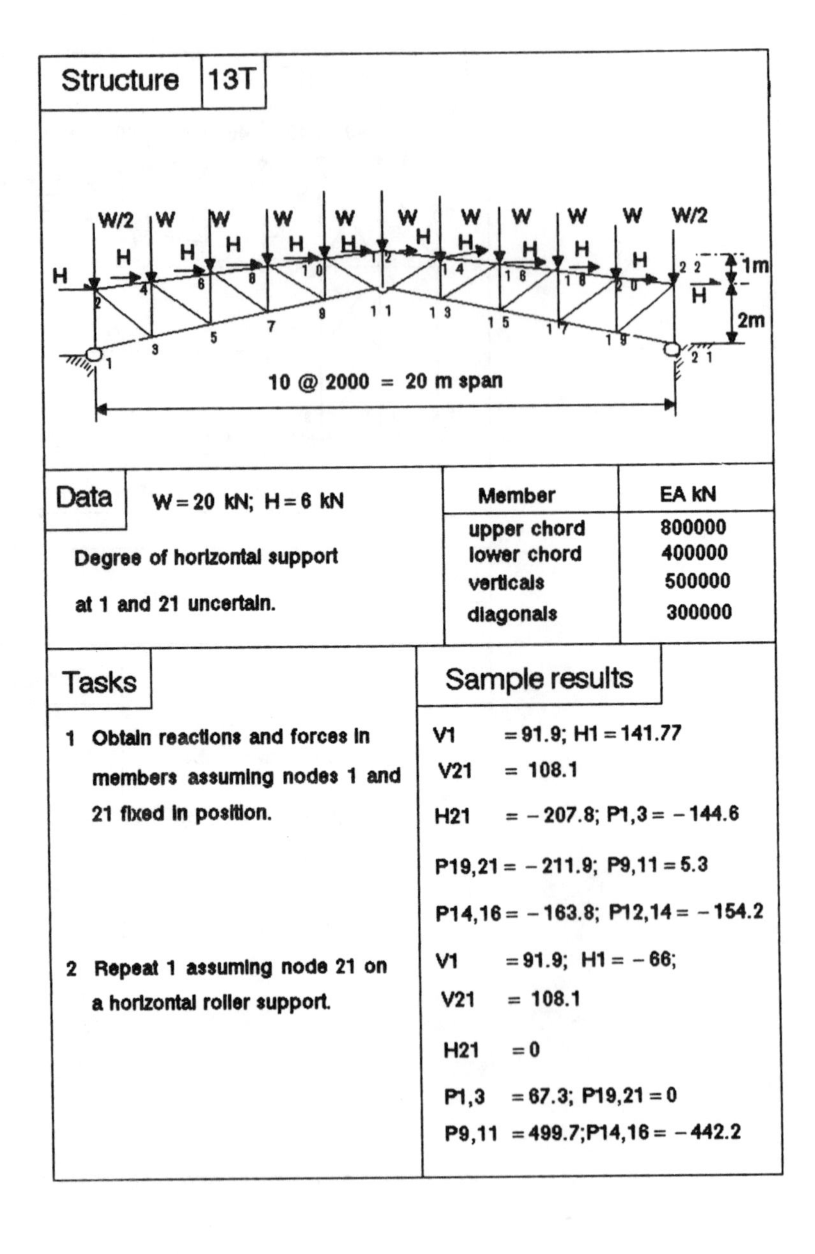

Structure 13T

Data W = 20 kN; H = 6 kN

Degree of horizontal support at 1 and 21 uncertain.

Member	EA kN
upper chord	800000
lower chord	400000
verticals	500000
diagonals	300000

Tasks	Sample results
1 Obtain reactions and forces in members assuming nodes 1 and 21 fixed in position.	V1 = 91.9; H1 = 141.77 V21 = 108.1 H21 = − 207.8; P1,3 = − 144.6 P19,21 = − 211.9; P9,11 = 5.3 P14,16 = − 163.8; P12,14 = − 154.2
2 Repeat 1 assuming node 21 on a horizontal roller support.	V1 = 91.9; H1 = − 66; V21 = 108.1 H21 = 0 P1,3 = 67.3; P19,21 = 0 P9,11 = 499.7;P14,16 = − 442.2

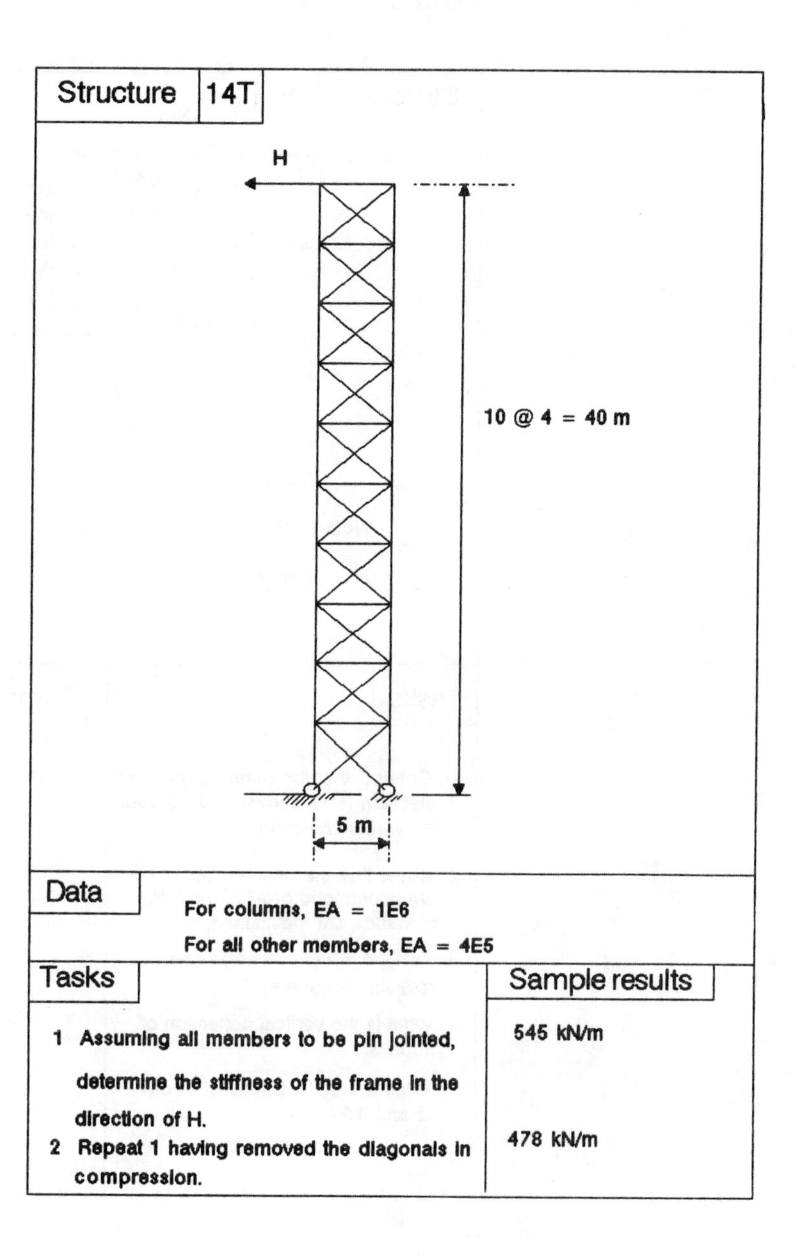

Structure 14T

Data

For columns, EA = 1E6

For all other members, EA = 4E5

Tasks	Sample results
1 Assuming all members to be pin jointed, determine the stiffness of the frame in the direction of H.	545 kN/m
2 Repeat 1 having removed the diagonals in compression.	478 kN/m

Frame structures

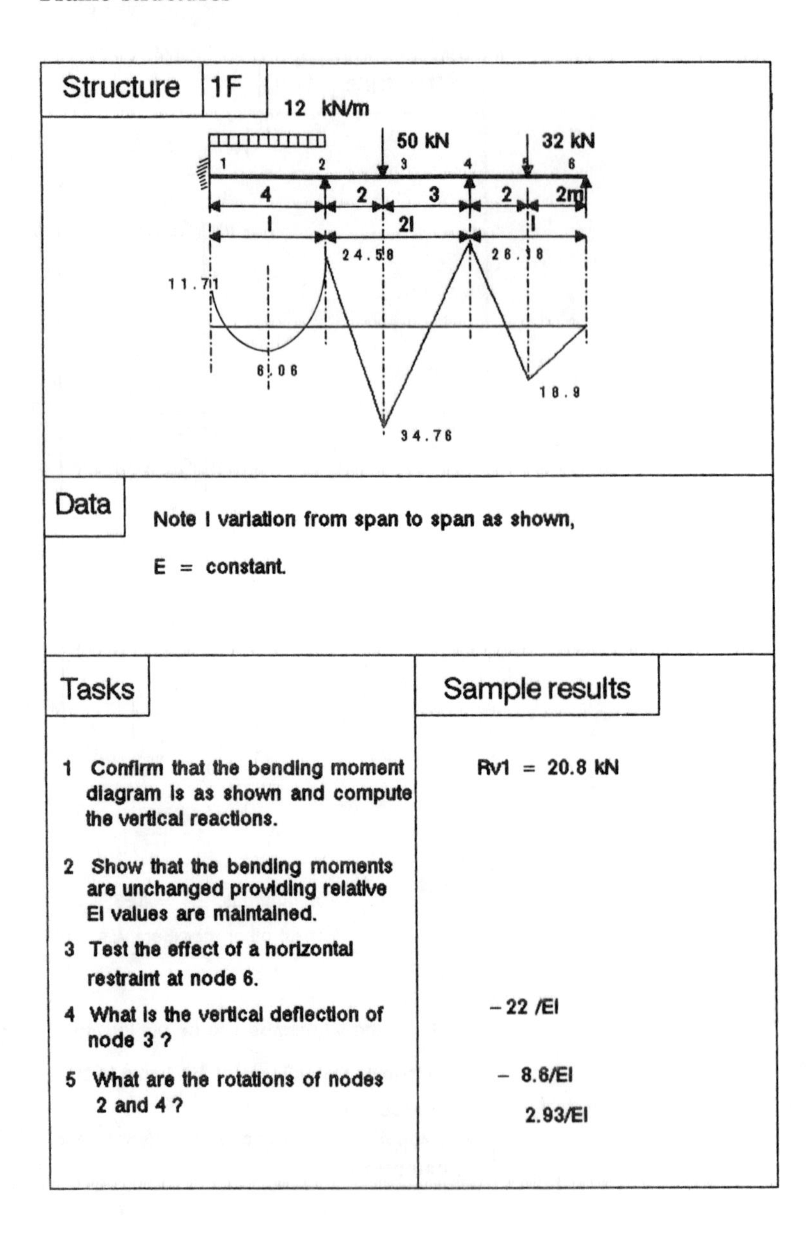

Data

Note I variation from span to span as shown,

E = constant.

Tasks	Sample results
1 Confirm that the bending moment diagram is as shown and compute the vertical reactions.	Rv1 = 20.8 kN
2 Show that the bending moments are unchanged providing relative EI values are maintained.	
3 Test the effect of a horizontal restraint at node 6.	
4 What is the vertical deflection of node 3 ?	−22 /EI
5 What are the rotations of nodes 2 and 4 ?	− 8.6/EI 2.93/EI

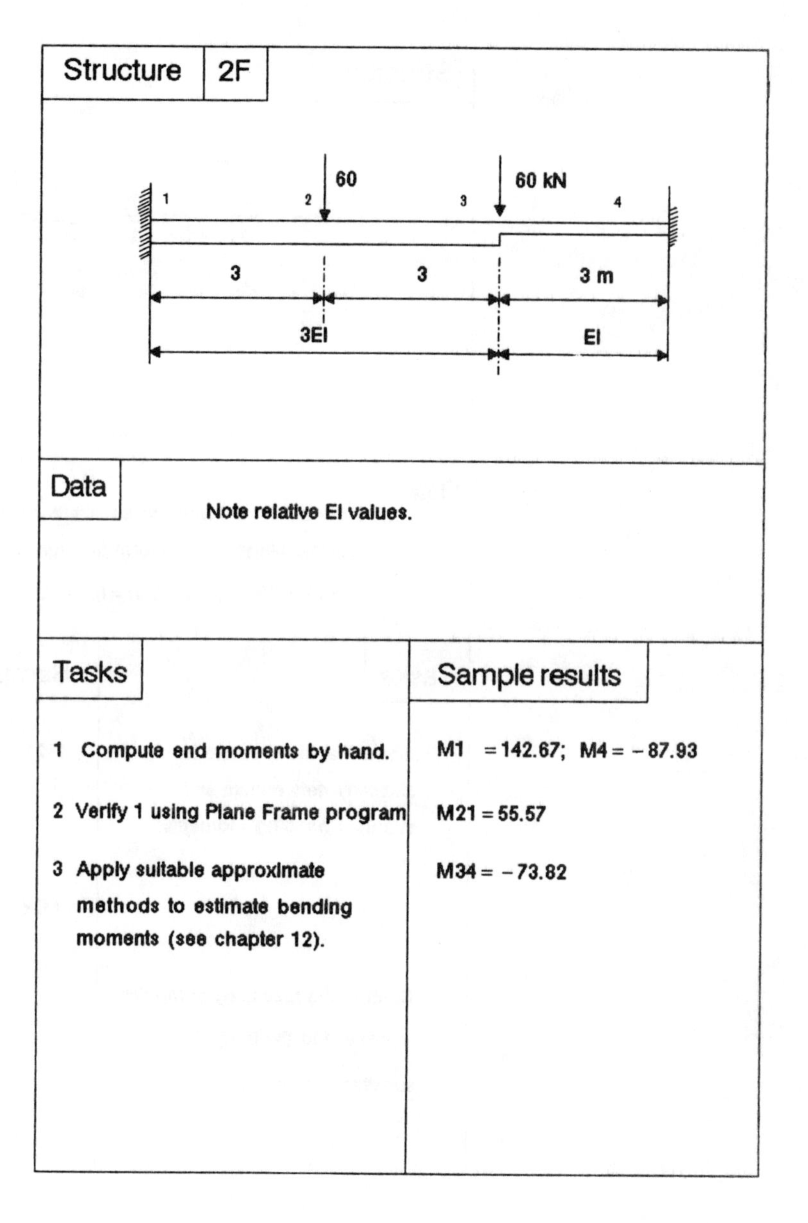

Structure 2F

Data

Note relative EI values.

Tasks

1 Compute end moments by hand.

2 Verify 1 using Plane Frame program

3 Apply suitable approximate methods to estimate bending moments (see chapter 12).

Sample results

M1 = 142.67; M4 = −87.93

M21 = 55.57

M34 = −73.82

Structure 3F

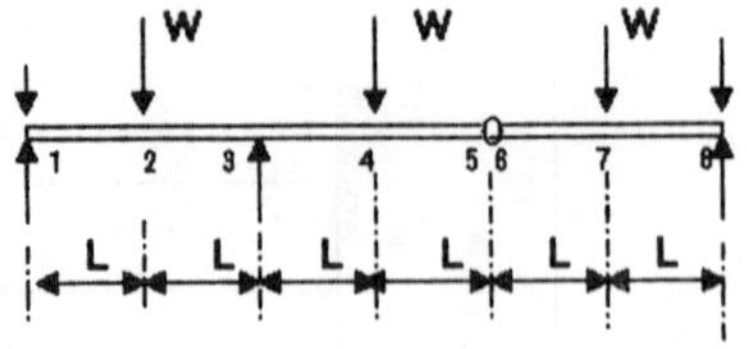

Data

Note additional node to constitute member 5,6 to represent internal hinge. EI uniform throughout.

I (5,6) = 1E − 7; L (5,6) = 0.001 L

Tasks

1 Show that the structure is statically determinate and evaluate bending moments.

2 Confirm the results by computer representing the hinge by a substitute member.

Sample results

M21 = −0.5 W

M34 = 2W

M43 = −0.5 W

M78 = −0.5 W

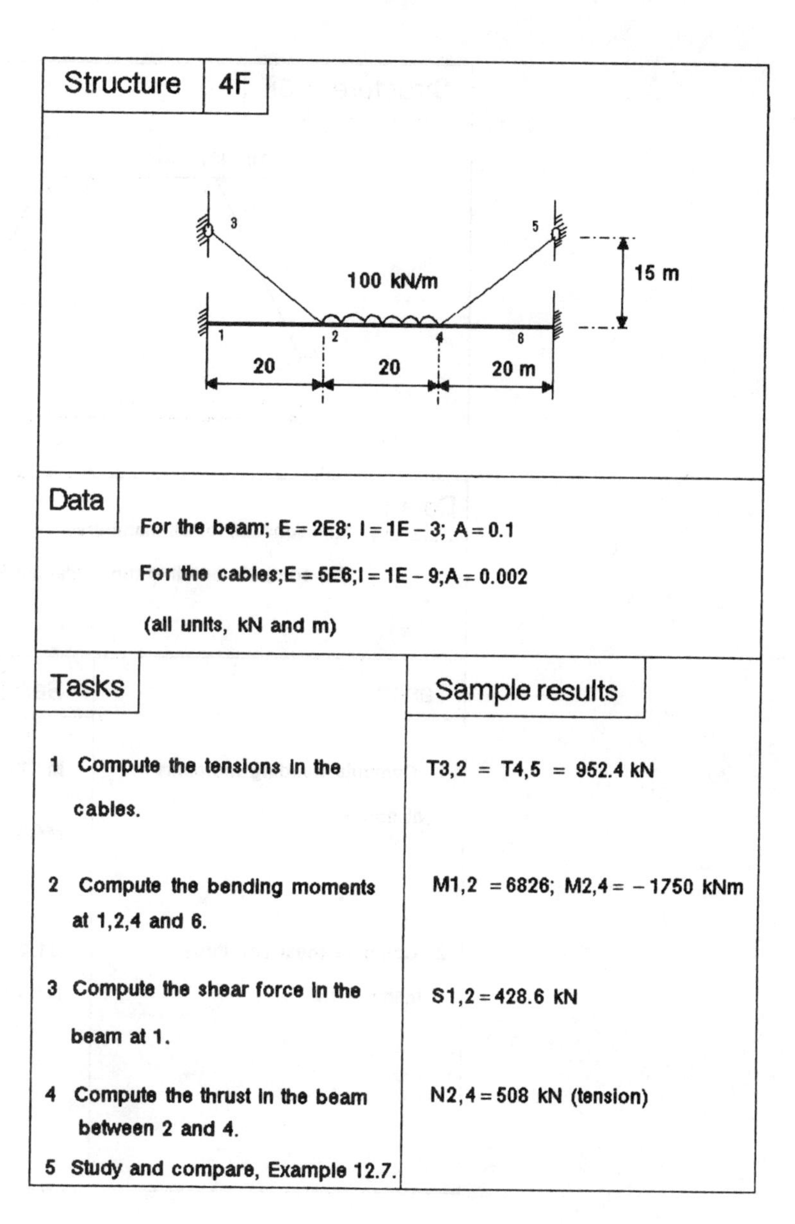

Structure	4F

Data

For the beam; E = 2E8; I = 1E − 3; A = 0.1

For the cables; E = 5E6; I = 1E − 9; A = 0.002

(all units, kN and m)

Tasks	Sample results
1 Compute the tensions in the cables.	T3,2 = T4,5 = 952.4 kN
2 Compute the bending moments at 1,2,4 and 6.	M1,2 = 6826; M2,4 = −1750 kNm
3 Compute the shear force in the beam at 1.	S1,2 = 428.6 kN
4 Compute the thrust in the beam between 2 and 4.	N2,4 = 508 kN (tension)
5 Study and compare, Example 12.7.	

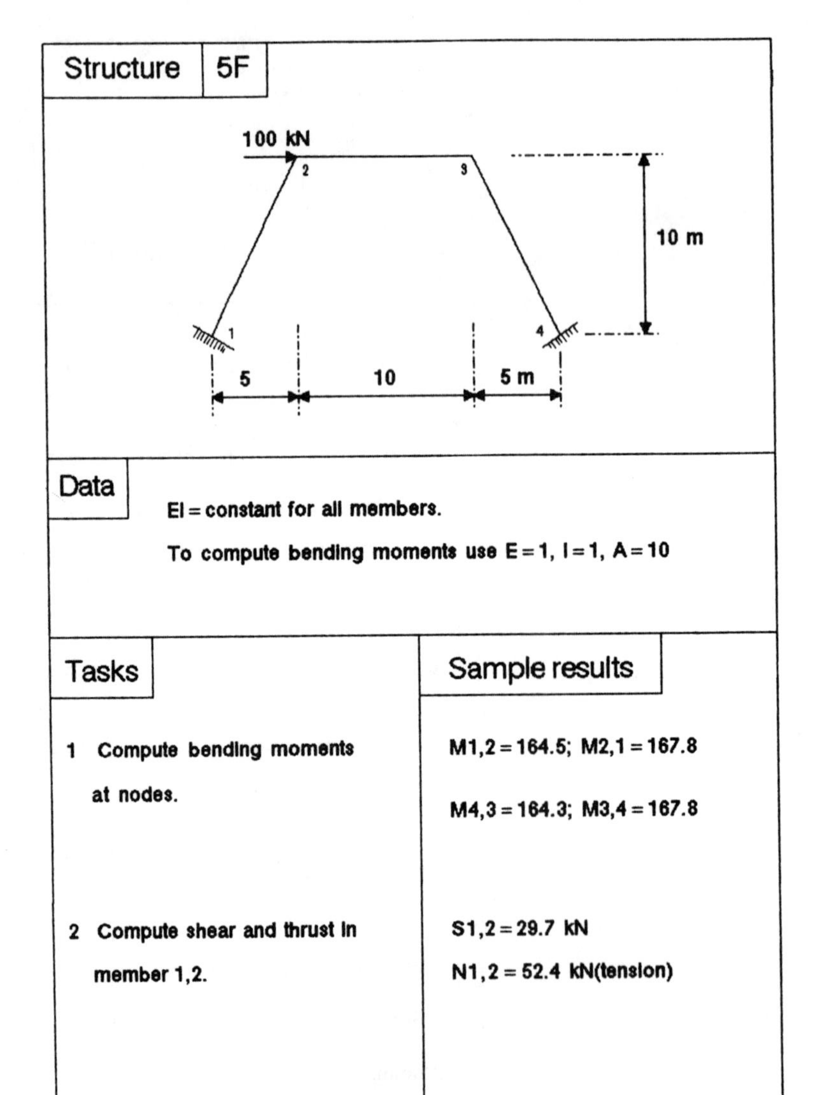

Structure | **5F**

Data

EI = constant for all members.

To compute bending moments use E = 1, I = 1, A = 10

Tasks	Sample results
1 Compute bending moments at nodes.	M1,2 = 164.5; M2,1 = 167.8 M4,3 = 164.3; M3,4 = 167.8
2 Compute shear and thrust in member 1,2.	S1,2 = 29.7 kN N1,2 = 52.4 kN(tension)

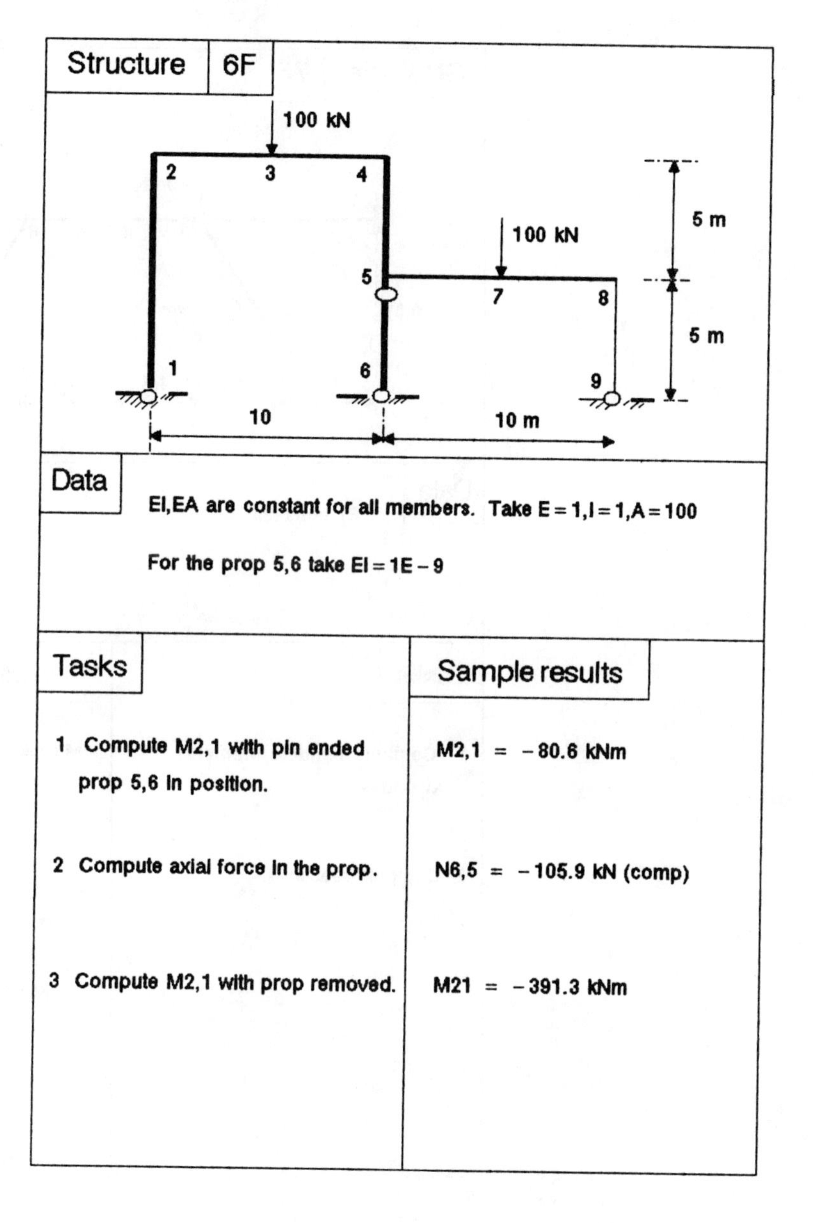

Structure 6F

Data

EI,EA are constant for all members. Take E = 1, I = 1, A = 100

For the prop 5,6 take EI = 1E − 9

Tasks	Sample results
1 Compute M2,1 with pin ended prop 5,6 in position.	M2,1 = −80.6 kNm
2 Compute axial force in the prop.	N6,5 = −105.9 kN (comp)
3 Compute M2,1 with prop removed.	M21 = −391.3 kNm

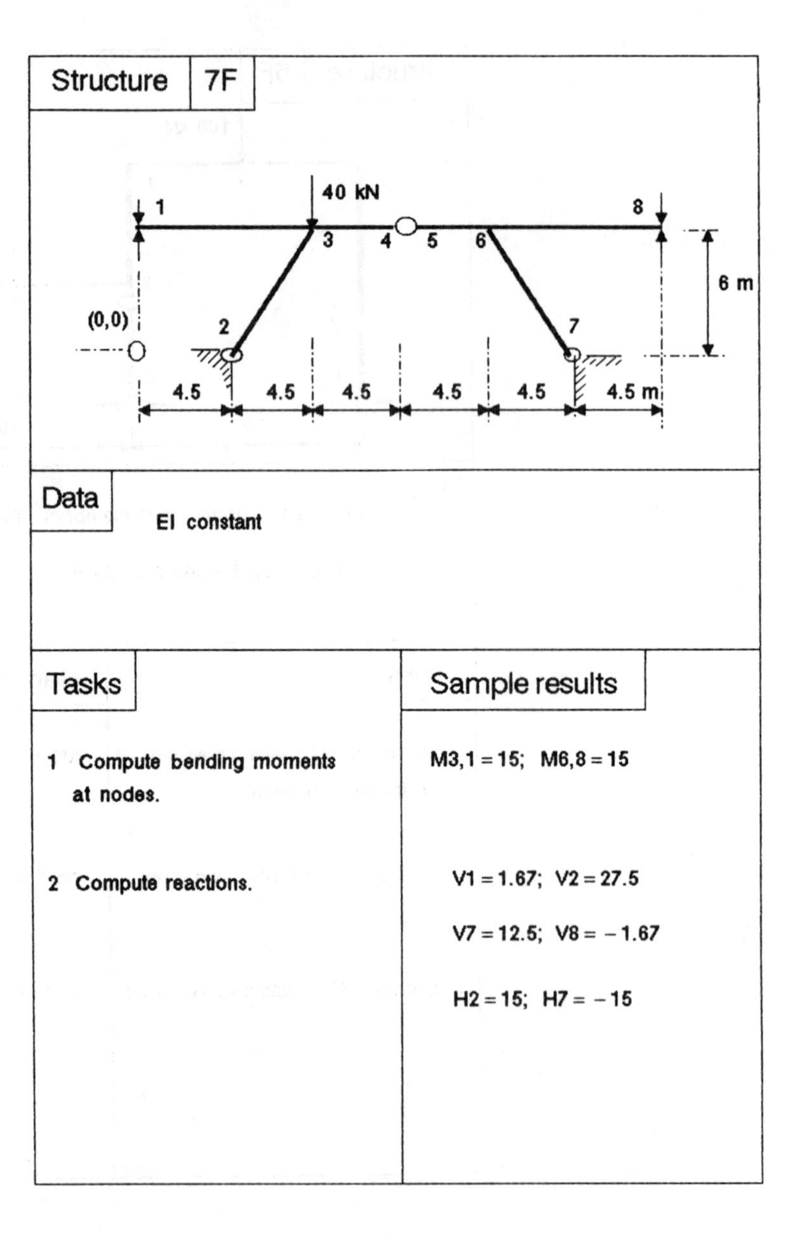

Data

EI constant

Tasks

1 Compute bending moments at nodes.

2 Compute reactions.

Sample results

M3,1 = 15; M6,8 = 15

V1 = 1.67; V2 = 27.5

V7 = 12.5; V8 = −1.67

H2 = 15; H7 = −15

Structure | 8F

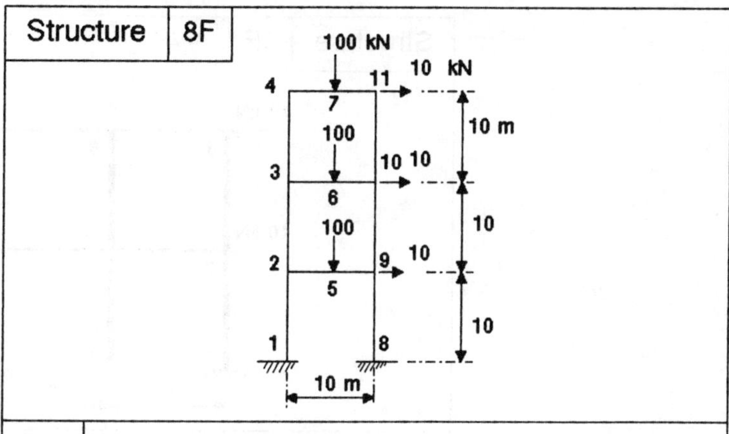

Data

EI and EA constant for all members.

Tasks	Sample results
1 Compute bending moment, shear and thrust distributions for all members.	M1,2 = 69.8 kNm M8,9 = 114.5 S1,2 = 8.283 kN S8,9 = 21.716 N1,2 = −108.4 N8,9 = −191.6
2 Compare the results with those obtained using the approximate method from Section 12.5.	

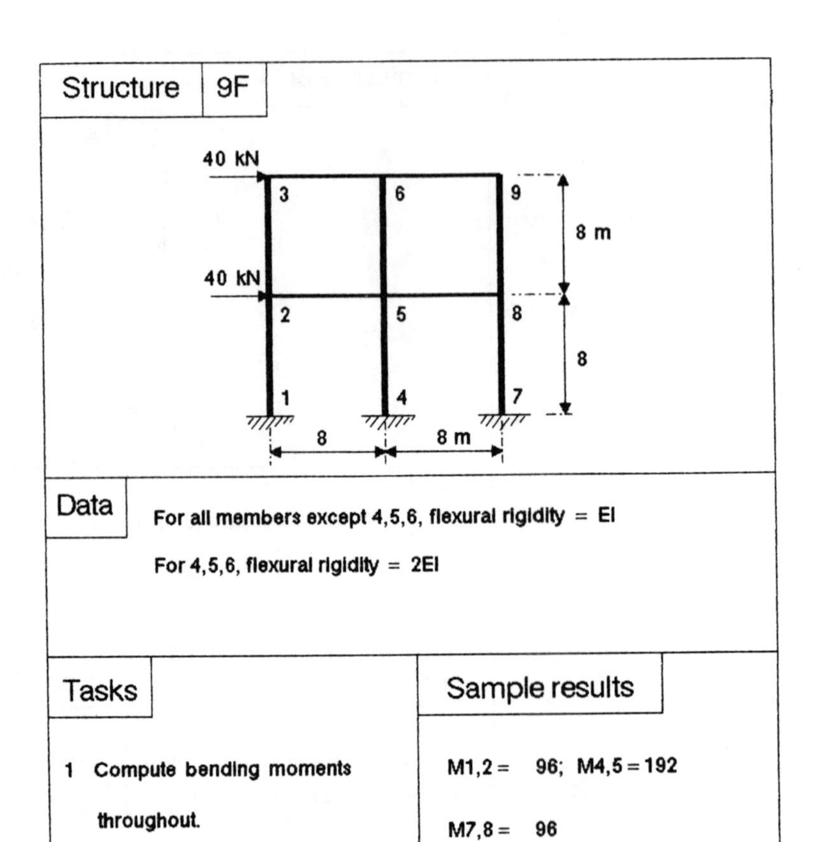

Data

For all members except 4,5,6, flexural rigidity = EI

For 4,5,6, flexural rigidity = 2EI

Tasks	Sample results
1 Compute bending moments throughout.	M1,2 = 96; M4,5 = 192 M7,8 = 96
2 Compute thrusts in columns	N1,2 = 36; N4,5 = 0 N7,8 = −36
3 Compute shears in columns.	S1,2 = 20; S4,5 = 40 S7,8 = 20

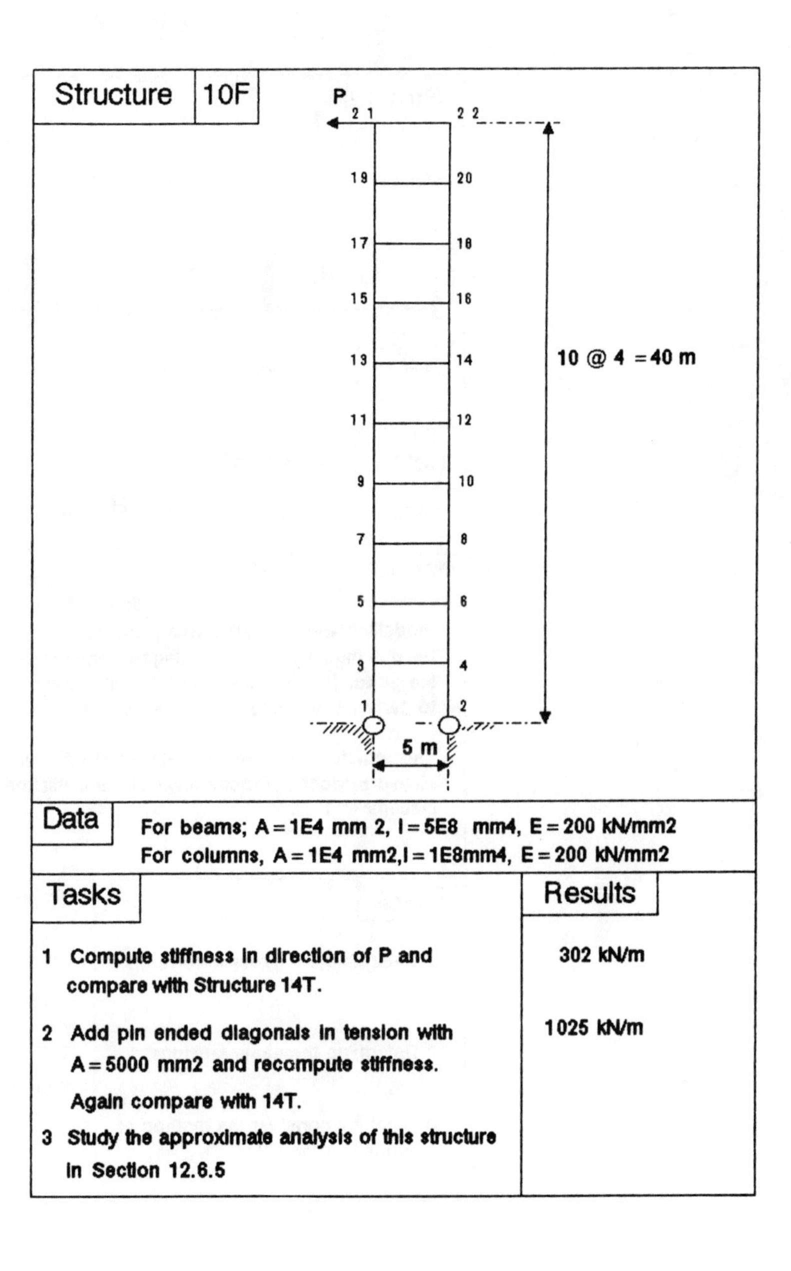

Data

For beams; A = 1E4 mm 2, I = 5E8 mm4, E = 200 kN/mm2
For columns, A = 1E4 mm2, I = 1E8mm4, E = 200 kN/mm2

Tasks	Results
1 Compute stiffness in direction of P and compare with Structure 14T.	302 kN/m
2 Add pin ended diagonals in tension with A = 5000 mm2 and recompute stiffness. Again compare with 14T.	1025 kN/m
3 Study the approximate analysis of this structure in Section 12.6.5	

Structure	11F

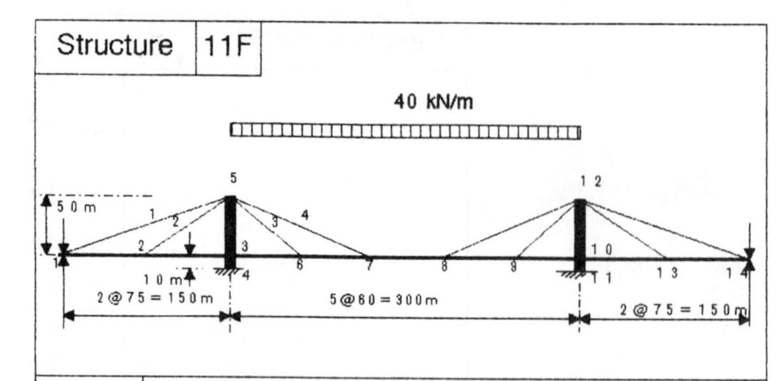

Data E = 200E6(kN/m2);EI(towers) = 8E9(kNm2);EI(girder) = 1.5E9

EA for cables (kN)			
1	2	3	4
79E4	45E4	78E4	130E4

Neglect axial strains in towers and girder by allocating large values of A.

The girder is simply – supported at the tower positions and this situation needs careful modelling with a plane frame program. A very good approximation can be obtained by adopting hinged connections between the tower and the girder (I = 0 for 4,3 and 3,5) and adding a member between 4 and 5 to carry the bending (EI = 8E9, A = 0).

This structure is taken from Hegab H I A, 1986, 'Static analysis of cable – stayed bridges', Proceedings of the Institution of Civil Engineers, Vol 81 December 1986.

Tasks	Sample results
1 Determine the girder deflections.	v(2) = 0.1219 m; v(6) = – 0.349 v(7) = – 0.631 m
2 Determine the cable tensions.	T(1) = 150.32; T(2) = – 205.8 T(3) = 1985; T(4) = 2134 kN
3 Carefully consider the method of modelling the tower.	

Structure 12F

From the same source as 11F.

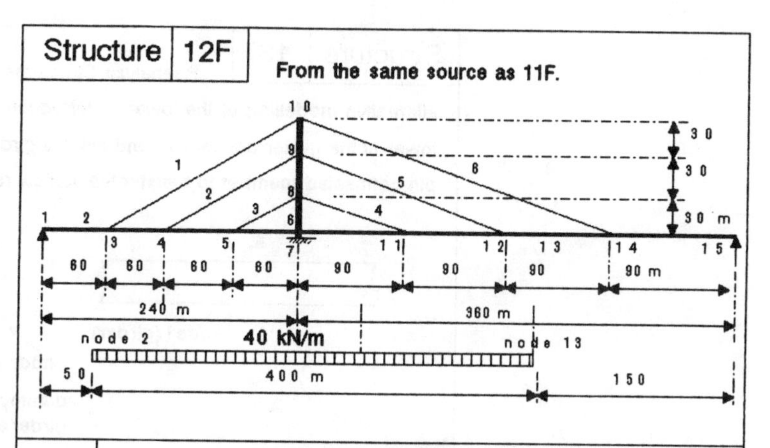

Data

E = 200 kNm2; EI(girder) = 4E9 kNm2; EI(tower) = 8E9 kNm2

cable	length(m)	EA (kN)
1	201.2	301.8E4
2	134.2	201.3E4
3	67.1	100.7E4
4	94.9	142.4E4
5	189.7	284.6E4
6	284.6	426.9E4

Neglect axial strains in tower and girder by allocating large values of A.

The modelling of the tower to represent the simply – supported condition of the deck at the tower is carried out in a similar way to Structure 11F but here the flexural modelling of the tower will connect with the thrust model at the cable connecting points. The flexural model of the tower must not be connected with the girder at node 6 and this can be achieved by setting the fixed tower support a small distance below node 6 at additional node 7.

Tasks

1 Determine the girder deflections.

2 Determine the cable tensions.

Sample results

v(13) = 0.472 m

T(1) = 858.8; T(2) = 679.8

T(3) = 117.4; T(4) = 1301.2

T(5) = 1976.4; T(6) = 1073.8 kN

Structure	13F

Reanalyse Structures 11F and 12F with an alternative modelling of the tower. Introduce an additional node at the tower at the girder connection and link the girder and tower with a dummy pin-connected member to transfer the vertical reaction.

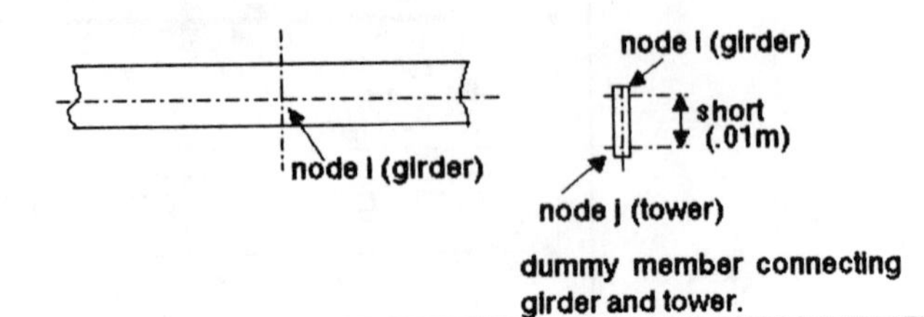

dummy member connecting girder and tower.

Data

For dummy member take, A = large, I = 0, (pin connected) length = small (say 0.01 m).

Tasks

Reanalyse both 11F and 12F using this alternative modelling of the tower. Compare the two approaches.

Sample results

The results will be the same as those for 11F and 12F.

PART III

Computer programs

The programs presented in this part are written in BASIC 2 for the Amstrad PC. It is unlikely that they will be immediately transportable to other microcomputers. However, each program is structured in subroutines with a considerable annotation in the form of dialogue, which should make adaptation to another form of BASIC relatively straightforward. The programs are also available in FAST BASIC for the Atari computer and in QUICK BASIC (MSDOS) and in TURBO PASCAL. These alternative versions all provide for data storage and editing in disc files. Discs (5.25 in or 3.5 in) may be obtained from the author, who may also be contacted about the availability of the programs in other forms of BASIC.

Editing facilities are written into the programs to try to simplify the editing of data and to avoid awkward disc file handling routines. Validation exercises are included so that the programs, or similar programs, can be tested. It is recommended that all the validation tests be carried out; even so it is possible that a program may produce unreliable results in certain unusual circumstances not covered by the validation tests. This is considered to be unlikely.

The programs are not intended to be comprehensive in scope. Neither are they particularly sophisticated or 'robust', and they will not therefore compete with commercial packages. They should be useful for the reader with limited or no access to large-scale structural computing facilities, or for study 'at home'.

In BASIC 2, the use of line numbers is optional. In PLATRUSS.BAS line numbers are used throughout, whereas in PLANFRAM.BAS they are used only occasionally, the subroutines being labelled with suitable titles. It is found that this approach makes the program structure easier to follow.

It will be appreciated that both programs demonstrate the general rule that 'improvement is always possible'!

PLATRUSS.BAS

This program uses the stiffness method to analyse plane, pin-jointed trusses. An interactive style is used for both the input of data and the output of results, the purpose being to allow the user to gain experience of data handling in the stiffness method and to control the output to suit the learning purposes.

Specification

The program is controlled by menu selection at all stages. The menu is as follows.

1. Enter data

The principal parameters – the number of nodes *nn* and the number of members *nm* – are input and the arrays are then dimensioned automatically. The nodal coordinates are typed in the form x, y, with x positive to the right and y positive upwards. It is suggested that the node nearest the bottom left is numbered 1. Any node numbering arrangement can be used. Similarly, the members are numbered from 1 to *nm* in any convenient order. The menu contains an option (10) to display a line drawing of the frame with the node numbers shown, so that the user can check that the geometry of the structure has been correctly represented. For each member, the node numbers and *EA* values are then typed in the form i, j, EA. Next, the applied loads are input. For each loaded node the required data is: node number, X direction load, Y direction load. Then the supported nodes are represented by typing in first the number of supported nodes, then, for each supported node, 'x' for support in the X direction, 'y' for support in the Y direction, or 'xy' for support in both directions. This concludes the entry of data. Care is needed to input data in the correct order and according to the format requested by the computer.

2. Edit nodal coordinate data

This option allows the user to verify and, if necessary, edit the nodal coordinates already input. In the dialogue, the user is requested to type in any node number, is given the current nodal coordinates, and is then given an opportunity to re-input changed values.

3. Edit member data

It is of course likely that typing errors will be made on entering data. This option allows the stored data for any member to be displayed and changed if desired. The member data is displayed in the format i, j, x, y, EA, where x and y are the horizontal and vertical projections of the member. The construction of the program is such that the support constraints will need to be re-entered if any member data is altered.

4. Edit applied load data

This option allows the stored loading data to be displayed and edited if necessary. The user simply follows the instructions on the screen, re-inputting data according to the given format if found necessary. Note the sign convention for the applied loads; horizontal loads are positive to the right and vertical loads are positive if downwards.

5. Re-input support constraint data

The user can re-input nodal support constraint data if there is any uncertainty as to the current state or if any changes have been made in the structural or loading data. Under certain circumstances the program will ask for this to be done.

6. Process data and analyse frame

Before choosing this option we should be satisfied that the computer has stored the correct data for our structure, since the data will now be processed into the form of the structure stiffness matrix and applied load vector. The equilibrium equations are then solved for the nodal displacements and the member forces are computed and retained for later access.

7. Display joint displacements

In this option, the user is asked for a joint number and the appropriate nodal displacements are displayed in the form X displacement and Y displacement.

8. Display reactive components

The reactions are not computed unless this option is chosen, when they are obtained by resolving member forces meeting at the joint in the X and Y directions. The user is asked to input a joint number and the horizontal and vertical components of the reaction are then displayed.

9. Display member forces

On entering this option, the user is asked for a member number and the computer then displays the force according to the convention of tension positive. This approach has some educational advantage if the user is carrying out a hand calculation and wishes to check progress step by step.

10. Display line drawing

Here the computer clears the screen and draws a simple line drawing of the frame with the node numbers shown. The origin (0, 0) will be placed towards the bottom left-hand corner of the screen, and the diagram will be suitably scaled to maximize its size either vertically or horizontally whichever overall dimension controls. The purpose of this option is simply to check that the frame geometry has been correctly input.

11. Quit, takes program control to END

Coordinate axes and sign conventions

The global coordinate axes X, Y are as shown in Fig. III.1. The ordering of the nodal forces is

$$\begin{bmatrix} P_{xi} \\ P_{yi} \\ P_{xj} \\ P_{yj} \end{bmatrix}$$

and the ordering of the nodal displacements is

$$\begin{bmatrix} d_{xi} \\ d_{yi} \\ d_{xj} \\ d_{yj} \end{bmatrix}$$

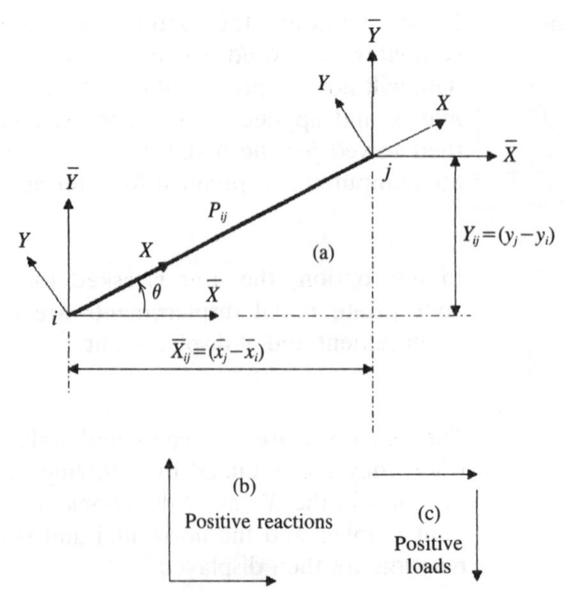

Figure III.1 General plane truss member

Sign conventions for displacements and stress resultants are as defined in Fig. 8.9. In the two-dimensional $X-Y$ plane with the positive Z axis directed towards the viewer, positive moments and rotations appear anticlockwise.

Sign conventions for applied loads and reactions are shown in Fig. III.1(c) and (b) respectively.

Array definitions

The program uses the following arrays:

Name	*Dimensions*	*Contents*
k	n, n	stiffness matrix
w	n	applied loads
d	n	nodal displacements
m	nm, 9	member properties
s	nm	sin
c	nm	cos
p	nm	member forces
v	n	temporary
x	nn	X coordinates of nodes
y	nn	Y coordinates of nodes

NB: $n = 2nn$

Listing: PLATRUSS.BAS

```
 10CLS #1
 15CLOSE WINDOW 2:CLOSE WINDOW 3:CLOSE WINDOW 4:WINDOW #1 FULL ON
 20PRINT AT(1;1) "this program is PLATRUSS, and it will"
 30PRINT AT(2;2) "analyse plane trusses using the"
 40PRINT AT(10;3) "stiffness method"
 50PRINT AT(1;5) "the program was written by W M Jenkins in 1987"
 60PRINT AT(1;6) "for his book STRUCTURAL ANALYSIS using COMPUTERS"
 70PRINT:PRINT "type m to display the menu"
 80IF INKEY$<>"m" THEN 80

 90CLS: PRINT" ************ M E N U **************"
 95PRINT "type"
100PRINT "1 to enter data"
105PRINT "2 to edit nodal coordinate data"
110PRINT "3 to edit member data"
120PRINT "4 to edit applied load data"
125PRINT "5 to re-input support constraint data"
130PRINT "6 to process data and analyse frame"
135PRINT "7 to display joint displacements"
140PRINT "8 to display reactive components"
145PRINT "9 to display member forces"
150PRINT "10 to display line drawing of frame"
170PRINT "12 to quit"
180PRINT
190PRINT "which do you require?"
200INPUT a
210IF a = 1 THEN 500
215IF a = 2 THEN 6000
220IF a = 3 THEN 1000
230IF a = 4 THEN 1500
240IF a = 5 THEN 2000
250IF a = 6 THEN 3000
260IF a = 7 THEN 4000
270IF a = 8 THEN 4200
280IF a = 9 THEN 4400
290IF a = 10 THEN 5000
320IF a = 12 THEN PRINT "          *******   C H E E R S   *******"
330FOR i = 1 TO 10000:NEXT i
400END

500CLS:PRINT "this is the subroutine to enter data"
510PRINT "please type in the number of nodes,nn":INPUT nn
520n = 2*nn
530PRINT "please type in the number of members, nm":INPUT nm
540DIM k(n,n),w(n),d(n),m(nm,9),s(nm),c(nm),p(nm),v(n),x(nn),y(nn),xg(nn)
545DIM yg(nn)
550PRINT "please type in the nodal coordinates in the form x,y"
560PRINT "x positive to the right and y positive upwards":PRINT
570 FOR i = 1 TO nn
580PRINT"type in x,y coordinates for node";i:INPUT x(i),y(i)
590NEXT i
600PRINT "now please type in node numbers and EA values for each"
610PRINT "member in the form I,J,EA":PRINT
620FOR i = 1 TO nm
630PRINT "type in I,J,EA for member";i
640INPUT m(i,1),m(i,2),m(i,5):NEXT i
650FOR k=1 TO nm
660i=m(k,1):j=m(k,2)
670m(k,3)=x(j)-x(i)
680m(k,4)=y(j)-y(i)
690NEXT k
700CLS:PRINT"type in number of loaded nodes":INPUT nwn
```

```
710FOR i=1 TO nwn
720PRINT "type in node number, applied horizontal load (positive to"
725PRINT "the right), applied vertical load (positive downwards)"
730PRINT "in the form;  node, horizontal load,  vertical load"
750INPUT node,wx,wy:wy=-wy
760q=2*node: w(q) = wy: w(q-1) = wx
770NEXT i
775GOSUB 5500
780CLS:PRINT "type in number of supported nodes":INPUT nsn
790FOR i=1 TO nsn
800PRINT "type in node number of supported node"
810INPUT node
820PRINT "type in x for support in x-direction, y for support in"
830PRINT "y- direction or xy for support in both directions"
860INPUT a$
870IF a$="x" THEN k(2*node-1,2*node-1)=10E20
890IF a$="y" THEN k(2*node,2*node)=10E20
900IF a$="xy" THEN k(2*node-1,2*node-1)=10E20
910IF a$="xy" THEN k(2*node,2*node)=10E20
920NEXT i
930CLS:PRINT "end of subroutine":GOTO 70


1000 PRINT"this routine allows you to verify and edit member data"
1020PRINT "please type in member number":INPUT p
1030PRINT "this is the current data for member";p
1040SET ZONE 5
1050PRINT m(p,1),m(p,2),m(p,3),m(p,4),m(p,5)
1060INPUT "do you want to edit this data? y/n";yn$
1070IF yn$ = "n" THEN 1110
1080PRINT "please type in revised data in the order  i,j,x,y,ea"
1100INPUT m(p,1),m(p,2),m(p,3),m(p,4),m(p,5)
1110PRINT "do you wish to verify/edit further member data? y/n"
1130INPUT yn$
1140IF yn$="y" THEN 1020
1150GOSUB 5500
1190PRINT "end of subroutine"
1200PRINT "if you wish to continue, please"
1210PRINT "re-enter support constraints using"
1220PRINT "the edit facility (menu option 5)"
1230GOTO 70


1500PRINT "this routine allows you to verify and edit applied load data"
1520PRINT "please type in node number":INPUT node
1530PRINT "this is the current record of loads on node";node
1550PRINT "x direction load =";w(2*node - 1)
1560PRINT "y direction load =";-w(2*node)
1570INPUT "do you wish to edit this data? y/n";yn$
1580IF yn$ = "n" THEN 1620
1590PRINT "please type in revised data in the form  x load, y load"
1610INPUT w(2*node -1),w(2*node):w(2*node)=-w(2*node)
1620PRINT "do you wish to verify/edit further applied load data? y/n"
1640INPUT yn$
1650IF yn$ = "y" THEN 1520
1660PRINT "end of subroutine.  If you wish to continue, please"
1670PRINT "re-enter support constraints using menu option 5"
1680GOSUB 5500
1690GOTO 70


2000PRINT"this routine allows you to re-input nodal constraint data"
2010PRINT"when you wish to change this or have changed other data"
```

```
2020GOSUB 5500
2030PRINT "     please type-in number of supported nodes":INPUT nsn
2040FOR i = 1 TO nsn
2050PRINT "type in node number of supported node":INPUT node
2060PRINT "type in x for support in the x-direction, y for support in"
2070PRINT "the y direction or  xy for support in both directions"
2080INPUT a$
2090IF a$ = "x" THEN k(2*node-1,2*node-1)=10E20
2100IF a$ = "y" THEN k(2*node,2*node)=10E20
2110IF a$ = "xy" THEN k(2*node-1,2*node-1)=10E20
2120IF a$ = "xy" THEN k(2*node,2*node)=10E20
2130NEXT i
2140PRINT
2150PRINT"  if you make further changes in data you will need to"
2160PRINT"  return to this subroutine to re-input support constraints"
2170PRINT
2180PRINT "end of subroutine":GOTO 70


3000CLS: PRINT "this subroutine processes the data"
3010PRINT "and analyses the frame, please wait."
3020FOR i = 1 TO nm
3030l=SQR(m(i,3)^2+m(i,4)^2)
3040m(i,9)=m(i,5)/l
3050s(i)=m(i,4)/l:c(i)=m(i,3)/l
3060m(i,0)=l:m(i,8)=s(i)*c(i)*m(i,9)
3070m(i,6)=s(i)^2*m(i,9):m(i,7)=c(i)^2*m(i,9)
3080NEXT i
3110FOR k = 1 TO nm
3120i=m(k,1):j=m(k,2)
3130r=2*(i-1)+1
3140k(r,r)=k(r,r)+m(k,7)
3150k(r,r+1)=k(r,r+1)+m(k,8)
3160k(r+1,r)=k(r+1,r)+m(k,8)
3170k(r+1,r+1)=k(r+1,r+1)+m(k,6)
3180r=2*(j-1)+1
3190k(r,r)=k(r,r)+m(k,7)
3200k(r,r+1)=k(r,r+1)+m(k,8)
3210k(r+1,r)=k(r+1,r)+m(k,8)
3220k(r+1,r+1)=k(r+1,r+1)+m(k,6)
3230r=2*(i-1)+1
3240p=2*(j-1)+1
3250k(r,p)=k(r,p)-m(k,7)
3260k(p,r)=k(p,r)-m(k,7)
3270k(r,p+1)=k(r,p+1)-m(k,8)
3280k(p+1,r)=k(p+1,r)-m(k,8)
3290k(r+1,p)=k(r+1,p)-m(k,8)
3300k(p,r+1)=k(p,r+1)-m(k,8)
3310k(r+1,p+1)=k(r+1,p+1)-m(k,6)
3320k(p+1,r+1)=k(p+1,r+1)-mk,6)
3330NEXT k
3340FOR i = 1 TO n
3350v(i)=w(i)
3360NEXT i
3370FOR i = 1 TO n-1
3380FOR j = i+1 TO n
3390d = k(j,i)/k(i,i)
3400FOR h = i TO n
3410p = k(i,h)*d:q=k(j,h)
3420k(j,h)=q-p
3430NEXT h
3440q=w(j):p=w(i)*d
3445w(j)=q-p
```

```
3450NEXT j
3455PRINT "              x";i;"of";n;"eliminated"
3460NEXT i
3470d(n)=w(n)/k(n,n)
3480FOR i = 1 TO n-1
3490j=n-i
3500d=0
3510FOR h=j+1 TO n
3520p=k(j,h)*d(h):d=d+p
3530NEXT h
3540q=w(j)/k(j,j):p=d/k(j,j)
3545d(j)=q-p
3550NEXT i
3560FOR i = 1 TO n
3570w(i)=v(i)
3580NEXT i
3590FOR k = 1 TO nm
3600i=2*(m(k,1)-1)+1:j=2*(m(k,2)-1)+1
3610px=(d(j)-d(i))*m(k,7)+(d(j+1)-d(i+1))*m(k,8)
3620py=(d(j+1)-d(i+1))*m(k,6)+(d(j)-d(i))*m(k,8)
3630p(k)=px*c(k)+py*s(k)
3640NEXT k
3650PRINT "end of subroutine":GOTO 70


4000CLS:PRINT "this routine lists joint displacements"
4010PRINT "please type in node number"
4020INPUT j
4030PRINT "x displacement at joint";j;"=";d(2*j-1)
4040PRINT "y displacement at joint";j;"=";d(2*j)
4050PRINT "do you want displacements at another"
4060PRINT "joint? y/n"
4070INPUT yn$
4080IF yn$ = "y" THEN 4010
4090PRINT "end of subroutine": GOTO 70


4200CLS:PRINT "this routine displays reactive components"
4210PRINT "type in node number":INPUT k:h=0:v=0
4220FOR i=1 TO nm
4230IF m(i,1)<>k THEN 4250
4240hk=p(i)*c(i):vk=p(i)*s(i)
4245h=h-hk:v=v-vk
4250NEXT i
4260FOR i=1 TO nm
4270IF m(i,2)<>k THEN 4290
4280hk=p(i)*c(i):vk=p(i)*s(i)
4285h=h+hk:v=v+vk
4290NEXT i
4300PRINT "horizontal reaction at joint";k;"=";h
4310PRINT "vertical reaction at joint";k;"=";v
4320PRINT "do you want further reactions? y/n"
4330INPUT yn$
4340IF yn$ = "y" THEN 4210
4350PRINT "end of subroutine":GOTO 70


4400CLS:PRINT "this routine displays member forces"
4410PRINT "type in member number":INPUT k
4415IF k > nm THEN 4410
4420PRINT "force in member";k;"=";p(k)
4430PRINT "do you want further forces? y/n"
4440INPUT yn$
```

```
4450IF yn$ = "y" THEN 4410
4460PRINT "end of subroutine":GOTO 70


5000REM **   commencement of draw routine   **
5020 CLS
5022xm=0:ym=0
5024FOR i=1 TO nn
5026IF x(i)>xm THEN xm=x(i)
5028IF y(i)>ym THEN ym=y(i)
5030NEXT i
5040IF xm>ym THEN 5070
5050s=4000/ym
5060GOTO 5080
5070s=7000/xm
5080FOR i=1 TO nn
5090xg(i)=x(i)*s:yg(i)=y(i)*s
5095xg(i)=xg(i)+500:yg(i)=yg(i)+500
5100NEXT i
5110FOR k=1 TO nm
5120i=m(k,1):j=m(k,2)
5130LINE xg(i);yg(i),xg(j);yg(j)
5140NEXT k
50FOR i=1 TO nn
5160MOVE xg(i);yg(i)
5170PRINT i;
5180NEXT i
5190MOVE 0;0
5200PRINT"Press any key to continue";
5210IF INKEY$=""THEN 5210
5220CLS
5230GOTO 70


5500 REM **    subroutine to clear stiffness matrix    **
5510FOR i = 1 TO n
5520FOR j = 1 TO n:k(i,j) = 0
5530NEXT j
5540NEXT i
RETURN

6000CLS:PRINT "this routine allows you to verify/edit nodal"
6010PRINT "coordinate data"
6020PRINT:PRINT"    Please type-in node number":INPUT node
6030PRINT"these are the current coordinates for node";node
6040PRINT"             x";node;"=";  x(node)
6050PRINT"             y";node;"=";  y(node)
6060PRINT:INPUT "do you wish to change these? y/n"; yn$
6070IF yn$ = "n" THEN 6110
6080PRINT:PRINT"  please type-in revised coordinates at node";node
6090PRINT"  in the form  x , y"
6100INPUT x(node),y(node)
6110PRINT"      do you wish to verify/edit further nodal"
6120PRINT"      coordinate data? y/n"
6130INPUT yn$
6140IF yn$ = "y" THEN 6020
6150FOR k=1 TO nm:i=m(k,1):j=m(k,2)
6160m(k,3)=x(j)-x(i):m(k,4)=y(j)-y(i)
6160NEXT k
6170PRINT"          end of subroutine.  If you wish to continue"
6180PRINT"          please re-enter support constraints using"
6190PRINT"          menu option number 5"

6200GOSUB 5500
6210GOTO 70
```

Validation exercises for PLATRUSS.BAS

Structure	Data	Sample results
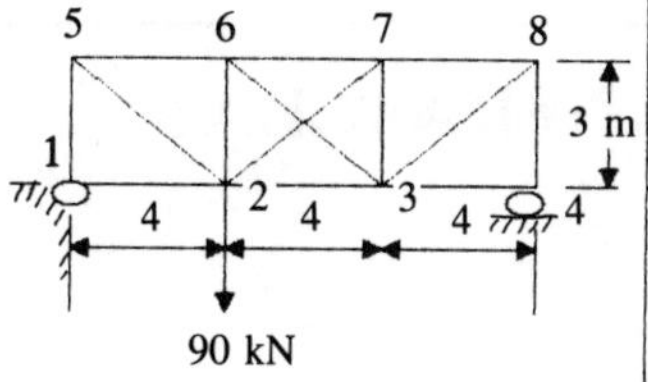 5 6 7 8; 1 2 3 4; 3 m; 4 4 4; 90 kN	Length (m) / EA (kN): 3 / 75 000; 4 / 100 000; 5 / 125 000	$v_2 = -15$ mm $P_{2,6} = 15$ $P_{6,3} = -25$ $P_{2,7} = 25$ $P_{3,8} = 50$
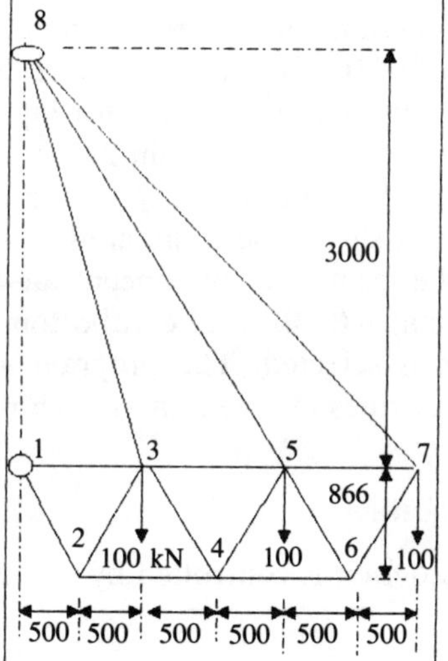 8; 3000; 1 3 5 7; 866; 2 4 6; 100 kN 100 100; 500 500 500 500 500 500	EA (truss members) $= 1000$ EA (cable 8–3) $= 791$ EA (cable 8–5) $= 902$ EA (cable 8–7) $= 1061$	$P_{8,3} = 110.8$ $P_{8,5} = 135.8$ $P_{8,7} = 126.8$ $P_{1,3} = -195.5$ $P_{1,2} = -8.93$ $P_{5,7} = -83.7$ $H_1 = 200$ $V_1 = -7.773$ $H_8 = -200$ $V_8 = 307.73$

PLANFRAM.BAS

This program analyses plane frames with rigid joints having three degrees of freedom at each node. As with PLATRUSS.BAS, an interactive style is adopted for input and output. The program is 'menu driven' with a main menu and a subsidiary menu for data editing. As a preliminary to the input of data, the nodes and members should be numbered in any convenient order. A node must be located at all concentrated load positions. The origin of coordinates (0, 0) will be at bottom left of the screen if the 'draw' option is selected. The program uses the stiffness method with three degrees of freedom at each node.

Specification

The program is controlled by menu selection as follows.

1. Input main parameters (LABEL maindat)

The number of nodes *nn* and the number of members *nm* are typed in and all the arrays are then dimensioned automatically. It is not possible to redimension arrays, so if *nn* or *nm* are to be changed the program must be restarted.

2. Input nodal coordinates (LABEL noddat)

This subroutine controls the input of nodal coordinate data in the form *x, y* for each node. The sign convention is that *x* is positive to the right and *y* positive upwards.

3. Input member data (LABEL memdat)

This data consists of five numbers for each member, the format for member *ij* being: node *i*, node *j*, *E*, *A*, *I*. All the numbers are typed on one line separated by commas. If a mistake is made and only noticed after RETURN has been keyed, the editing menu can be used at a later stage to effect a correction. The conventions are shown in Fig. III.2; the general member *ij* is shown in (a). Difficulties in interpretation of signs can be avoided if ends *i* and *j* for horizontal and vertical members are oriented as shown at (b) and (c) respectively.

4. Input load data (LABEL loaddat)

Provision is made for two types of applied load: nodal point loads in the global *X* and *Y* directions; and uniformly distributed loads applied to members in either a vertical or a horizontal direction. The format for load data input for nodal point loads is: node number; magnitude of load; 'h' for horizontal loads or 'v' for vertical loads. The sign convention for nodal loads is shown in Fig. III.2(d). Horizontal loads to the right are positive, and

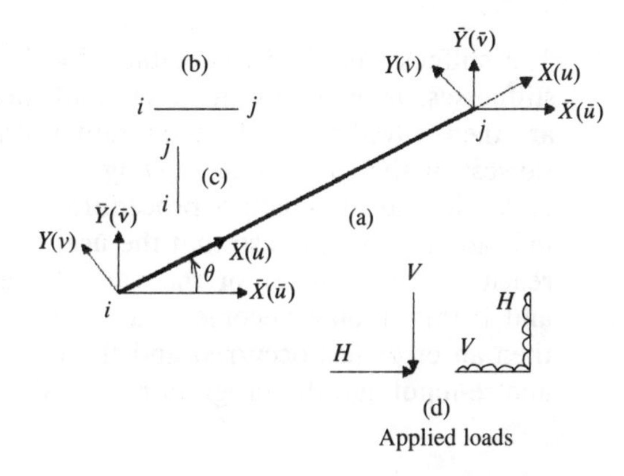

Figure III.2 General plane frame member

vertically downwards loads are positive. The format for input of distributed loads is: member number; 'udl' intensity; 'v' or 'h'.

This provides for the majority of practical situations. Note that a node must be located at all concentrated load positions. Nodes can be located anywhere within the lengths of members in addition to those at the joints of the strucural frame.

5. To introduce springs (LABEL springs)

Linear and rotational springs can be introduced at any of the nodes using this option. The springs must have a known stiffness and can be linear horizontal (h), linear vertical (v) or rotational (r). The data input format is: node number; 'r', 'h' or 'v'; spring stiffness. The units of the spring stiffness are assumed to be consistent with units used elsewhere in the program, usually kN/m or kN m/rad. If springs are introduced at a node there is no need to specify that the node is supported in the spring direction. Support in other directions may, however, be specified.

6. Input support data (LABEL suppdat)

The structural supports, constrained coordinates, are represented mathematically by planting a large number on the corresponding diagonal of the stiffness matrix. This is done automatically on input of the following data:

Number of supported node
'hs' or 'nhs' (for horizontal or no horizontal support)
'vs' or 'nvs' (for vertical or no vertical support)
'rs' or 'nrs' (for rotational or no rotational support).

It is important to ensure that at least one restraint (either spring or full restraint) is provided in each of the three coordinates (x, y, θ). At this stage all data has been input. It is important to do this in the correct order as already indicated.

7. Process data and analyse frame (LABEL procdat)

The stiffness matrix is assembled by the successive transfer of stiffnesses, member by member, and the equilibrium equations are then solved to produce the nodal displacements. This is the slowest of the subroutines, owing to the 'three-deep' nesting of cycles in the elimination procedure. The program displays an indication of progress so that the user has some idea of the stage reached. The number of the variables eliminated is displayed, and if this display becomes stationary before the last equation then an error has occurred and the data should be re-examined and re-input and the program restarted.

8. Display line drawing of frame (LABEL draw)

This subroutine displays a simple line drawing of the structure with the node numbers shown. This allows the user to verify that the members and nodes have been represented correctly. The origin of coordinates (0, 0) should be chosen towards the bottom left-hand part of the structure to maximize the scale of the line drawing. If an error is revealed, the edit menu can be accessed (option 11).

9. Output displacements (LABEL disps)

The nodal displacements are listed by keying in the node number where displacements are required. In practice it is unlikely that all displacements will be required, so time is saved by allowing the user to select as required. The units are those adopted in the data, e.g. metres and radians.

10. Output member forces (LABEL memforces)

The member forces are listed by keying in the member number. The format is:

Moment at end i (anticlockwise positive)
Moment at end j (anticlockwise positive)
Thrust at end i (tension positive)
Shear at end i (positive when in the positive direction of Y: Fig. III.2(a)).

Thrust and shear at end j are not listed but can easily be deduced from the results above and the applied loading (care is needed when this is uniformly distributed between nodes).

11. To edit data (LABEL changedata)

On selecting this option the user is given a submenu as follows:

EDIT MENU
1. Edit nodal coordinates
2. Edit member data
3. Verify loading data
4. Verify support data
5. Return to main menu.

Options 1 and 2 allow verification/editing of nodal coordinates and member data according to the dialogue provided. If changes are required it will be necessary to re-input data as requested and to re-input all applied load and support data (this is usually very quickly done). Options 3 and 4 allow verification of loading and support data but do not allow editing; if this is required, fresh data has to be input via the appropriate options in the main menu.

12. To quit (LABEL exit)

This displays a goodbye message and takes the program control to END.

Coordinate axes and sign conventions

These are as defined in Fig. III.2. The ordering of the nodal forces is

$$\begin{bmatrix} M_{zi} \\ N_i \\ S_{yi} \\ M_{zj} \\ N_j \\ S_{yj} \end{bmatrix}$$

and the nodal displacements are

$$\begin{bmatrix} \theta_{zi} \\ u_i \\ v_i \\ \theta_{zj} \\ u_j \\ v_j \end{bmatrix}$$

The sign convention for M_z needs a little care. The Z axis is directed towards the viewer and the moment M_z is positive when clockwise as viewed in the positive direction of the Z axis. This means that a positive M_z will appear anticlockwise when taking the normal view of the X–Y plane.

Array definitions

Name	*Dimensions*	*Contents*
k	n, n	stiffness matrix
w	n	applied loads
d	n	nodal displacements
m	nm, 9	member data
ms	nm, 9	member stiffnesses
s	nm	sin
c	nm	cos
xn	nn	X coordinates of nodes
yn	nn	Y coordinates of nodes
f	nm, 4	member forces

NB: $n = 3nn$

Simulations

The introduction of springs at nodes has been explained under option 5. There is no need to increase the number of nodes when springs are introduced (see the validation exercises after the program listing). The spring is an 'elastic' support in contrast to the rigid supports initiated by option 6.

Whilst the program is written primarily for rigid-jointed frames, it is possible with care to insert hinges at the ends of members. Some practical situations are shown in Fig. III.3, and

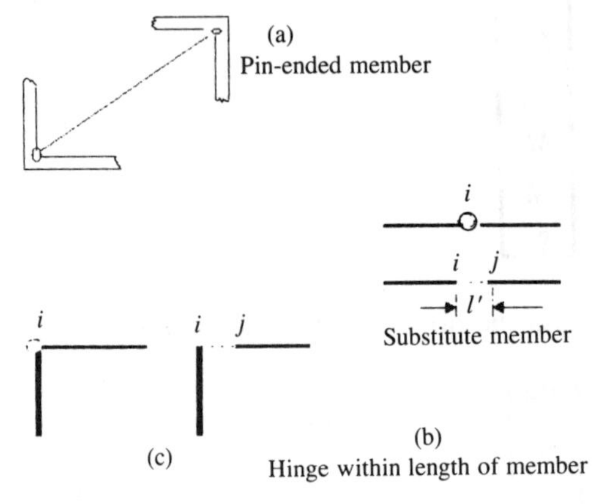

Figure III.3 Structural hinges represented by substitute members

the action via the program data for each is as follows:

(a) For a simple member pinned at both ends, put $I = 0$.
(b) If a hinge exists within the length of a member at say node i, an additional node is introduced, say j, adjacent to i and a

small distance l from i. A 'substitute' member ij is inserted. The requirements for the substitute member are:

Infinite transverse stiffness: EI'/l^3 large
Infinite axial stiffness: EA'/l large
Zero rotational stiffness: EI'/l small

This situation can be closely approximated mathematically if: $l = 0.0001L$, where L is a representative member length; $A' = 10^3A$, where A is a representative cross-sectional area; and $I' = 10^{-8}I$, where I is a representative second moment of area.

(c) If a hinge is introduced at a node as in Fig. III.3(c), the treatment is similar to that in (b). The substitute member can be aligned with either axis.

The substitute member device is effective even if a point load is placed at the hinge; the load can be applied at either end of the substitute member. The device can also be used if members connected by the hinge carry uniformly distributed loads.

Listing: PLANFRAM.BAS

```
REM **  plane frame analysis program  **
CLS #1
CLOSE WINDOW 3:CLOSE WINDOW 4:CLOSE WINDOW 2
WINDOW #1 FULL ON
PRINT AT (4;1) "this program is PLANFRAM and it will"
PRINT AT (6;2) "analyse plane frames using the"
PRINT AT (10;3) "stiffness method"
PRINT AT (4;5) "the program was written by W M Jenkins in 1988"
PRINT AT (4;6) "for his book STRUCTURAL ANALYSIS using COMPUTERS"
PRINT AT (4;8) "PRESS ANY KEY WHEN READY TO START"

ON ERROR GOTO 2000
REPEAT
presskey$ = INKEY$
UNTIL presskey$<>""
10CLS:PRINT "      key m to display the menu"
REPEAT
presskey$=INKEY$
UNTIL presskey$="m"
15CLS:PRINT "          **********  M  E  N  U  **********"
PRINT "       key....."
PRINT "       1 to input basic parameters"
PRINT "       2 to input nodal coordinates"
PRINT "       3 to input member data"
PRINT "       4 to input load data"
PRINT "       5 to introduce springs"
PRINT "       6 to input support data"
PRINT "       7 to process data and analyse frame"
PRINT "       8 to display line drawing of frame"
PRINT "       9 to output displacements"
PRINT "       10 to output member forces"
PRINT "       11 to edit data"
PRINT "       12 to quit"
20PRINT "     which do you require?"
INPUT c: IF c < 1 THEN 20
IF c > 12 THEN 20
IF c = 1 THEN GOSUB maindat
IF c = 2 THEN GOSUB noddat
IF c = 3 THEN GOSUB memdat
IF c = 4 THEN GOSUB loaddat
IF c = 5 THEN GOSUB springs
IF c = 6 THEN GOSUB suppdat
IF c = 7 THEN GOSUB procdat
IF c = 8 THEN GOSUB draw
IF c = 9 THEN GOSUB disps
IF c = 10 THEN GOSUB memforces
IF c = 11 THEN GOSUB changedata
IF c = 12 THEN GOSUB exit
PRINT:PRINT:PRINT
PRINT "              please key m to display the menu"
REPEAT
presskey$ = INKEY$
UNTIL presskey$ = "m"
GOTO 15

LABEL maindat
CLS:PRINT "     this is the subroutine to enter the main parameters"
PRINT "    please type in the number of nodes":INPUT nn
n = 3*nn
PRINT "    please type in the number of members":INPUT nm
DIM k(n,n), w(n), d(n), m(nm,9), ms(nm,9), s(nm), c(nm)
DIM xn(nn), yn(nn), f(nm,4), xg(nn), yg(nn)
```

```
RETURN


LABEL noddat
CLS:PRINT "    this is the subroutine to enter nodal coordinates"
PRINT "    the coordinates for each node are typed-in in the"
PRINT "    form,  x,y   where x is positive to the right and"
PRINT "    y is positive upwards"
FOR i = 1 TO nn
PRINT:PRINT "    type in x,y coordinates for node";i:INPUT xn(i),yn(i)
NEXT i
RETURN


LABEL memdat
CLS:PRINT "   this is the subroutine to enter member data"
PRINT "   please number the members 1 to";nm;"if you have not"
PRINT "   done so.    If you make a mistake in typing in, leave"
PRINT "   it since you will get an opportunity to edit data later"
PRINT:PRINT "            PRESS ANY KEY WHEN READY TO START"
REPEAT
presskey$=INKEY$
UNTIL presskey$ <>"":CLS
PRINT "   the member data will be typed-in in the order,"
PRINT "   node number at end 1, node number at end 2, E, A, I"
PRINT "   on one line in the form N1, N2, E, A, I"
FOR k = 1 TO nm
PRINT "   please type in N1, N2, E, A, I for member";k
INPUT m(k,1),m(k,2),m(k,5),m(k,6),m(k,7)
i = m(k,1): j=m(k,2)
m(k,3)=xn(j)-xn(i): m(k,4)=yn(j)-yn(i)
a=ATAN2(m(k,3),m(k,4))
s(k)=SIN(a): c(k)=COS(a)
m(k,8)=SQR(m(k,3)^2 + m(k,4)^2)
ms(k,1)=m(k,5)*m(k,7)/m(k,8)
ms(k,2)=ms(k,1)*c(k)/m(k,8)
ms(k,3)=ms(k,1)*s(k)/m(k,8)
ms(k,4)=ms(k,2)*c(k)/m(k,8)
ms(k,5)=ms(k,3)*s(k)/m(k,8)
ms(k,6)=ms(k,2)*s(k)/m(k,8)
ms(k,7)=m(k,5)*m(k,6)*c(k)*c(k)/m(k,8)
ms(k,8)=m(k,5)*m(k,6)*s(k)*s(k)/m(k,8)
ms(k,9)=m(k,5)*m(k,6)*s(k)*c(k)/m(k,8)
NEXT k
RETURN


LABEL loaddat
GOSUB cleardata
CLS: PRINT "   this is the subroutine to enter load data"
PRINT "   we shall apply nodal point loads first and then any"
PRINT "   distributed loads on the members.    If you have a"
PRINT "   point load between nodes you should have put a node at"
PRINT "   this position."
PRINT:PRINT "   please type in number of nodal point loads"
PRINT "   to be applied":INPUT nnw
IF nnw = 0 THEN 50
PRINT "   for each nodal load, type;  node number, magnitude of"
PRINT "   load, h (for horizontal load) OR v (for vertical load)"
FOR k = 1 TO nnw
30 PRINT "   type in data for load number"; k
INPUT node,w,hv$
IF hv$ = "v" THEN 40
```

```
IF hv$ <> "h" THEN 30
w(3*node -1) = w(3*node -1) + w
GOTO 45
40 w(3*node) = w(3*node) - w
45NEXT k
50CLS: PRINT "  now we input data for uniformly distributed loads if any"
PRINT "   please type-in number of such loads":INPUT nudl
IF nudl = 0 THEN 85
PRINT "   please type-in member number, udl intensity, v  or  h"
PRINT "   for vertical OR horizontal loading
FOR k = 1 TO nudl
60PRINT "    please type-in data for udl";k
INPUT m,udl,vh$
i = m(m,1): j = m(m,2)
i=3*i - 2 : j=3*j - 2
c=c(m):s=s(m)
IF vh$="h" THEN 70
IF vh$<>"v" THEN 60
m1=-udl*c^2*m(m,8)^2/12:m2=-m1
v1=-udl*c*m(m,8)/2:v2=v1
w(i)=w(i)+m1:w(j)=w(j)+m2
f(m,1)=f(m,1)-m1:f(m,2)=f(m,2)-m2
f(m,3)=f(m,3) + v1*s
f(m,4)=f(m,4)-v1*c
w(i+2)=w(i+2)+v1:w(j+2)=w(j+2)+v2
GOTO 80
70m1=-udl*s^2*m(m,8)^2/12:m2=-m1
h1=udl*s*m(m,8)/2:h2=h1
w(i)=w(i)+m1:w(j)=w(j)+m2
f(m,1)=f(m,1)-m1:f(m,2)=f(m,2)-m2
f(m,3)=f(m,3) + h1*c
f(m,4)=f(m,4)+h1*s
w(i+1)=w(i+1)+h1:w(j+1)=w(j+1)+h2
80NEXT k
85CLS:PRINT "    that concludes the input of loading data"
PRINT
RETURN


LABEL springs
CLS:PRINT "    this subroutine will allow you to introduce springs"
PRINT "    of given stiffness into the structure: the springs"
PRINT "    may be rotational (r), linear horizontal (h), or linear"
PRINT "    vertical (v)"
PRINT:PRINT "      please type-in the number of springs you wish to use"
INPUT sp:FOR i = 1 TO sp
PRINT:PRINT "    please type-in; node number, r/h or v, spring stiffness"
PRINT"   for spring number";i
INPUT node,rhv$,ss: node = 3*node
IF rhv$ ="r" THEN k(node-2,node-2)=k(node-2,node-2) + ss
IF rhv$ ="h" THEN k(node-1,node-1)=k(node-1,node-1) + ss
IF rhv$ ="v" THEN k(node,node)=k(node,node) + ss
NEXT i
PRINT "     that concludes the input of spring data"
RETURN


LABEL suppdat
CLS:PRINT "   this is the subroutine to input support data"
PRINT "   please type in number of supported nodes":INPUT nsn
FOR k = 1 TO nsn
PRIN   please type in number of supported node":INPUT node
coord = 3 * node
```

```
90PRINT "    please type in hs for horizontal support OR nhs for"
PRINT "    no horizontal support at node";node:INPUT a$
IF a$ = "hs" THEN k(coord-1,coord-1) = 10E20:GOTO 95
IF a$<>"nhs" THEN 90
95PRINT "    please type in vs for vertical support OR nvs for"
PRINT "    no vertical support at node";node:INPUT a$
IF a$ = "vs" THEN k(coord,coord)= 10E20:GOTO 100
IF a$<>"nvs" THEN 95
100PRINT "    please now type-in rs for rotational support OR nrs for"
PRINT "    no rotational support at node";node:INPUT a$
IF a$ = "rs" THEN k(coord-2,coord-2) = 10E20:GOTO 105
IF a$<>"nrs" THEN 100
105NEXT k
CLS:PRINT "this concludes the input of support data"
PRINT
RETURN


LABEL procdat
CLS:PRINT "    this is the subroutine to process the data and"
PRINT "    analyse the frame:   this may take a little time with"
PRINT "    a big structure, please wait for prompts."
FOR k = 1 TO nm
i=m(k,1):j=m(k,2)
r=3*(i-1)+1:p=3*(j-1)+1
120k(r,r)=k(r,r) + 4*ms(k,1)
k(r,r+1)=k(r,r+1) - 6*ms(k,3)
k(r,r+2)=k(r,r+2) + 6*ms(k,2)
k(r+1,r)=k(r+1,r) - 6*ms(k,3)
k(r+1,r+1)=k(r+1,r+1) + 12*ms(k,5) + ms(k,7)
k(r+1,r+2)=k(r+1,r+2) - 12*ms(k,6) + ms(k,9)
k(r+2,r)=k(r+2,r) + 6*ms(k,2)
k(r+2,r+1)=k(r+2,r+1) - 12*ms(k,6) + ms(k,9)
k(r+2,r+2)=k(r+2,r+2) + 12*ms(k,4) + ms(k,8)

k(p,p)=k(p,p) + 4*ms(k,1)
k(p,p+1)=k(p,p+1) + 6*ms(k,3)
k(p,p+2)=k(p,p+2) - 6*ms(k,2)
k(p+1,p)=k(p+1,p) + 6*ms(k,3)
k(p+1,p+1)=k(p+1,p+1) + 12*ms(k,5) + ms(k,7)
k(p+1,p+2)=k(p+1,p+2) - 12*ms(k,6) + ms(k,9)
k(p+2,p)=k(p+2,p) - 6*ms(k,2)
k(p+2,p+1)=k(p+2,p+1) - 12*ms(k,6) + ms(k,9)
k(p+2,p+2)=k(p+2,p+2) + 12*ms(k,4) + ms(k,8)

k(r,p)=k(r,p) + 2*ms(k,1)
k(r,p+1)=k(r,p+1) + 6*ms(k,3)
k(r,p+2)=k(r,p+2) - 6*ms(k,2)
k(r+1,p)=k(r+1,p) - 6*ms(k,3)
k(r+1,p+1)=k(r+1,p+1) - 12*ms(k,5) - ms(k,7)
k(r+1,p+2)=k(r+1,p+2) + 12*ms(k,6) - ms(k,9)
k(r+2,p)=k(r+2,p) + 6*ms(k,2)
k(r+2,p+1)=k(r+2,p+1) + 12*ms(k,6) - ms(k,9)
k(r+2,p+2)=k(r+2,p+2) - 12*ms(k,4) - ms(k,8)

k(p,r)=k(p,r) + 2*ms(k,1)
k(p+1,r)=k(p+1,r) + 6*ms(k,3)
k(p+2,r)=k(p+2,r) - 6*ms(k,2)
k(p,r+1)=k(p,r+1) - 6*ms(k,3)
k(p+1,r+1)=k(p+1,r - 12*ms(k,5) - ms(k,7)
k(p+2,r+1)=k(p+2,r+1) + 12*ms(k,6) - ms(k,9)
k(p,r+2)=k(p,r+2) + 6*ms(k,2)
k(p+1,r+2)=k(p+1,r+2) + 12*ms(k,6) - ms(k,9)
```

```
k(p+2,r+2)=k(p+2,r+2) - 12*ms(k,4) - ms(k,8)
NEXT k

130PRINT "   the stiffness matrix is now completed and the"
PRINT "   equilibrium equations are being solved:   please wait"


FOR i=1 TO n-1
FOR j=i+1 TO n
d=k(j,i)/k(i,i)
FOR h=i TO n
p=k(i,h)*d:q=k(j,h)
k(j,h)=q-p
NEXT h
q=w(j):p=w(i)*d:w(j)=q-p
NEXT j
PRINT"          x";i;"of";n;"eliminated"
NEXT i
d(n)=w(n)/k(n,n)
FOR i=1 TO n-1
j=n-i:d=0
FOR h=j+1 TO n
p=k(j,h)*d(h):d=d+p
NEXT h
q=w(j)/k(j,j):p=d/k(j,j):d(j)=q-p
NEXT i
FOR k = 1 TO nm
i=3*m(k,1)-2:j=3*m(k,2)-2
f=-6*ms(k,3)*(d(i+1)-d(j+1)) + 6*ms(k,2)*(d(i+2)-d(j+2))
f(k,1)=f(k,1) + 4*ms(k,1)*d(i) + 2*ms(k,1)*d(j)
f(k,1)=f(k,1) + f
f(k,2)=f(k,2) + 2*ms(k,1)*d(i) + 4*ms(k,1)*d(j)
f(k,2)=f(k,2) + f
f=6*ms(k,3)*(d(i) + d(j))
f=f + (ms(k,7) + 12*ms(k,5))*(d(j+1) - d(i+1))
f=f + (ms(k,9) - 12*ms(k,6))*(d(j+2) - d(i+2))
f(k,3)=f(k,3) + f*c(k)
f=-6*ms(k,2)*(d(i) + d(j))
f=f + (ms(k,9) - 12*ms(k,6))*(d(j+1) - d(i+1))
f=f + (ms(k,8) + 12*ms(k,4))*(d(j+2) - d(i+2))
f=f*s(k)
f(k,3)=f(k,3) + f
f=6*ms(k,2)*(d(i)+d(j)) + (ms(k,9) - 12*ms(k,6))*(d(i+1) - d(j+1))
f=f + (ms(k,8)+12*ms(k,4))*(d(i+2)-d(j+2)):f = f*c(k)
f(k,4)=f(k,4) + f
f=6*ms(k,3)*(d(i)+d(j)) - (ms(k,7)+12*ms(k,5))*(d(i+1)-d(j+1))
f=f - (ms(k,9)-12*ms(k,6))*(d(i+2)-d(j+2)):f = f*s(k)
f(k,4)=f(k,4) + f
NEXT k
RETURN


LABEL draw
CLS
xm=0:ym=0
FOR i = 1 TO nn
IF xn(i)>xm THEN xm=xn(i)
IF yn(i)>ym THEN ym=yn(i)
NEXT i
IF xm/7500 > ym/4500 THEN 200
s=4000/ym
```

```
GOTO 210
200 s=7000/xm
210FOR i=1 TO nn
xg(i)=xn(i)*s + 500:yg(i)=yn(i)*s + 500
NEXT i

FOR k = 1 TO nm
i=m(k,1):j=m(k,2)
LINE xg(i);yg(i),xg(j);yg(j)
NEXT k

FOR i=1 TO nn
MOVE xg(i);yg(i)
PRINT i
NEXT i

MOVE 0;0:PRINT "             PRESS ANY KEY TO CONTINUE";
REPEAT
presskey$=INKEY$
UNTIL presskey$<>""
RETURN


LABEL disps
CLS:PRINT "   thsubroutine lists nodal displacements"
140PRINT "   please type-in node number":INPUT j
PRINT "   horizontal displacement of node";j;"=";d(3*j-1)
PRINT "   vertical displacement of node";j;"=";d(3*j)
PRINT "   rotation of node";j;"=";d(3*j-2)
PRINT:PRINT "      do you want displacements at another node? y/n"
INPUT yn$
IF yn$="y" THEN 140
RETURN


LABEL memforces
CLS:PRINT"   this subroutine displays member forces."
150PRINT"    please type-in the number of the member for which"
PRINT"    you need the forces":INPUT m
PRINT"           in member";m
PRINT"    moment at end";m(m,1);"=";f(m,1)
PRINT"    moment at end";m(m,2);"=";f(m,2)
PRINT"    thrust at end";m(m,1);"(tension positive) = ";f(m,3)
PRINT"    shear at end";m(m,1);" =";f(m,4)
160PRINT:PRINT"    do you want forces in another member? y/n"
INPUT yn$
IF yn$ = "y" THEN 150
IF yn$ <>"n" THEN 160
RETURN


LABEL changedata
CLS:PRINT"    this subroutine will allow you to verify/edit the"
PRINT"    existing data:   you cannot however change the main"
PRINT"    parameters, ie numbers of nodes and members"
PRINT"    if you wish to do this you will need to restart"
PRINT"    the program"
PRINT"    PRESS ANY KEY WHEN READY TO START"
REPEAT
presskey$=INKEY$
UNTIL presskey$ <> ""
170CLS:PRINT"         ********  E D I T   M E N U  ********"
PRINT"          key ....."
```

```
PRINT"       1 to edit nodal coordinates"
PRINT"       2 to edit member data"
PRINT"       3 to verify loading data"
PRINT"       4 to verify support data"
PRINT"       5 to return to main menu"
PRINT"   which do you require?"
INPUT choice: IF choice < 1 THEN 170
IF choice > 5 THEN 170
IF choice = 1 THEN GOSUB chnodcoords
IF choice = 2 THEN GOSUB chmemdat
IF choice = 3 THEN GOSUB chloaddat
IF choice = 4 THEN GOSUB chsuppdat
IF choice = 5 THEN 180
GOTO 170
180GOSUB cleardata
190RETURN


LABEL chnodcoords
CLS:PRINT"    this subroutine will allow you to edit nodal coordinates"
1000PRINT"    please type in node number":INPUT node
PRINT:PRINT"    the existing coordinates of node";node;"are"
PRINT "X";node;"=";xn(node)
PRINT "Y";node;"=";yn(node)
1010INPUT " do you want to change these";yn$
IF yn$ = "n" THEN 1020
IF yn$ <> "y" THEN 1010
PRI  please type in revised data in the form x,y"
INPUT xn(node),yn(node)
1020INPUT "   do you wish to edit further nodal coordinate data? y/n";yn$
IF yn$ = "y" THEN 1000
IF yn$ <> "n" THEN 1020
PRINT "    please re-input applied load, support and any spring data"
PRINT"    BEFORE analysing this structure"
PRINT:PRINT"      PRESS ANY KEY WHEN READY TO CONTINUE"
REPEAT
presskey$ = INKEY$
UNTIL presskey$ <> ""
1040 RETURN


LABEL chmemdat
CLS:PRINT"    this subroutine will allow you to edit member data"
1050PRINT:PRINT"    please type-in member number":INPUT m
PRINT"    this is the current data for member";m
SET ZONE 5
PRINT m(m,1),m(m,2),m(m,5),m(m,6),m(m,7)
1060INPUT "    do you wish to edit this data? y/n";yn$
IF yn$ = "n" THEN 1070
IF yn$ <> "y" THEN 1060
PRINT"    please type-in revised data in the order N1,N2,E,A,I"
INPUT m(m,1),m(m,2),m(m,5),m(m,6),m(m,7)
i = m(m,1): j = m(m,2)
m(m,3)=xn(j)-xn(i): m(m,4)=yn(j)-yn(i)
a=ATAN2(m(m,3),m(m,4))
s(m)=SIN(a):c(m)=COS(a)
m(m,8)=SQR(m(m,3)^2+m(m,4)^2)
ms(m,1)=m(m,5)*m(m,7)/m(m,8)
ms(m,2)=ms(m,1)*c(m)/m(m,8)
ms(m,3)=ms(m,1)*s(m)/m(m,8)
ms(m,4)=ms(m,2)*c(m)/m(m,8)
ms(m,5)=ms(m,3)*s(m)/m(m,8)
ms(m,6)=ms(m,2)*s(m)/m(m,8)
```

```
ms(m,7)=m(m,5)*m(m,6)*c(m)*c(m)/m(m,8)
ms(m,8)=m(m,5)*m(m,6)*s(m)*s(m)/m(m,8)
ms(m,9)=m(m,5)*m(m,6)*s(m)*c(m)/m(m,8)
1070INPUT "     do you want to edit further member data? y/n";yn$
IF yn$ = "y" THEN 1050
IF yn$ <> "n" THEN 1070
PRINT "    please re-input all applied loadsupport and any spring data"
PRINT"     BEFORE  analysing this structure"
PRINT:PRINT "    PRESS ANY KEY WHEN READY TO CONTINUE"
REPEAT
presskey$ = INKEY$
UNTIL presskey$ <>""
1080RETURN


LABEL chloaddat
CLS: PRINT "     this subroutine will allow you to verify applied"
PRINT "    load data"
1100PRINT"      please type-in the number of the node where you"
PRINT"     wish to edit loading data":INPUT node
PRINT"     this is the current record of loads at node";node
PRINT"     (please note that it includes udl's if any)"
PRINT"       applied moment   ="; w(3*node - 2)
PRINT"       horizontal load ="; w(3*node - 1)
PRINT"       vertical load    ="; w(3*node)

1120PRINT"     do you wish to edit further nodal loads?  y/n"
INPUT yn$
IF yn$ = "y" THEN 1100
IF yn$ <> "n" THEN 1120
PRINT"     please re-input all applied load, support and any spring data"
PRINT"     BEFORE analysing this structure"
PRINT:PRINT"       PRESS ANY KEY WHEN READY TO CONTINUE"
REPEAT
presskey$=INKEY$
UNTIL presskey$<>""
1130RETURN


LABEL chsuppdat
CLS:PRINT"     this subroutine allows you to verify nodal support"
PRINT"     data:  if editing is required you will need to re-input"
PRINT"     all nodal support constraints using option 5 in the"
PRINT"     main menu"
1900PRINT"     please type-in node number of supported node":INPUT node
1IF k(3*node-2,3*node-2) > 10E19 THEN 1200
PRINT"       Node";node;"is NOT supported rotationally"
GOTO 1210
1200PRINT"       Node";node;"IS supported rotationally"
1210IF k(3*node-1,3*node-1)>10E19 THEN 1220
PRINT"       Node";node;"is NOT supported horizontally"
GOTO 1230
1220PRINT"       Node";node;"IS supported horizontally"
1230IF k(3*node,3*node)>10E19 THEN 1240
PRINT"       Ne";node;"is NOT supported vertically"
GOTO 1250
1240PRINT"       Node";node;"IS supported vertically"
PRINT
1250PRINT"     do you wish to verify support data at another node?y/n"
INPUT yn$
IF yn$ = "n" THEN 1260
IF yn$<>"y" THEN 1250
GOTO 1900
```

```
1260PRINT:PRINT"    please return to the main menu and re-input all"
PRINT"    applied load, support and any spring data BEFORE"
PRINT"    re-analysing this structure"
PRINT:PRINT"            PRESS ANY KEY TO CONTINUE"
REPEAT
presskey$ = INKEY$
UNTIL presskey$ <>""
RETURN


LABEL cleardata
FOR i = 1 TO n
w(i)=0
FOR j = 1 TO n
k(i,j) = 0
NEXT j
NEXT i
FOR i = 1 TO nm
FOR j = 1 TO 4
f(i,j) = 0
NEXT j
NEXT i
RETURN


2000RESUME

LABEL exit
CLS:PRINT AT (15;10) "* * * * * *  C H E E R S * * * * * * "
FOR i=1 TO 10000:NEXT i
END
```

Validation exercises for PLANFRAM.BAS

Structure	Data	Sample results
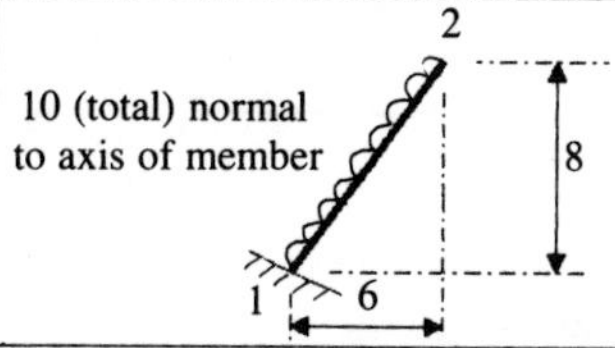 10 (total) normal to axis of member; nodes 1, 2; dimensions 6, 8	$E=200$ $I=500$ $A=1000$ $w(v)=1$ $w(h)=1$	$v_2=-0.0075$ $u_2=0.01$ $\theta_2=-0.00167$ $M_{1,2}=50$ $N_{1,2}=0$ $S_{1,2}=10$
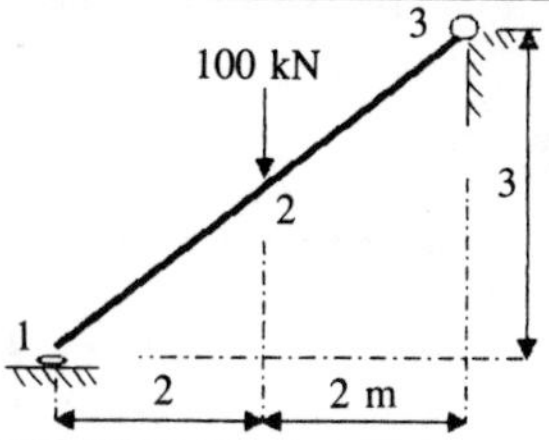 100 kN; nodes 1, 2, 3; dimensions 2, 2 m, 3	$E=200$ $A=1000$ $I=100$	$M_{2,1}=100$ $S_1=40$ $N_{1,2}=-30$ (compression) $N_{2,3}=30$ (tension) $M_{2,3}=-100$
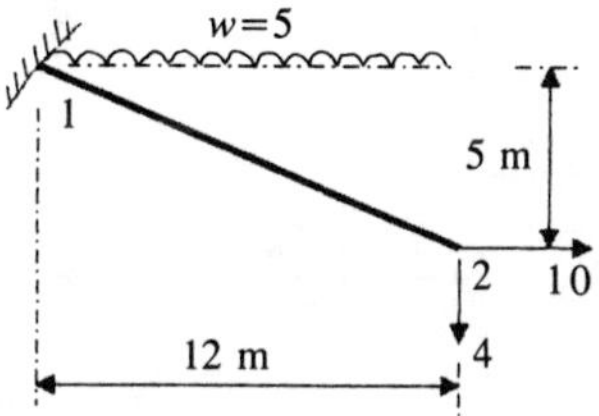 $w=5$; nodes 1, 2; loads 10, 4; dimensions 12 m, 5 m	$E=1$ $A=1000$ $I=1$	$M_{1,2}=358$ $S_1=55.23$ $N_1=33.85$
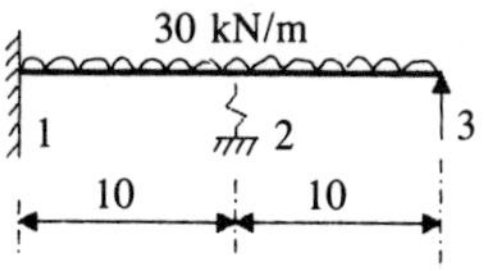 30 kN/m; nodes 1, 2, 3; dimensions 10, 10	$E=100$ $A=1000$ $I=1000$ k(spring) $=2400$	$\theta_2=-0.00568$ $v_2=-0.0909$ $\theta_3=0.02273$ $M_{1,2}=681.8$ $M_{2,1}=68.2$ $S_1=225$ $S_2=143.2$

Structure	Data	Sample results
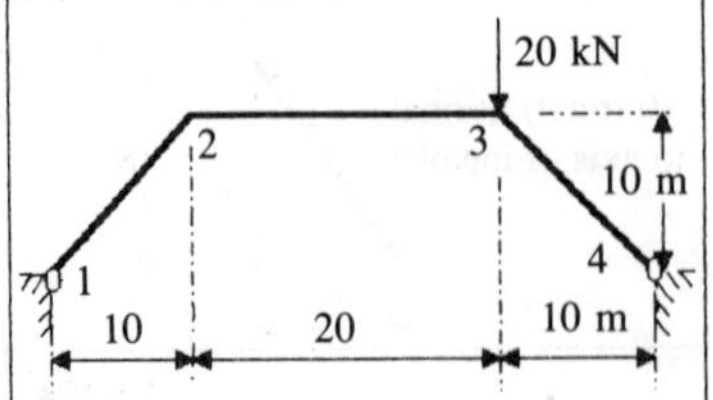 20 kN; 1, 2, 3, 4; 10 m; 10, 20, 10 m	$E=1$ $A=1000$ $I=1$	$M_{2,1}=-50$ $M_{3,4}=-50$ $N_{1,2}=-10.6$ $S_{1,2}=-3.54$ $N_{3,4}=-17.7$ $S_{3,4}=-3.54$
30 kN/m; 1, 2, 3, 4; 4 m; 4; 8	$E=1$ $A=1000$ $I=1$	$M_{2,1}=125.2$ $M_{3,4}=417.4$ $N_{1,2}=67.8$ $S_{1,2}=135.7$ $N_{4,3}=-67.8$ $S_{4,3}=104.3$

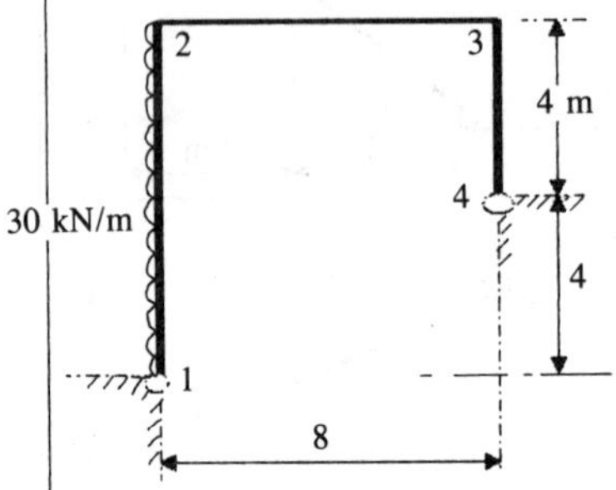

APPENDIXES

Appendix 1 Section properties of structural members

Solid sections

Cross-section	A	I_z	I_y	I_{zy}	J
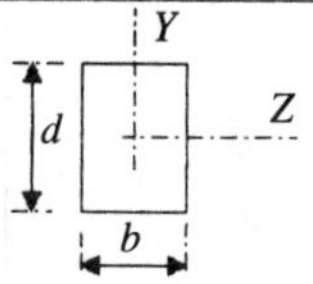	bd	$bd^3/12$	$db^3/12$	0	*
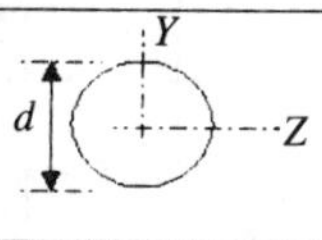	$\pi d^2/4$	$\pi d^4/64$	$\pi d^4/64$	0	$\pi d^4/32$
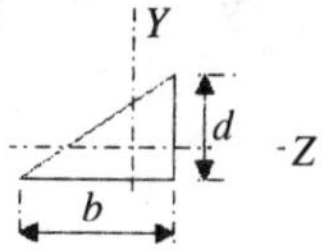	$bd/2$	$bd^3/36$	$db^3/36$	$d^2b^2/72$	
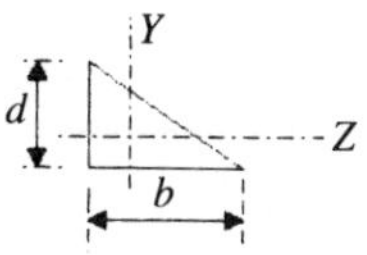	$bd/2$	$bd^3/36$	$db^3/36$	$-d^2b^2/72$	
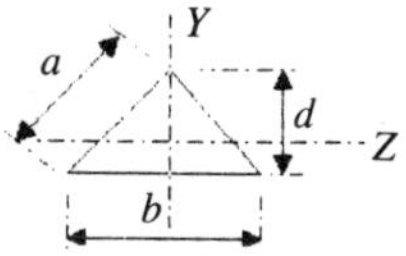	$bd/2$	$bd^3/36$	$db^3/48$		$\sqrt{(3)}b^4/80$ for $a=b$

$$* \; db^3\left\{\frac{1}{3}-0.21\,\frac{b}{d}\left(1-\frac{b^4}{12d^4}\right)\right\};\; d \geqslant b$$

Thin-walled sections

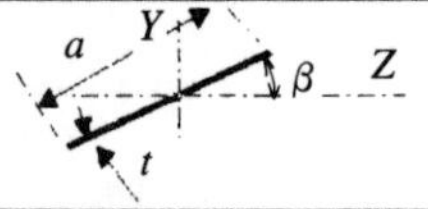	at	$\frac{a^3t}{12}\sin^2\beta$	$\frac{a^3t}{12}\cos^2\beta$	$\frac{a^3t}{24}\sin 2\beta$	
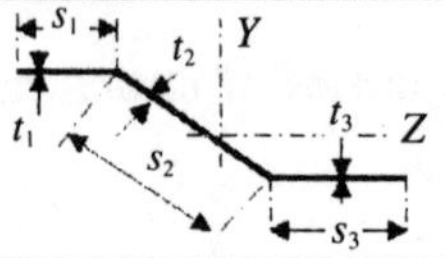	$\int \mathrm{d}A$	$\int y^2 \mathrm{d}A$	$\int z^2 \mathrm{d}A$		$\Sigma\,(st^3/3)$ (open section)
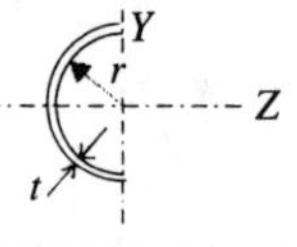	πrt	$\pi r^3t/2$			
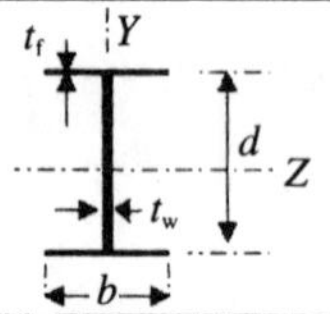	$2bt_f+dt_w$	$\frac{d^2}{12}(6bt_f+dt_w)$	$t_f b^3/6$		$\frac{1}{3}(dt_w^3+2bt_f^3)$
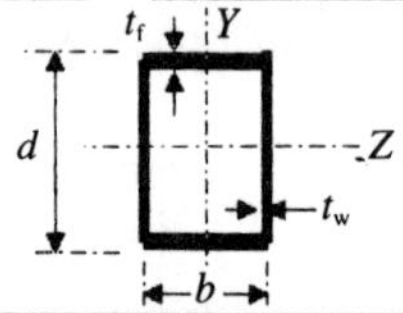	$2(bt_f+dt_w)$	$\frac{d^2}{6}(dt_w+3bt_f)$	$\frac{b^2}{6}(bt_f+3dt_w)$		$\frac{2b^2d^2t_wt_f}{bt_w+dt_f}$
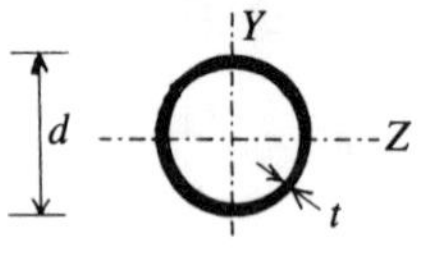	πdt	$\pi d^3t/8$	$\pi d^3t/8$		$\pi d^3t/4$

Appendix 2 Standard displacements in prismatic beams

Beam and loading	$EI\theta_1$	EIy_{max}	$EI\theta_2$
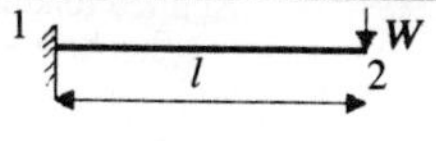 1 W l 2	0	$\frac{1}{3}Wl^3$	$\frac{1}{2}Wl^2$
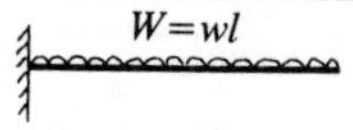 $W=wl$	0	$\frac{1}{8}Wl^3$	$\frac{1}{6}Wl^2$
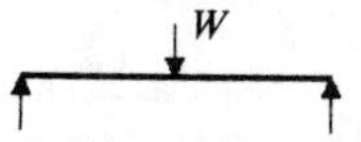 W	$\frac{1}{16}Wl^2$	$\frac{1}{48}Wl^3$	$\frac{1}{16}Wl^2$
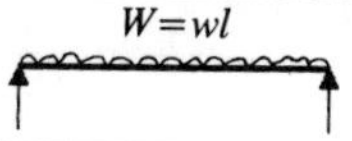 $W=wl$	$\frac{1}{24}Wl^2$	$\frac{5}{384}Wl^3$	$\frac{1}{24}Wl^2$
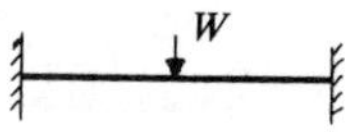 W	0	$\frac{1}{192}Wl^3$	0
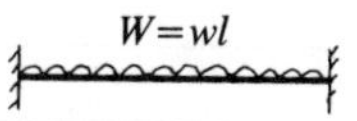 $W=wl$	0	$\frac{1}{384}Wl^3$	0
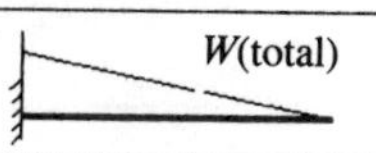 W(total)	0	$\frac{1}{15}Wl^3$	$\frac{1}{12}Wl^2$
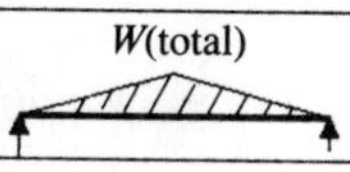 W(total)	$\frac{5}{96}Wl^2$	$\frac{1}{60}Wl^3$	$\frac{5}{96}Wl^2$
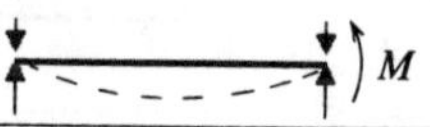 M	$\frac{Ml}{6}$		$\frac{Ml}{3}$
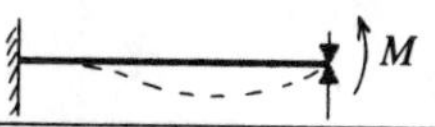 M	0		$\frac{Ml}{4}$
 $6EI\Delta/l^2$ Δ $6EI\Delta/l^2$	0	$EI\Delta=\dfrac{Ml^2}{6}$	0
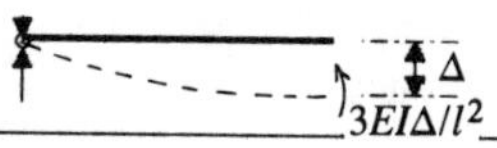 Δ $3EI\Delta/l^2$	$\dfrac{3\Delta}{2l}$	$EI\Delta=\dfrac{Ml^2}{3}$	0

Appendix 3 Fixed-end moments for prismatic beams

M_{f1}	Applied loading	M_{f2}
$(Wab/l)\,b/l$	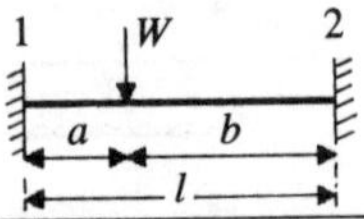 1, W, 2; a, b, l	$-(Wab/l)a/l$
$\dfrac{W}{12l^2}\left\{12ab^2+c^2(a-2b)\right\}$	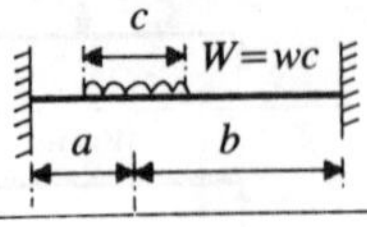 c, $W=wc$; a, b	$-\dfrac{W}{12l^2}\left\{12a^2b+c^2(b-2a)\right\}$
$Wl/12$	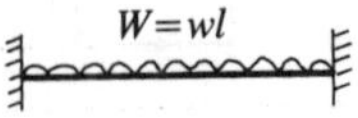 $W=wl$	$-Wl/12$
$Wl/15$	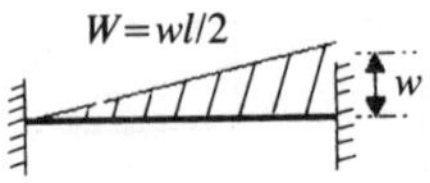 $W=wl/2$; w	$-Wl/10$
$5Wl/48$	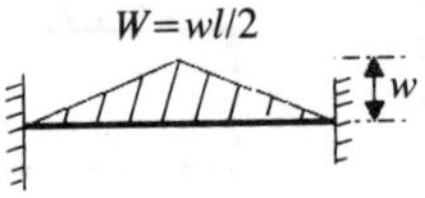 $W=wl/2$; w	$-5Wl/48$
$Mb(2a-b)/l^2$	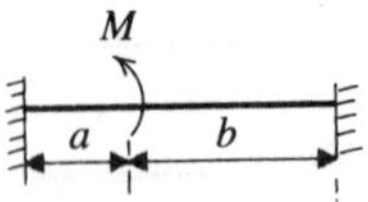 M; a, b	$Ma(2b-a)/l^2$

Appendix 4 Product integrals

$\int m_i m_j \, ds$

m_j \ m_i	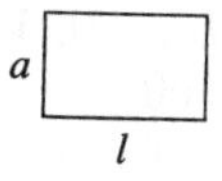 a, l	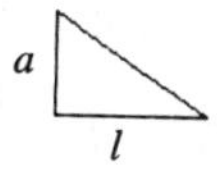 a, l	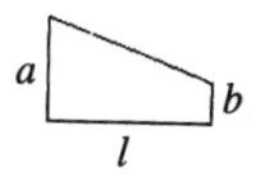 a, b, l
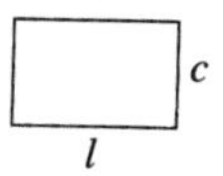 c, l	lac	$\frac{l}{2}ac$	$\frac{l}{2}(a+b)c$
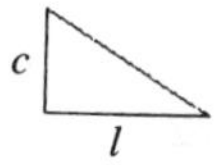 c, l	$\frac{l}{2}ac$	$\frac{l}{3}ac$	$\frac{l}{6}(2a+b)c$
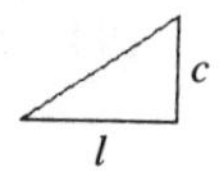 l, c	$\frac{l}{2}ac$	$\frac{l}{6}ac$	$\frac{l}{6}(a+2b)c$
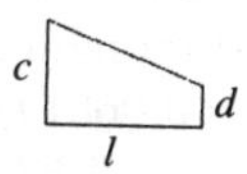 c, d, l	$\frac{l}{2}a(c+d)$	$\frac{l}{6}a(2c+d)$	$\frac{l}{6}\left\{a(2c+d) + b(2d+c)\right\}$
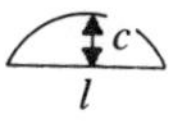 c, l	$\frac{2l}{3}ac$	$\frac{l}{3}ac$	$\frac{l}{3}(a+b)c$

Appendix 5 Summary of equations in the stiffness method

Operation	*Equations*	
Displacement function	$\boldsymbol{d} = \boldsymbol{L\alpha}$	[8.2]
Nodal displacements	$\boldsymbol{s} = \boldsymbol{A\alpha}$	[8.3]
Internal displacements	$\boldsymbol{d} = \boldsymbol{LA}^{-1}\boldsymbol{s}$	[8.5]
	$= \boldsymbol{Ns} = \sum \boldsymbol{N}_i \boldsymbol{s}_i$	[8.11]
Internal strains	$\boldsymbol{\epsilon} = \boldsymbol{Bs}$	[4.46]
Internal stresses	$\boldsymbol{\sigma} = \boldsymbol{D\varepsilon}$	[4.47]
	$= \boldsymbol{DBs}$	[4.48]
Element stiffness	$\boldsymbol{k} = \int \boldsymbol{B}^{\mathrm{T}}\boldsymbol{DB} \, dvol$	[4.49]
Consistent nodal forces	$\boldsymbol{W}_{\mathrm{e}} = \boldsymbol{WN}^{\mathrm{T}}$	[8.40]
	$\boldsymbol{W}_{\mathrm{e}} = \int_0^l \boldsymbol{N}^{\mathrm{T}} q \, dx$	[8.42]

Appendix 6 Answers to exercises

Chapter 1

1.4 (a) $H_A = 0$, $V_A = 6.75$; (b) $H_E = 10$, $V_E = 19.25$; (c) -28; (d) 100.
1.5 (a) 2; (b) 9; (c) 2; (d) 3.
1.6 (a) 9; (b) 18 (9 without axial straining); (c) 10 (without axial straining); (d) 12 (7 without axial straining).

Chapter 5

5.1 $0.211\,l$.
5.2 $M_A = WL$; $N_{BC} = 0.6W$; $S_{BC} = 0.8W$.
5.3 37.5 w.
5.4 (a) $M_C = 141.9$; (b) $M_C = 80$.
5.5 $M_A = Wl/2$; $T_A = Wl/2$.

Chapter 6

6.1 682 mm; 224 E8 mm^4; 194.3 E8 mm^4.
6.2 100, 50 N/mm^2.
6.3 51.7 kN.
6.4 −19 mm; −13 mm.
6.7 222 N/mm^2; 14.8 N/mm^2; 0.014 15°/m; 6.67 kN m.
6.9 153.6 mm above bottom flange.
6.10 $f_{11} = 0.193\, l/EI_0$; $f_{22} = 0.0681\, l/EI_0$; $f_{12} = 0.0567\, l/EI_0$; $k_{11} = 6.81\, EI_0/l$; $k_{22} = 19.3\, EI_0/l$; $k_{12} = 5.66\, EI_0/l$

Chapter 7

7.2 1.75.

Chapter 8

8.1 (ii) $EA/4l\begin{bmatrix} 16 & -18 & 2 \\ -18 & 27 & -9 \\ 2 & -9 & 7 \end{bmatrix}$; (iii) $w_0 l\begin{bmatrix} -1/12 \\ 3/8 \\ 5/24 \end{bmatrix}$

8.2

$$
EI\begin{bmatrix}
\frac{8}{l}+\frac{4}{h} & & & & & & & & \\
\frac{4}{l} & \frac{16}{l}+\frac{4}{h} & & & & & \text{Symmetric} & & \\
0 & \frac{4}{l} & \frac{8}{l}+\frac{4}{h} & & & & & & \\
-\frac{6}{h^2} & -\frac{6}{h^2} & -\frac{6}{h^2} & \frac{36}{h^3} & & & & & \\
\frac{2}{h} & 0 & 0 & -\frac{6}{h^2} & \frac{8}{l}+\frac{8}{h} & & & & \\
0 & \frac{2}{h} & 0 & -\frac{6}{h^2} & \frac{4}{l} & \frac{16}{l}+\frac{4}{h} & & & \\
0 & 0 & \frac{2}{h} & -\frac{6}{h^2} & 0 & \frac{4}{l} & \frac{8}{l}+\frac{8}{h} & & \\
\frac{6}{h^2} & \frac{6}{h^2} & \frac{6}{h^2} & -\frac{36}{h^3} & 0 & \frac{6}{h^2} & 0 & \frac{60}{h^3} & \\
-\frac{12}{l^2} & 0 & \frac{12}{l^2} & 0 & -\frac{12}{l^2} & 0 & \frac{12}{l^2} & 0 & \frac{96}{l^3}
\end{bmatrix}
$$

8.4

$$
\begin{bmatrix}
\frac{EA}{l} & & & & & \\
0 & \frac{3EI}{l^3} & & & \text{Symmetric} & \\
0 & 0 & 0 & & & \\
-\frac{EA}{l} & 0 & 0 & \frac{EA}{l} & & \\
0 & -\frac{3EI}{l^3} & 0 & 0 & \frac{3EI}{l^3} & \\
0 & \frac{3EI}{l^3} & 0 & 0 & -\frac{3EI}{l^2} & \frac{3EI}{l}
\end{bmatrix}
$$

8.5

$\frac{GJ}{l}$				Symmetric	
0	$\frac{4EI}{l}$				
0	$\frac{6EI}{l^2}$	$\frac{12EI}{l^3}$			
$-\frac{GJ}{l}$	0	0	$\frac{GJ}{l}$		
0	$\frac{2EI}{l}$	$\frac{6EI}{l^2}$	0	$\frac{4EI}{l}$	
0	$-\frac{6EI}{l^2}$	$-\frac{12EI}{l^3}$	0	$-\frac{6EI}{l^2}$	$\frac{12EI}{l^3}$

8.6

$\frac{EI}{37l^3}$ ×

$56l^2$		sym
$34l^2$	$92l^2$	
$-30l$	$-42l$	24

8.7 $k_{11} = k_{22} = 0.673\ EI$; $k_{12} = k_{21} = 0.272\ EI$.

Chapter 9

9.4 $R_A = 1$ at A; $-1/2$ at C
$R_B = 1$ at B; 1.5 at C
$M_B = 5$ at C
$S_C = 1$ at C.

Chapter 10

10.1 109.8, 140.2 kN m.
10.3 65.6 kN m.
10.4 2.
10.5 48 kN m.
10.6 1.32.

Chapter 11

11.2 5.29 Hz.
11.3 (a) 0.0707 s
(b) $y = 10 \cos 88.9\, t$
(c) 6.35 mm
(d) 88.26 rad/s; 0.0712 s.
11.4 $\omega^2 = 97.5\, EI/ml^4$.

Chapter 12

12.2 0.0918 m; 5.8 per cent.

BIBLIOGRAPHY

Bhatt P 1981 *Problems in Structural Analysis by Matrix Methods.* Construction Press

British Standards Institution BS 5950 1985 *Structural Use of Steel in Building.*

Brotton D M 1962 *The Application of Digital Computers to Structural Engineering Problems.* Spon

Budynas R G 1977 *Advanced Strength and Applied Stress Analysis.* McGraw-Hill

Cheung Y K, Yeo M F 1979 *A Practical Introduction to Finite Element Analysis.* Pitman

Civil Engineer's Reference Book. Butterworth

Desai C S 1979 *Elementary Finite Element Method.* Prentice-Hall

Elias Z M 1986 *Theory and Methods of Structural Analysis.* Wiley

Ghali A, Neville A M 1978 *Structural Analysis – A Unified Classical and Matrix Approach.* Chapman and Hall

Gordon J E 1978 *Structures, or Why Things Don't Fall Down.* Penguin

Horne M R 1971 *Plastic Theory of Structures.* Nelson

Horne M R, Merchant W 1965 *The Stability of Frames.* Pergamon Press

Jenkins W M 1969 *Matrix and Digital Computer Methods in Structural Analysis.* McGraw-Hill

Jennings A 1966 A compact storage scheme for the solution of linear simultaneous equations. *Computer Journal* **9**: 281–5

Jennings A 1977 *Matrix Computation for Engineers and Scientists.* Wiley

Jennings A, Gilbert S 1988 Where now the teaching of structures? *The Structural Engineer* **66**(1)

Kopal Z 1961 *Numerical Analysis.* Chapman and Hall

Livesley R K 1964 *Matrix Methods of Structural Analysis.* Pergamon

McMinn S J 1962 *Matrices for Structural Analysis.* Spon

Majid K I 1978 *Theory of Structures with Matrix Notation.* Newnes-Butterworth

Megson T H G 1974 *Linear Analysis of Thin-Walled Elastic Structures.* Surrey University Press

Morice P B 1959 *Linear Structural Analysis.* Thames and Hudson

Moy S J 1981 *Plastic Methods for Steel and Concrete Structures.* Macmillan

Przemieniecki J S 1968 *Theory of Matrix Structural Analysis.* McGraw-Hill

Rubinstein M F 1966 *Matrix Computer Analysis of Structures*. Prentice-Hall

Taniguchi T, Shivaishi N, Soga A 1985 Numerical error of elimination method for large sparse sets of linear equations. Proceedings Civil-Comp 85. Civil-Comp Press, Edinburgh

Timoshenko S P, Gere J M 1961 *Theory of Elastic Stability*. McGraw-Hill

Timoshenko S P, Woinowsky-Kreiger S 1959 *Theory of Plates and Shells*. McGraw-Hill

Zbirohowski-Koscia K 1967 *Thin-Walled Beams*. Crosby-Lockwood

Zienkiewicz O C 1967 *The Finite Element Method in Structural and Continuum Mechanics*. McGraw-Hill

INDEX

Computer programs

To obtain further information on the programs PLATRUSS and PLANFRAM the reader is invited to detach (or photocopy) this slip, complete it and send it to the author . A stamped addressed envelope (A5 size) should be enclosed.

To
Professor W M Jenkins
6 Bridge Road
MALTON, North Yorkshire, YO17 0JX

Please send information regarding alternative versions of the computer programs PLATRUSS and PLANFRAM.

(Please tick version required)

FAST BASIC ☐
QUICK BASIC ☐
TURBO PASCAL ☐
Other (please specify) ....

Name: ....
Address: ....
....
....